THE APE THAT

HOW THE MIND

UNDERSTOOD

AND CULTURE EVOLVE

THE UNIVERSE

Steve Stewart-Williams

CAMBRIDGE
UNIVERSITY PRESS

CAMBRIDGE
UNIVERSITY PRESS

University Printing House, Cambridge CB2 8BS, United Kingdom

One Liberty Plaza, 20th Floor, New York, NY 10006, USA

477 Williamstown Road, Port Melbourne, VIC 3207, Australia

314–321, 3rd Floor, Plot 3, Splendor Forum, Jasola District Centre,
New Delhi – 110025, India

79 Anson Road, #06-04/06, Singapore 079906

Cambridge University Press is part of the University of Cambridge.

It furthers the University's mission by disseminating knowledge in the pursuit of
education, learning, and research at the highest international levels of excellence.

www.cambridge.org
Information on this title: www.cambridge.org/9781108425049
DOI: 10.1017/9781108348140

© Steve Stewart-Williams 2018

First published 2018

Printed in the United Kingdom by TJ International Ltd. Padstow Cornwall

A catalogue record for this publication is available from the British Library.

Library of Congress Cataloging-in-Publication Data
Names: Stewart-Williams, Steve, 1971– author.
Title: The ape that understood the universe : how the mind and culture evolve /
Steve Stewart-Williams, University of Nottingham.
Description: New York: Cambridge University Press, [2018] |
Includes bibliographical references and index.
Identifiers: LCCN 2018003789 | ISBN 9781108425049 (hardback)
Subjects: LCSH: Evolutionary psychology. | Human behavior. |
Culture. | Interpersonal relations.
Classification: LCC BF698.95.S745 2018 | DDC 155.7–dc23
LC record available at https://lccn.loc.gov/2018003789

ISBN 978-1-108-42504-9 Hardback

For Jane, Darwin, and India

Contents

Acknowledgments

Thanks go, above all, to my wife, Jane Stewart-Williams. I can say without a shadow of a doubt that, without Jane, this book would not exist. Thanks also to our kids, Darwin and India, who've changed a lot since I first put pen to paper, but who've always been supportive and fun. Thanks to my former colleagues at Swansea University, Mark Blagrove and Michelle Lee, for allowing me to take a sabbatical to begin working on the book, and thanks to my current colleagues at the University of Nottingham Malaysia Campus, in particular David Keeble, for giving me the freedom to continue working on the book and for supporting my work in general. Thanks to Martin Daly and the late Margo Wilson, who've had a huge impact on my thinking about evolutionary psychology. Thanks to everyone who's read drafts of individual chapters, or of the entire manuscript, including Mike Alb, Amy Alkon, John Archer, Susan Blackmore, Baba Brinkman, Neil Carter, Helen Fisher, Rob Lowe, Robert King, Geoffrey Miller, Will Reader, Matt Ridley, Todd Shackelford, Michael Shermer, Beatrice Stewart, Jolyon Stewart, Andrew Thomas, Ben Winter, Lance Workman, and three anonymous reviewers for Cambridge University Press. Thanks to my editor at Cambridge University Press, Janka Romero, who's been a pleasure to work with from day one and who's made numerous suggestions that have greatly improved the book. And thanks to everyone else who's helped out in one way or another, either with the book itself or with my thinking on these topics, including Pat Barclay, Chloe Bradley, Andrew Clark, Jerry Coyne, Oliver Curry, Greg Dingle, Céline Durassier, Martie Haselton, Adam Hooper, Stephanie Huitson, Toko Kiyonari, Danny Krupp, Claire Lehmann, Andrew Loughnan, James

McKellar, Stewart McWilliam, Randy Nesse, Nikki Owen, Adam Perrott, Steven Pinker, John Podd, David Schmitt, Delia Shanly, Christina Hoff Sommers, Phil Tucker, Alison Walker, Abigail Walkington, Lee White, Barbara Williams, and Brian Williams. Last but not least, thanks to John Anderson for the plant genitals joke (see the Alien's Report).

I

The Alien's Challenge

This book is about the strangest animal in the world – the animal that's reading these words and the animal that wrote them: the human animal. Because we're so used to being human, and to living with humans, we sometimes don't notice what a peculiar creature we are. As a corrective, I want to begin by looking at our species from a new perspective. This perspective might initially seem somewhat alien to you... but so it should because that's the perspective we'll be using. We'll be looking at our species through the eyes of a hypothetical, hyperintelligent alien – an anthropologist from the planet *Betelgeuse III* – as it visits the Earth on an intergalactic *Beagle* and studies us "as someone with a microscope studies creatures that swarm and multiply in a drop of water."[1] But this isn't just any old hyperintelligent alien. It's a gender-neutral, asexual, asocial, amoral, areligious, and amusical alien. It is, in other words, a stranger to many elements of human life that are so familiar to us that we simply take them for granted. And that's why its perspective is useful. The alien's uncomprehending eyes will make the familiar seem strange, waking us to aspects of humanity that we normally overlook and which are so deeply ingrained that we don't even notice require an explanation.

Before going any further, I should probably make clear that I don't believe for a minute that extraterrestrials have actually visited Earth. It's certainly possible that intelligent life has evolved elsewhere in the universe. But there's no good evidence that any of it has traversed the interstellar darkness to spy on us, probe us, or otherwise interact with our civilization. (As Stephen Hawking once said, "I am discounting reports of UFOs. Why would they appear only to cranks and weirdos?")[2] Nonetheless, it is still valuable to ask: If an alien *did* drop in on us, how would it view our species? What would it make of our sex differences, our sexual behavior, our child-rearing patterns, our social behavior, our religions, our languages, music, and science? By way of an answer, here's how I imagine that an alien's report on our unusual species might read...

The Alien's Report

Excerpted from the prestigious journal, *Proceedings of the Betelgeusean Academy of Sciences.*

Earth is a small and inauspicious planet orbiting an average star on the outskirts of our galaxy. It first came to the attention of the Great Galactic Counsel around twenty-five cycles ago, when one of our top otherworldologists, Seer Ram Tin, detected some strange happenings in and around the planet.[3] For most of its four-and-a-half-billion-year history, the Earth changed only slowly, other than an occasional shakeup caused by asteroid impacts or volcanic activity. A few thousand years ago, however, the pace of change suddenly and dramatically quickened. Forests began disappearing, to be replaced with fields of "corn," and "rice," and "wheat." The ground started sprouting strange structures, now known as "cities," which spread across the planet's surface like bacterial molds. The land surrounding the cities divided itself into rectangles and other geometrical shapes. Then, in the last century, the process went into overdrive. The Earth suddenly became a major emitter of radio waves. Weird metallic objects jumped into orbit around the Earth, or left the Earth altogether and traveled to other worlds. These lonely wanderers sent a continuous stream of information back to their home planet, leading some Betelgeuseans to suggest that the Earth was growing a sensory system, thereby becoming aware of itself and its surroundings. It was this possibility that first drew the attention of the Great Galactic Council.

After three nanoseconds of in-depth discussion, the Council decided that a senior otherworldologist should be dispatched to the Earth to investigate more closely. As the most brilliant and distinguished otherworldologist available on short notice, this great honor went to me, and I charted a course for the Earth. Upon arrival, I quickly ascertained that some of the metal machines orbiting the planet contained meat creatures. This confirmed what many already suspected: that evolution by natural selection was taking place on Earth. At first, it was unclear what role the meat creatures (or "humans") played in the strange events we'd detected. One suggestion was that they were the sex organs of the cities or maybe of the metal machines – the means by which the cities and the machines reproduced themselves.[4] Another suggestion was that the meat creatures were slaves of the corn, the rice, and the wheat. Perhaps, these plants had somehow tricked the humans into launching a genocidal culling spree on their main competitors – the trees – and figured out how to get the humans to sow and spread them instead.[5]

Although we haven't completely ruled out any of these hypotheses, most otherworldologists now believe that the humans themselves were the driving force behind the strange events on Earth. The most plausible computer simulations suggest that this fragile planet recently suffered a plague of technology-wielding humans. By accident or by design, these crazy meat robots discovered ways to convert more and more of the matter of the Earth's biosphere into more and more human beings. In doing so, they rapidly overran the Earth. They enslaved the plants, built the cities, and ordered the metal sense organs into orbit. Based on this assessment, the humans became the primary focus of my investigation. What follows is a summary of my observations to date.

Trigger-Warning! Human beings are so extreme and bizarre that some Betelgeuseans were initially unwilling to accept that they genuinely existed. Perhaps, they suggested, the whole thing was a hoax set up by the Grimmies of Sector Four. Alternatively, perhaps humans were selectively bred as a practical joke by the Narbo people of the D-Star region, well known for their overenthusiastic application of the techniques of selective breeding to inferior life forms. In light of these allegations, I must stress that everything here reported has now been verified by the Great Galactic Council. Human beings really do exist, and there's not a shred of evidence that they've been tampered with by any extraterrestrial intelligence.

The Strangest Animal

Let's start at the beginning. Human beings are collections of atoms organized into self-replicating systems or "life forms." Most life on Earth consists of single-celled organisms: free-living, solitary cells. Every human being starts out this way as well – as a cell so small it would barely be visible to its own adult eyes. But this original cell soon begins to divide and multiply, and before too long, the expanding society of cells arranges itself into the shape of a human being. An individual human organism, therefore, is actually a vast colony of single-celled organisms. Human cell colonies belong to a family of cell colonies known as "animals." Periodically, these cell colonies establish new cell colonies – in other words, they reproduce. Like most animals, humans reproduce in a strange and inefficient manner. Rather than simply cloning themselves, pairs of humans merge their hereditary material to create offspring that are mere "half-clones" of each parent. Not just any pairing will do, though. For some reason, humans come in two basic morphs: male and female. One of each is required to establish a new human cell colony. Most animals live in the sea. Humans, in contrast, live at the bottom of the sky on slow-moving rafts called "land masses." Like all land animals, however, humans are the descendants of sea creatures. To be precise, humans are modified fish – land fish, one might say. Their arms are modified fins; their jaws are modified gills.[6] Like most land fish, and indeed like most animals, humans hibernate once or twice a day, ultimately spending around a third of their brief lives in this plant-like state. No wonder it took them so long to establish their civilization, such as it is.

The more closely I looked at these tiny monsters, the more confused I became. By this time, I'd established that humans are products of natural selection, not some weird lab experiment. As any well-educated life-form knows, natural selection produces entities designed to keep themselves alive and create more of their kind. In many ways, humans fit this description: They suck up bits of the environment as fuel; they run away from threats; they make and care for new human cell colonies. In many other ways, though, humans defy the ancient biological imperative to survive and reproduce. For one thing, they crave substances that are bad for them. People everywhere have a strong, almost irresistible appetite for foods that make them unhealthy, and which drastically shorten their lives. This makes about as much sense as having an appetite for poison. How could it evolve? Another mystery is that human fears are poorly matched to the dangers in their environment. Many humans are afraid of animals such as snakes and spiders, even though most humans live in cities where these creatures pose little threat. Similarly, many miniature humans (or "children") are terrified of the dark, even though most humans live in boxes that keep them safe from night-lurking dangers. As well as fearing things that pose little threat to their survival or reproductive prospects, humans show no natural fear of many things that genuinely do pose a threat. This includes items as diverse as "junk food," "cigarettes," and "driving too fast without a seat belt." It also includes items such as "condoms" and "the pill." Not only are most humans unafraid of these fertility-harming contraptions, many use them deliberately to block their own fertility. Why on Betelgeuse would natural selection design them to do *that*?

As mentioned, humanoids come in two main varieties: male and female. Some don't fit easily into these categories, but the vast majority do. Males and females not only look somewhat different, they behave somewhat differently too. The larger ones (males) tend to be more aggressive, more sexually reckless, and more willing to take life-threatening risks. The smaller ones (females) tend to be more selective about their sexual partners, more involved in childcare, and somewhat longer lived. On spotting these differences, I immediately became curious about how they'd come to exist. Are they programmed into the humans' necktop computers (or "brains") by their families and the people around them? Or do the differences go deeper than that? Are they part of the bedrock nature of these upright, furless apes?

With these questions lodged firmly in mind, I quickly scanned the entire animal kingdom with my Animascope 5000, and made two profound discoveries. The first was that the same kinds of sex differences I'd spotted in the humans are found in many other animals as well, including most mammal-fish. This led me to suspect that the human sex differences are not purely products of cultural programming. The second discovery, however, was that the human sex differences are not nearly as pronounced or polarized as those found in most other species. In most mammal-fish, for instance, males alone compete for mates, and females alone choose from the males on offer. Among humans, on the other tentacle, *both* sexes compete for mates and *both* are choosy about their mates, at least when it comes to long-term pairings. Similarly, in most mammal-fish, the females alone care for the young and the males are mere sperm donors. Among humans, in contrast, *both* sexes commonly care for their hideous young – an arrangement more often found in bird-fish than in mammal-fish. Certainly, male humans tend to compete more fervently, and female humans tend to be choosier and more parental. But these differences are much more modest than those seen in most other species. What in the name of Kurtron the Destroyer is going on?

(Incidentally, should you ever have the misfortune of meeting a human being, don't mention any of these sex differences. Many humans find them strangely upsetting. At the risk of being vaporized, I must admit that I was unable to work out why.)

The human reproductive process is exceedingly bizarre, even by Earthling standards. Humans size up potential mates by visually inspecting the outermost layers of their bodies, and especially the front parts of their heads. To Betelgeuseans, all humans look the same. To one another, however, the subtlest differences in appearance – a slight deviation from symmetry, for instance, or the slightest hint of a wrinkle – can make the difference between being "beautiful" and being "aesthetically challenged." Why do they have these seemingly arbitrary preferences?

Once a suitable mate is located, humans engage in various peculiar mating rituals. The male, for example, may give the female a bundle of plant genitals (or "flowers"), or the pair may take turns making noises at each other while imbibing fermented plant juice. Every now and then, these mating rituals trigger an intense two-person madness called "love." Humans in the grip of the love-madness become obsessed with each other, and idolize each other in a way that seems unlikely to survive rational scrutiny. Stranger still, if a love affair collapses, this can usher in months or even years of misery and moping. It's hard to imagine that this is a biologically useful way for a human to spend its time. And love can even be actively dangerous. Humans sometimes kill over love – themselves, a former lover, a rival for a lover's affection. At first, I wondered whether the love-madness is a maladaptive glitch or perhaps the product of a mind-controlling virus. But after discreetly scanning a sample of humans with my trusty Animascope 5000, it became clear that love is built into the basic structure of the human animal. Why would natural selection favor such an irrational and potentially debilitating syndrome?

Every now and then, the humans' strange mating antics lead to the production of half-clones – or "babies," as the humans more often call them. To Betelgeuseans, human babies are gruesome and frightening. For the humans themselves, however, babies are the cutest, most intrinsically important aggregations of matter in the entire visible universe. Indeed, I'm not exaggerating when I say that the bond between parents and their offspring is stronger than a Martian mud wrestler. Parents will risk life-and-modified-fin to protect their offspring, and should one of their offspring die, parents will regularly leak salt water from their eyeballs for the rest of their natural lives. To be clear, I'm not suggesting that humans act this way toward *all* the young they encounter. People are vastly more likely to feed, clothe, and love their own offspring than they are the offspring of others. For an impartial Betelgeusean like myself, this makes little sense. Wouldn't it be better for the species if everyone just cared for everyone?

I know what you're thinking. You're thinking: "Listen here, you multi-tentacled fool! Human beings are in the pre-singularity phase of their evolution. During that phase, which mercifully we passed through eons ago, the only individuals that survive and leave descendants are those that ruthlessly pursue their own interests. So *of course* humans care more about their own offspring than they care about anyone else's!" It's a fair point (although the "multi-tentacled fool" comment was a little out of line). However, when we look at things from this perspective, it simply flips the mystery on its head. Humans clearly care a lot about themselves and their offspring. But that's not *all* they care about. To begin with, they care as well about other relatives, including siblings, cousins, nieces, and nephews, none of whom could continue their line. Why would that be? Whatever the answer, it's not unique to humans. *Most* animals are nicer to their relatives than to unrelated life forms. In fact, compared to other animals, the thing that really stands out about human beings is how nice they can be to *non*-relatives. Humans cooperate with non-relatives to a degree unmatched by any other species on Earth. And not only do they cooperate, they often care deeply about them. Sometimes they leak from their eyeballs if they spot images of non-relatives suffering or starving half-a-world away. Sometimes they help non-relatives even when the non-relatives could never return the favor, and when nobody else is watching. And sometimes they risk life-and-modified-fin to help unre-lated humans – or even members of other species. This selfless behavior once again makes humans a puzzling exception to every known rule. Why don't they just look out for themselves and their nearest relatives, like any other self-respecting Earthling?

I should add that, nice though they can be, humans are also quite capable of treating each other abysmally. Young male humans period-ically coalesce into groups and start maiming or killing members of other groups. People have been known to incinerate each other, slice off each other's body parts, and call each other hurtful names. They imprison other animals by the millions, torture them, and then heat and eat their corpses. And yet still it has to be said: Despite their many,

many flaws, human beings are among the most cooperative and altruistic carbon-based life forms in this neck of the galaxy.

Infested with Ideas

They're also among the most baffling. Human beings devote extraordinary quantities of time and energy to activities that, as far as this Betelgeusean can tell, do nothing to help them survive or reproduce. I'll give you some examples. First, humans spend many hours of every day making noises at each other through the holes in their faces. Indeed, except when they're hibernating, humans just don't shut up! It took me quite some time to determine what function their funny face noises served – a full nanosecond, to be precise. For the most part, they use the noises to transmit ideas between their tiny brains. But here's the peculiar thing: Most of these ideas are entirely unrelated to matters of survival or reproduction. Often they're designed simply to elicit the weird "ha-ha" noises that humans regularly emit. Otherwise, they tend to be about the weather or other people's failings. Strange.

Second, most humans believe in invisible beings called "ghosts," or "spirits," or "gods." Many spend vast amounts of time thinking about these beings and attempting to communicate with them telepathically, and many gather together regularly to perform costly and elaborate rituals, aimed at persuading the beings to be nice to them. On top of that, many make strenuous efforts to persuade *others* to believe in the invisible beings and to engage in the costly rituals. And they do all of this despite a relative paucity of evidence that the invisible beings exist or that the rituals actually work.

Third and finally, humans everywhere devote significant chunks of their lives to stimulating their brains in various inexplicable ways. Many, for instance, while away the hours making or staring at colored splodges on canvass or cave walls – splodges that resemble items in the world such as nude humans, other animal cell colonies, or plant genitalia. Many deliberately hypnotize themselves with rhythmically patterned noises – simple, repetitive sounds which, for some strange reason, evoke strong emotional responses in humans, and produce weird side effects such as foot tapping, head bobbing, and even full-body rhythmical spasms. And many spend hour after hour and day after day sitting in a peculiar trance-like state, staring at flickering images on flat screens – images, for example, of simulated events they know full well never happened, of other human beings mating, or of baby mammals behaving incompetently.

Consider how bizarre this all is. If humans devoted, say, 20 percent of their time to the chattering, the beliefs and rituals, or the splodges and rhythmical noises, this would mean that every fifth mouthful of food they ate would be siphoned off to fund these seemingly pointless activities. Why don't they devote this time and energy instead to manufacturing as many babies and grandbabies as possible, like every other species on Earth? At this stage, I have only one firm conclusion: I'm stumped. Truly, humans are strange fish.

And I've yet to discuss the strangest thing of all! As you've probably gathered, humans are far from mental giants. Their brains take years to master even simple ideas, such as relativity theory and quantum electrodynamics. Their memories are leaky, such that they rapidly forget most of the moments of their short and troubled lives. And they need intellectual prostheses, such as calculators and computers, to solve mathematical problems that Betelgeuseans solve unaided in mere picoseconds. Yet for all their intellectual foibles and frailties, these odd bio-robots have somehow come to possess knowledge far in advance of their powers. Needless to say, their knowledge is primitive compared to the glories of our own. But it still vastly outstrips their paltry intellects. How did a species whose brightest members can't even do differential calculus in their heads come to acquire such an intricate knowledge of reality? How did meat robots come to glimpse, however dimly, the nature of the Big Bang that birthed the entire universe, the evolutionary process that created them from dust, and the physical laws that govern all matter and energy? You're probably thinking: "It's obvious! These lowly creatures were visited by the Lokies of the Third Quadrant, long suspected of breaking Galactic Law and giving the sacred knowledge to primitive life forms." But the Lokies have watertight alibis spanning the last three thousand years, so that can't be it. Thus, the question remains: How did this hapless creature come to possess such an impressive body of knowledge? How did a mere ape come to understand the vast universe of which it is but a tiny, fleeting fragment?

Meeting the Alien's Challenge

As you can see, the alien scientist would be entirely confused by our species. Hopefully, having read its report, you're confused about a few things too. It might sound strange to say that we could be confused by ourselves; after all, we *are* ourselves, and surely if anyone's going to understand human beings, it's going to be us. But a moment's reflection reveals that this can't be right. History presents us with an endless parade of conflicting ideas about human nature and the human condition: the traditional Christian view that humans are fallen angels tainted by original sin and bound for heaven or hell; the Hindu view that we're karma-collecting souls living out one life after another; the Cartesian view that we're immaterial minds piloting mechanistic animal bodies; the materialist view that we're soulless arrangements of matter; Hobbes's view that we're violent and selfish creatures tamed by civilization; Rousseau's view that we're kind and peaceful creatures *corrupted* by civilization; Freud's view that we're conflicted beings driven by childhood traumas and powerful unconscious urges; Skinner's view that we're learning machines shaped almost entirely by reward and punishment – and so on. Logic dictates that not all of these ideas can be the whole truth and nothing but the truth. At the most, one can be. But none of them has been held by more than a fraction of the world's population. Thus, over the course of human history, *most* human beings must have held false beliefs about the nature of human nature and the causes of their own behavior. Being human is no guarantee that we'll understand human beings. Even in the age of science, we understand the motions of the distant, ancient stars better than we understand the organism observing them: ourselves.

Still, we may find ourselves at a unique moment in history: a moment in which it's possible for the first time to sketch out an explanation for human behavior and human culture that has at least a reasonable chance of being accurate. That's the aim of this book. Given that so many have failed before me – the bulk of humankind, to be precise – this might seem like a somewhat over-ambitious aim, if not a delusional one. But the answers I'll give are not mine alone. They're based on the cumulative efforts of thousands of philosophers, scientists, and psychologists working over many centuries. These thinkers and tinkerers haven't just been spinning their wheels. They've made some real progress. Since the mid-twentieth century in particular, science has made enormous strides toward answering our deepest questions about human nature – much more than most people realize. My job is to piece together some of the best answers, and to try to solve the mysteries highlighted in the Alien's Report.

The guiding assumption of the book is that the alien's questions can be answered using two broad theoretical frameworks, both of which are children of evolutionary theory. The first uses evolutionary theory to shed light on the human mind and behavior.[7] The second uses evolutionary principles to shed light on human culture.[8] The approaches go by various different names, but I'll refer to them as *evolutionary psychology* and *cultural evolutionary theory*. Between them, these approaches provide us with a toolkit of ideas and hypotheses to help us to answer the alien's questions.

The starting point for evolutionary psychology is the idea that human beings are animals, and that like all animals, we're products of natural selection. This is true of our bodies, but it's also true of our minds. To see what this means, we can start by looking at our pets: our cats, our dogs – even our pet snakes or tarantulas. As different as we are from these creatures, we also have a lot in common: We eat; we retreat from threats; we mate with members of our own species. Clearly, these are things we evolved to do. Equally clearly, however – at least in the case of the humans, cats, dogs, and snakes – conscious mental states play a crucial role in orchestrating these behaviors. Hunger motivates eating; fear and pain motivate retreat from threats; sexual desire motivates mating. These basic feelings, desires, and drives are just as much a product of natural selection as the behavior they help to generate. The same goes for the various general-purpose psychological abilities that we all possess and without which no behavior would be possible, including our basic sensory abilities, the capacity to learn and remember, and the capacity to move our bodies. Just like eyes and wings, teeth and claws, these elements of human psychology are adaptations crafted by natural selection.

At this basic level, no sane psychologist would deny that selection has helped shape the human mind. What distinguishes evolutionary psychologists from other psychologists is that they take this line of thought much further. Natural selection, in their view, doesn't just explain our most basic drives and abilities, or the things we obviously have in common with other animals. It also helps to explain many things that psychologists traditionally ascribed solely to learning, socialization, and culture. This includes various sex differences, a range of mate preferences, complex emotions such as love and jealousy, and the tendency to favor our relatives over unrelated individuals. For much of the twentieth century, psychologists failed to spot the evolutionary underpinnings of these phenomena, for two main reasons. First, they didn't know enough about other animals and thus didn't realize that similar traits and tendencies are found in other species. And second, they didn't know enough about evolution. To truly understand ourselves, argue evolutionary psychologists, we need to understand the process that created us. What did natural selection design us to do?

To cut a very long story very short, selection designed us and all other organisms to do one thing and one thing only: to pass on our genes to future generations. The rationale for this is simple: If you were a gene and you helped to build an organism that *didn't* pass on its genes, or that did so less competently than its neighbors, you would quickly disappear from the gene pool. The only genes that have staying power are those that help to build "gene machines" – organisms that act as if their one true goal in life is to make sure that their genetic endowment is passed on to as many new organisms as possible. The three main ways they do this are by staying alive, having lots of offspring, and helping their relatives to stay alive and have lots of offspring as well. Obviously, things get a lot more complicated in the human case. But the complications don't nullify the importance of the gene machine perspective. It's the foundation of the evolutionary approach to psychology.

As we'll see in later chapters, evolutionary psychology answers many long-standing questions about human nature and human behavior. Why are men the main consumers of action movies, pornography, and prostitutes, whereas women are the main consumers of romance novels and health advice? Why, when choosing sexual partners, are close relatives almost always off the menu? Why do we fall in love and why doesn't it always go smoothly? Why do women do the lion's share of child rearing? Why do men do any at all? (In most species, they don't.) Why is blood thicker than water? And why are people so much more cooperative than virtually any other animal on Earth? In each case, these tendencies are not merely products of learning or culture. They're woven into the very fabric of human nature.

Evolutionary psychology represents a profound shift in our view of ourselves and the wellsprings of our behavior. Many argue that it's the biggest paradigm shift in psychology since the cognitive revolution of the 1950s. Indeed, the biologist Richard Alexander once went as far as to describe the application of evolutionary principles to social behavior as "the greatest intellectual revolution of the century," on a par with relativity theory and quantum mechanics.[9]

But it's not enough!

It's enough for goats and jellyfish and hummingbirds, but it's not enough for us. The reason is simple: culture. Culture is our real gimmick as a species. It's not our big brains, our opposable thumbs, or our facility for finding new and improved ways of killing each other. It's our capacity for culture. Humans are as dependent on culture as we are on oxygen; without it, we're as naked and vulnerable as a crab without a shell. True, we're not the only animals with culture. But no other animal does it quite like us. Ten thousand years ago, the pinnacle of chimpanzee culture was using twigs to extract termites from termite mounds. Today, the pinnacle of chimpanzee culture is... using twigs to extract termites from termite mounds. Humans, on the other hand, went from Stone Age technology to Space Age technology in less than 10,000 years. The gaping chasm between our cultural achievements and chimps' has a lot to do with the fact that we're smarter than they are. But it has at least as much to do with the fact that human culture is *cumulative*: It advances through the accretion of a thousand little improvements, and then a thousand more. As soon as we evolved the capacity for cumulative culture – as soon as we opened that Pandora's Box – culture began to evolve in its own right, independently of biology.

To illustrate, consider Plato and Aristotle. Plato and Aristotle were almost certainly more intelligent than most people living today. And yet most people living today have a vastly more accurate view of the universe than these Ancient Greek philosophers. Indeed, most *preschool children* have a more accurate view, because most know that the world is a spinning rock orbiting a great ball of fire (to paraphrase the physicist Richard Feynman). In a certain sense, then, today's preschoolers are smarter than the greatest sages of the ancient world. This has nothing to do with biological evolution, and everything to do with our ability to stockpile knowledge and add to the common pool of knowledge over time. Of course, human biology makes culture possible in the first place. Our dogs and cats are no better informed than they were in Plato or Aristotle's day, despite being at our sides throughout the great expansion of our knowledge. But although culture is, in that sense, a biological phenomenon, it's one that has come loose of its biological moorings. Culture is a semi-autonomous evolving system, made out of ideas.

What drives the evolution of culture? How did we get from hunting and gathering in small nomadic bands to building skyscrapers and catapulting ourselves to the moon? Intelligence no doubt played a large role. But intelligence turns out to be a lot less important than we'd always assumed. Culture is shaped as well by the blind, mindless process of natural selection. As we'll see later, this applies to everything from languages to businesses and even to teddy bears.

To be clear, the claim is not that cultural evolution is a result of natural selection operating on genes. The claim instead is that cultural evolution is a result of natural selection operating on what the evolutionary biologist Richard Dawkins called *memes*: ideas, beliefs, practices, tools, and anything else that gets passed on via social interaction.[10] To understand cultural evolution, we need to understand which memes natural selection favors and which it discards – in other words, which memes survive in the culture and which go extinct. Often the memes that survive are those that benefit the individuals who hold them or the groups to which they belong. Sometimes, however, memes survive even when they *don't* benefit the individual or the group, purely because they're good at surviving. Buzzwords and irritatingly catchy tunes (also known as *earworms*) are innocuous examples of memes that survive despite not being good for us. As I'll argue in Chapter 6, other examples are not so innocuous.

The parallels between biological and cultural evolution run deep. I mentioned earlier that natural selection operating on genes gives rise to gene machines: organisms designed to pass on their genes. Natural selection operating on *memes* doesn't give rise to "meme machines" exactly, but it does do something similar. It gives rise to ideas and ideologies that, in effect, convert human gene machines *into* meme machines – that is, into beings that devote their time and energy to passing on their memes: their values, their religions, their love of modern art.[11] Again, the rationale for this is simple: Any memes that *didn't* somehow motivate their hosts to pass them on would quickly vanish from the culture, leaving only those that did.

What are we, then – gene machines or meme machines? The philosopher Daniel Dennett put it best. Human beings, argued Dennett, are *gene–meme hybrids*: morphs of the sometimes conflicting agendas of our genes and our memes. This is the perspective we'll adopt in this book in our quest to answer the alien's questions. To get the ball rolling, let's turn to the much loved, much maligned field of evolutionary psychology.

2

Darwin Comes to Mind

Man or *Homo sapiens*, as he somewhat arrogantly calls himself, is the most interesting, and also the most irritating, of animal species on the planet Earth.

—*Alien scientist from Bertrand Russell's*
Has Man a Future? *(1961), p. 7*

Super Freaks

To say that human beings are interesting is an understatement. We're freaks of nature! We're blobs of matter that fall in love with each other. We're mammals with the child-rearing patterns of birds. We're mortal beings that, alone among the animals, know that we're going to die one day and flee in terror from this knowledge. We're bald apes that can think each other's thoughts simply by making noises at each other. We're creatures designed by a cruel, amoral process which invent moral codes for ourselves and sometimes even live up to them. We're carnivores that sympathize with our food. We're biological mechanisms designed to pass on our genes, but which fritter away our time playing games and weaving a web of fantasy around ourselves. We're clusters of chemical reactions that contemplate deep truths about the nature of reality. And we're little pieces of the Earth that can get outside our mother planet and venture to other worlds.

How can we explain how such a bizarre creature came to exist? How, in other words, can we solve the riddles posed in the Alien's Report? This is a complex problem, and complex problems call for complex solutions. To make any real progress, we're going to need a range of tools from a range of disciplines. But we've got to start somewhere, and it makes sense to begin our exploration with the thing that makes love and knowledge and culture possible in the first place: the human mind. Where did this strange contraption come from? To answer this question, we need to start by making a lateral move. We need to leave the psychologists to their own devices for a while and pay a visit to the biology department. We need to get a firm grasp on one of the most important theories in all of science: evolutionary theory.

The Greatest Idea Anyone Ever Had

The year 1838 represents an important milestone for the planet Earth. For this was the year in which a tiny fragment of the Earth – a fragment known as Charles Darwin – answered an ancient and bewildering question: Why are we here? As the answer crystallized in his young mind, Darwin became the first life form in the history of the planet, and perhaps even the history of the universe, to understand how life had come to exist. The answer, said Darwin, is evolution. Humans didn't just pop into existence at the whim of an intelligent designer. We exist because we evolved.[1]

The basic claims of evolutionary theory are simple to state and as well-established as anything else in science. Species are not static; they change and evolve over time. Every species we share the planet with today evolved from an earlier species, which itself evolved from an earlier species, and so on and so on, back into the mists of time. As we trace back the genealogies of existing species, they soon begin to converge like the branches of a tree, traced from tip to trunk. Chimpanzees and bonobos share a common ancestor around a million years ago. Chimpanzees, bonobos, and human beings share a common ancestor around seven million years ago. And we're only just scratching the surface! *Any* two species share a common ancestor if you trace it back far enough: humans and tigers, tigers and goldfish, goldfish and toadstools. Ultimately, all life on Earth traces back to a simple, self-replicating molecule that appeared on the planet around four billion years ago.[2]

The implications of this view of life are astounding. One implication is that all life on Earth is one big, not-so-happy family. A trip to the zoo is literally a family reunion. In fact, even a trip to the *fridge* is a family reunion of sorts. Another implication – a particularly striking one, to my mind – is that any individual who chooses not to have children, or who has childlessness thrust upon them, brings to an end an unbroken chain of life that stretches back four billion years.

No one should feel bad about this, though. The fact is that the vast majority of organisms throughout all of Earth's history died before they ever reproduced, and in doing so, brought to an end their personal chains. And this stark fact turns out to be critical to understanding the process of evolutionary change. One of Darwin's central observations was that many more organisms are born, hatched, or seeded than could ever possibly survive to adulthood and reproduce. The question is, therefore: Which organisms become links in the great chain of life, and which become end points? It's partly a matter of luck; one puppy or grizzly bear gets struck by lightning, while its neighbor escapes unscathed. But it's not just luck. Some organisms happen to have traits which boost their chances of surviving and reproducing. Lions that have sharper teeth get more food and survive for longer than their blunt-toothed comrades; gazelles that have faster hooves evade the clutches of sharp-toothed lions more reliably than their sluggish running mates. In both cases, these higher achievers produce more offspring, and thus the traits that helped them to do so become more common in the population. As the generations fly by, lions' fangs get sharper and sharper; gazelles' hooves get faster and faster. In effect, nature functions as a giant animal breeder: It "chooses" which individuals get to reproduce and which don't, and over long periods of time, adapts organisms to their environments. Of course, no one actually chooses anything. Instead, some traits just happen to persist for longer than others, and these are the traits we tend to see around us in the world.

The process I'm describing, as I'm sure you know, is what Darwin dubbed *natural selection*. At first glance, it might not sound like a big deal. And over the short term, it really *isn't* a big deal. Evolution by natural selection usually involves imperceptible baby steps, rather than large leaps across the space of possibilities. But baby steps soon add up, and unlike an animal breeder, natural selection has all the time in the world. Just as a mere trickle of water, given sufficient time, can carve the Grand Canyon out of solid rock, so too natural selection, given sufficient time, can fashion new biological structures out of old. Not only does selection sharpen teeth and hasten hooves, it creates complex adaptations where once there were none. Indeed, natural selection explains where the teeth and hooves came from in the first place – along with eyes, and cocoons, and a billion other wonders. And not only does natural selection create adaptations, in the fullness of time, it carves out new species from the gene pools of existing ones.

The concept of natural selection is the single most important tool in our Darwinian toolkit. And the most important thing to remember about natural selection – a point we'll come back to again and again – is that natural selection creates an illusion of intelligent design, or what Richard Dawkins calls *design without a designer.*[3] Adaptations look as if they were invented by a conscious agent for a particular purpose: hands for grasping; eyes for seeing. But they weren't; Darwin stuck a pitchfork in that hypothesis. The design in nature comes not from a designer but from the mindless accumulation of favorable accidents over vast periods of time. The only genuine intelligent design found in the biological world comes from us, in the shape of our dogs and cows and other organisms which we've deliberately modified via selective breeding. Beyond that, the apparent intelligent design in nature is a forgery perpetrated by natural selection. Conscious intentions play no role.

In 1748, the philosopher Julien Offray de La Mettrie published a controversial book entitled *Man a Machine*. La Mettrie's thesis was that human beings, like all organisms, are soulless biological machines. Darwin and his intellectual descendants added the finishing touch to the thesis: We are soulless biological machines *designed by natural selection*. But what has natural selection designed us to do? Are we machines designed to perpetuate our species, or to survive, or reproduce… or what? This turns out to be a deceptively difficult question. The best way to answer it, I propose, is to adopt a hypothesis-testing approach. That's what we'll do over the next few sections. We'll work our way through a series of hypotheses, critiquing each in turn and moving closer with every step to the most accurate view on offer. We'll start at the bottom of the barrel.

What Are People For?

Our first hypothesis reflects a widespread view of evolution, according to which natural selection is all about perpetuating the species. People rarely offer an explicit argument for this position, but if they did, it might go something like this. Animals run away from predators, mate with members of their own species, and care for their own offspring. If they *didn't* do these things – if they fed themselves to lions, mated with rocks, or used their offspring for target practice – the species would go extinct. Thus, animals' normal behavior *prevents* the species from going extinct. In short, it's all for the good of the species. Here, then, is our first hypothesis.

Hypothesis #1: Evolution is about the survival of the species. Genes are selected if they help the species to survive. Adaptations, from sharp eyes to sharp teeth, are designed to protect and propagate the species. And organisms, considered as a whole, are species-survival machines.

I won't keep you in suspense. Evolution is *not* about the survival of the species. This is an important point, so I'll repeat it. Is evolution about the survival of the species? No, it's not. It is about something, but that something isn't the survival of the species. If you were to make a list of all the things that evolution is *not* about, the survival of the species would be on that list, probably near the top. Every year, I teach a course on evolutionary psychology, and before the final exam, I give tips for anyone who wants to *fail* the course. Tip number one? Say that evolution is about the survival of the species.

What's so wrong with the species-survival hypothesis? Simple: Natural selection works almost entirely *within* species. Lions with sharper teeth do better than other lions at capturing prey. Gazelles with faster hooves do better than other gazelles at evading lions. Sharp teeth and fast hooves evolved because they're good for their owners, not because they're good for the species. Certainly, these traits *are* also good for the species. If lions or gazelles didn't keep getting better at catching or evading each other, they could very well go extinct. But that's not why these traits were favored. We know it's not because, in some cases, traits are favored despite being *bad* for the species. The primatologist Jane Goodall discovered in the 1960s, much to her surprise, that male chimpanzees occasionally band together to kill members of other groups.[4] This gives the killers an advantage, gaining them territory and females. But it's hardly advantageous for the species. Chimps are endangered, and thus the last thing they ought to be doing is going around murdering each other. Nonetheless, as long as killer chimps do better than pacifists, violence will be the norm, even if it's suboptimal for the species as a whole. Bottom line: Evolution ain't about the good of the species.

But we've now got some clues to help us figure out what it *is* about. What do lions with sharper teeth, and gazelles with faster hooves, do better than other lions and other gazelles? Well, for one thing, they survive better. Perhaps, then, evolution is all about survival – not of the species but of the individual organism. If you think this sounds plausible, you're in good company. In his later years, Darwin often described natural selection as "the survival of the fittest" – a phrase he borrowed from the philosopher Herbert Spencer. And it's easy to think of adaptations whose sole function is to keep their owners alive and kicking. Cactuses and porcupines have defensive spikes. Nuts and turtles have protective shells. Monarch butterflies and Brussels sprouts harbor noxious chemicals that put off would-be predators. The list goes on. It's time to unveil our second hypothesis.

> **Hypothesis #2:** Evolution is all about the survival of the fittest. Genes are selected if they help their owners to survive. Adaptations are designed to facilitate the organism's survival, even sometimes at a cost to the species. And organisms themselves are survival machines.

This hypothesis is a hundred times better than the first. Nonetheless, if you gave this as an answer in my evolutionary psychology exam, you'd probably still fail. Survival is important, but it's not enough. Even if you lived for a thousand years, if you didn't have any offspring in that time, your genes would disappear from the gene pool the moment you died. If you want your genes to survive, you have to reproduce. And that's where things start to get interesting.

Darwin's Sexy Idea

If evolution were a movie, what kind of movie would it be? If you're like a lot of people, you might think it'd be a horribly violent action movie, its characters locked in a bloody struggle for existence. And to some extent, you'd be right. However, you would have overlooked another element which is at least as important as survival, and maybe even more so: sex. Nature isn't just a bloodbath; it's also a vast, unending orgy. To put it in more prosaic terms, reproduction is absolutely critical in evolution. Organisms have numerous adaptations that have nothing to do with survival and everything to do with making new organisms. The most obvious examples are the sex organs. Testicles and ovaries don't help us to ward off predators or find water in the desert, but they do help us to make babies. That's all they do; it's their sole function. The same is true of the uteruses of female mammals, the egg pouches of male seahorses, and the flowers of flowering plants. The genes contributing to the development of these various contraptions keep themselves in the gene pool purely by helping their owners to have offspring – offspring that have a good chance of inheriting those very genes.

This much is clear enough. But the idea that adaptations might be about reproduction rather than survival turns out to be a lot more illuminating than it first appears. For Darwin, it was the key to explaining certain biological structures that initially seemed to thumb their noses at his whole theory. The classic example is the peacock's tail. This oversized, colorful appendage is found only in the males, and has no clear survival value. On the contrary, rather than increasing the peacock's chances of survival, it increases its chances of death. The tail advertises the peacock's whereabouts to predators, like a giant billboard advertising fast food, and it then impedes the peacock's efforts to escape those predators, because it slows the peacock down. From a health-and-safety perspective, sporting this ostentatious organ is about as sensible as strapping dynamite to one's hind quarters. On top of that, it's a huge waste of resources. The energetic costs of growing a new set of rump feathers every year seem hard to justify. Why not devote that energy to foraging for food or looking for mates?

Darwin's brilliant solution was that the peacock's tail is not a survival organ; it's a *reproductive* organ.[5] It doesn't look like one, but it is. Its function is to attract females. Peahens prefer to mate with males whose tails are larger than average, more colorful than average, and more ornately patterned than average. Males fitting this description attract more female attention than less well-endowed males, and therefore have more offspring – offspring that tend to inherit their father's tail. The net effect is that the average peacock's tail gets larger, more colorful, and more ornately patterned with each new generation. Of course, the tail can't just keep growing forever. Instead, it reaches an upper limit at the precise point where the survival costs of having a large tail match the reproductive benefits of being popular with the ladies. Like every trait, the peacock's tail represents a trade-off between competing selection pressures.

The evolutionary explanation for the peacock's tail has an interesting implication, namely that the mind of one sex can help shape the body of the other. The peacock's tail was literally built by the peahen's sexual preferences – a strange kind of Darwinian mind over matter. But although this accounts for the tail, it raises a new question as well: Why did peahens evolve a preference for these unwieldy appendages in the first place? There are several theories, but the most popular proposes that the tail is a hard-to-fake signal of fitness.[6] To see what this means, consider the different ways that people can signal wealth. One is simply to say "I'm rich – really, really rich." But this signal is easy to fake; even a homeless person could utter those words. A more trustworthy signal would be one that *can't* be easily faked – owning a mansion, for instance, or lighting Cuban cigars with hundred dollar bills. Only the genuinely wealthy can do those things. According to many biologists, the peacock's tail is a Cuban cigar: Males can only grow a good one if they're genuinely "rich" – that is, if they're in good condition, have relatively few mutations, and are relatively immune to the local pathogens and parasites. In effect, males with the most impressive tails are announcing to the world that "I can afford to grow this useless ornament, and can keep myself alive despite having to lug it around. I am therefore a particularly fit and virtuous specimen of manhood."[7] And that, apparently, is what peahens most want to hear.

The evolution of the peacock's tail is an example of a process that Darwin called *sexual selection*.[8] Sexual selection occurs when a trait is selected not because it promotes an organism's survival, but because it promotes its reproductive success. Darwin saw sexual selection as something separate from natural selection; these days, however, most biologists view it as a subtype of natural selection. Either way, sexual selection is a powerful force in evolution. With their sexual preferences, females force males to dance or sing or hang upside down flapping their wings. Most of the beauty and color in nature comes from sexual selection: the scent and appearance of flowering plants; the rainbow plumage and melodious songs of many songbirds – and, according to some, the art, music, and humor of *Homo sapiens*.[9]

But sexual selection doesn't just manufacture gaudy sexual ornaments. It also manufactures weapons. These include everything from the antlers of deer and stag beetles to the flesh-seeking fangs of many male primates and the unicorn-like tusk of the narwhal. As with the sexual ornaments, sexually selected weapons are usually found only in males. Males use them to fight over females, or to fight over the territories and resources needed to win the females' affection. Thus, whereas sexual ornaments are products of mate choice (also known as *intersexual* selection), weapons are products of same-sex competition (also known as *intrasexual* selection).

Human beings are not exempt from sexual selection, but the situation is somewhat different for us. In most species, males compete for mates and females choose from among the competing males; as a result, males have ornaments and weapons, whereas females tend to be drab and less well armed. Occasionally, the roles are reversed, but it's not particularly common. In our species, however, and a handful of others, the asymmetries are far less pronounced. Both sexes are choosy about their mates, and both compete with members of the same sex for access to the best mates. As a result, human beings are a special case: a species in which males and females both have their equivalents of the peacock's tail, and their equivalents of the deer's antlers.[10] I'll say more about this later.

For now, though, the important point is this: The peacock's tail and the deer's antlers don't promote their owners' survival, but they do still have an evolutionary function. Their function is to boost their owners' reproductive success. With that in mind, let's return to our original question: What are organisms designed to do? An initial suggestion might be that they're designed not merely to survive, but to *survive and reproduce*. But that's not quite right, either. Selection often favors traits that increase the organism's reproductive success but lower its odds of survival; the peacock's tail is just one example. This suggests that survival isn't valuable in itself from an evolutionary perspective, but is valuable only to the extent that it makes reproduction possible. Reproduction, we might say, is the true currency in evolution. And with that, we're ready to frame our third hypothesis.

> **Hypothesis #3:** Natural selection isn't about the survival of the fittest; it's about the *reproduction* of the fittest. Genes are selected if and only if they help their owners to reproduce. Adaptations – including spikes, shells, and other survival devices – are ultimately designed to facilitate reproduction. And organisms, at their core, are baby-making machines.

You could probably pass my exam with this answer – but the hypothesis still needs tweaking. Having offspring is clearly important. However, even if you had a thousand offspring, if none of them survived to have offspring of their own, your genes would be culled from the gene pool just as surely as if you'd stayed childless. Technically, then, organisms are not designed simply to produce offspring; they're designed to produce offspring that *themselves* produce offspring. The way most species do this is simply by pumping out as many offspring as they can. It's a numbers game: Most of these offspring will perish long before reproducing, but a tiny fraction will make it to adulthood and successfully reproduce. Some species, however, take a different path to the same goal: They have fewer offspring but invest much more into each. Mammals are the masters of this strategy. Indeed, most mammals have specialized organs for investing in their progeny: nipples. But regardless of whether a species takes a quantity approach or a quality approach, the ultimate benchmark of evolutionary success is not the number of babies an organism has, but rather the number of *grandbabies*. Here's an amended version of our last hypothesis.

Hypothesis #4: Natural selection is all about grandchildren. Genes are selected if they increase their owner's chances of ending up with lots of grandchildren. Adaptations – from spikes to sex organs to nipples – are ultimately a means to the end of producing grandchildren. And organisms, in the final analysis, are grandchild-making machines.

Now we're getting somewhere! You could earn a reasonable mark with this answer. But you wouldn't be at the top of the class. There's still room to take the analysis several steps forward. To see how, though, we're going to need to dive into some deeper waters.

The Inclusive Fitness Revolution

According to our most recent hypothesis, natural selection favors traits that maximize the number of grandchildren that an organism ends up producing. This hypothesis probably explains most of the adaptations found in the natural world. But it doesn't explain them all. We know it doesn't because of a strange group of animals called the *eusocial* insects. Most eusocial insects belong to the order *Hymenoptera*, which includes ants, bees, and wasps. These diminutive creatures live in teeming metropolises consisting mainly of workers, plus a queen or sometimes two. The queens are fully fledged baby-making factories; for most of their lives, they do nothing but pump out eggs. The workers, in contrast, have no offspring of their own (except in rare circumstances), but instead spend each waking moment helping the queen to produce *her* offspring. Some workers even give up their lives in the process. Among honeybees, for instance, workers sting any creature that ventures too close to the hive, which protects the queen but also usually kills the worker. Likewise, among carpenter ants, soldiers sometimes blow themselves up to protect the colony, spraying a murderous gluey substance over any would-be invader. In both cases, these tiny insect martyrs have no offspring of their own, and thus no grandkids. How, then, does natural selection favor their self-sacrificial tendencies?

This question – an example of what later came to be called *the problem of altruism* – drove Darwin crazy. Like the peacock's tail, the eusocial insects threatened to topple his fledgling theory. Darwin did catch a glimpse of the solution, as did several other biologists over the course of the subsequent century. But the mystery wasn't solved in a rigorous way until the 1960s. The solution came from the great British biologist William D. Hamilton.[11] The key to the evolution of eusociality, argued Hamilton, is the fact that eusocial insect colonies are vast cities of relatives.

To see why this is important, consider the mammalian nipple. Nipples are designed to deliver a benefit to another organism at a cost to the nipple's owner. They're literally altruism machines. How could selection favor these self-sacrificial devices? The answer is that mammalian mothers usually breastfeed their own offspring. Because parents pass on half their genes to each offspring they have, their offspring are more likely than chance to share any given gene – including any gene that helped to build the mother's nipples. In effect, the nipple-building genes keep themselves afloat in the gene pool by probabilistically caring for copies of themselves in other bodies. Thus, the key to understanding the evolution of nipples, and parental care in general, is shared genes – or rather, *genes shared at above-chance rates*.

If that's what matters, though, why stop with offspring? Other relatives share genes at above-chance rates as well. In most species (though not, as it happens, ants, bees, or wasps), full siblings are exactly as closely related to each other as parents are to their offspring. They too are half-clones, to use the alien's lingo. (Identical twins are full-clones: Ignoring new, unique mutations, they possess identical genomes.) Because full siblings are so closely related, they're more likely than a randomly chosen individual to share any given gene. Thus, for the same reason that selection can favor traits that benefit offspring, it can also favor traits that benefit full siblings. Grandchildren, half-siblings, nieces, and nephews are only half as closely related as offspring and full siblings – they're quarter-clones, if you will. But they're still more likely than a randomly chosen individual to share genes, and thus selection can favor a tendency to benefit them as well. The same basic principle applies to all relatives.

This type of selection, in which genes proliferate by reaching out and caring for themselves in other bodies, is known as *kin selection*, and it's another important tool in our Darwinian toolkit. As I've hinted, it's also the solution to the puzzle of eusociality. Consider again the eusocial ants, bees, and wasps. The mystery, if you recall, is that workers in these species produce no offspring of their own, but instead help the queen to produce *her* offspring. How could the genes underpinning this behavior persist in the gene pool? To answer this, imagine that you yourself are a worker. The first thing to notice is that, when you help the queen, you're not helping just anyone. The queen, you see, is your mother. That means that all her offspring are your siblings. And because they're your siblings, they're more likely than chance to share any genes underpinning your altruistic tendencies. Now, most of the time, the queen's offspring are fellow workers, and thus they're ultimately evolutionary dead-ends, just like you. Sometimes, however, her offspring develop into reproductively active individuals – new queens or reproductively active males – which can go on to found new colonies. It's through these individuals that the genes underpinning your altruism can make their way into the next generation. The upshot is that, by helping the queen to reproduce, you and your fellow workers are indirectly helping to spread the genes that give rise to that tendency.

The logic applies even in extreme cases. When a soldier carpenter ant blows herself up to protect the colony, for instance, she's actually helping to propagate the genes that made her do it. Copies of those genes are sealed away inside the queen, either in the queen's own genome or in stored sperm from the soldier's long-dead father. They're never *expressed* in the queen, of course; it would be no good if the queen blew herself up. But the genes are still locked up inside her, like gold in Fort Knox, and thus when the soldier ant makes the ultimate sacrifice, she's protecting the very genes that made her do it. In effect, the queen sends those genes out in expendable emissaries – the soldiers – which blow themselves up on her behalf.

Exploding soldier ants, and non-reproductive workers in general, tell us something important, namely that selection doesn't just favor traits that result in grandchildren. What *does* it favor, though? Different thinkers have given different answers over the years, but one of the most influential came from William Hamilton. Selection, argued Hamilton, favors traits that maximize *inclusive fitness*. Inclusive fitness has two components: *direct fitness* and *indirect fitness*.[12] Direct fitness is the individual's reproductive success; it's classical Darwinian fitness as captured in Hypothesis #4. *Indirect* fitness is the individual's contribution to the reproductive success of *kin*, weighted for the degree of relatedness between individual and kin member. (Don't worry about the details at this stage; I'll come back to this in Chapter 5.) Inclusive fitness is the sum of direct and indirect fitness. According to Hamilton, selection favors traits that produce the highest inclusive fitness for their owners, whether through direct fitness, indirect fitness, or some combination of the two. This brings us to our next hypothesis.

> **Hypothesis #5:** Natural selection is all about inclusive fitness. Genes are selected if they enhance the inclusive fitness of their bearers. Adaptations are designed to maximize the organism's inclusive fitness. And organisms, taken as a whole, are inclusive-fitness machines.

This answer would earn you a high mark on my exam. Inclusive fitness is one of the most important ideas in evolutionary biology, and has brought to light much about the living world that we hadn't previously known. That said, however, inclusive fitness is also a somewhat cumbersome concept. It involves attributing to the organism a strange, abstract kind of fitness, consisting of its own Darwinian fitness plus little bits of the Darwinian fitness of others. There's a clearer, less awkward, and – to my mind – more accurate way to describe what's going on. To get to it, though, we're going to need a fundamental shift in our view of evolution.

The Gene's-Eye View

> Samuel Butler's famous aphorism, that the chicken is only an egg's way of making another egg, has been modernized: the organism is only DNA's way of making more DNA.
>
> —*E. O. Wilson (1975), p. 3*

> The individual himself regards sexuality as one of his own ends; whereas from another point of view he is an appendage to his germ-plasm, at whose disposal he puts his energies in return for a bonus of pleasure. He is the mortal vehicle of a (possibly) immortal substance – like the inheritor of an entailed property, who is only the temporary holder of an estate which survives him.
>
> —*Sigmund Freud (1914), p. 78*

Up until now, we've surreptitiously adopted what we might call an "organism-centered" view of evolution. We've asked: What property of the organism does natural selection maximize? Our best answer so far has been inclusive fitness. But in exploring this answer, we brushed up against an intriguing possibility: the idea that genes can reach out and look after themselves in other bodies. This suggests a whole new approach to the question. It suggests that, in the final analysis, natural selection is not about how well organisms do; it's about how well genes do. Hamilton himself expressed this view in 1963, a year before he outlined his inclusive fitness framework, writing that, for a gene, G, which helps shape a given behavior, "the ultimate criterion which determines whether G will spread is not whether the behaviour is to the benefit of the behaver but whether it is to the benefit of the gene G."[13] This was the first explicit statement of what's now called the *gene-centered* or *gene's-eye view* of evolution – my pick for the most important advance in our thinking about evolution since Darwin.

The gene's-eye view, like most good ideas, is the product of many minds working over many decades.[14] But it was Richard Dawkins who most lucidly laid out this view of life and the arguments for it.[15] Natural selection, argued Dawkins, acts on *replicators*. A replicator is any entity that makes copies of itself. Not all replicators are alike, and purely by chance, some are better at copying themselves than others. Those that copy themselves more rapidly and reliably than their neighbors tend to increase in frequency in the population of replicators, and ultimately to dominate it – that is, they tend to be selected. In contrast, those that copy themselves less rapidly and less reliably tend to decrease in frequency and ultimately to disappear – they tend to be selected against. Natural selection, from a Dawkinsian perspective, happens whenever a replicator increases or decreases in frequency, as a result of its intrinsic skill at copying itself (rather than just by chance). Through the relentless winnowing of replicator populations by selection's invisible hand, replicators evolve over time to get better and better at copying themselves.

When people first hear natural selection described this way, their immediate assumption is usually that the replicators in question are organisms: Organisms make copies of themselves, and organisms evolve to get better at doing so. But that can't be right, says Dawkins. Organisms *don't* make copies of themselves. Sure, humans make other humans. But Johns don't make Johns and Janes don't make Janes. If you have your mother's nose, it's not because her nose was copied directly onto your face. It's because you inherited the genes that produced your mother's noble nose, and those genes then built a new and equally noble nose for you – and would have done even if your mother had somehow lost her nose before you were conceived. Organisms are built from scratch, not copied. This has an important implication. As mentioned, natural selection happens whenever an entity increases or decreases in frequency by virtue of its ability to copy itself. As also mentioned, however, organisms don't copy themselves; each is a one-of-a-kind. This means that individual organisms can't increase or decrease in frequency (other than going from a frequency of zero to one when they're born, and of one to zero when they die). And that in turn means that organisms can't be the entities upon which natural selection acts: the immediate targets of selection. Nor can groups, species, or ecosystems, none of which makes copies of themselves, either.

If organisms aren't the immediate targets of selection, what is? What, in other words, is the replicator in terrestrial evolution? The answer, as perhaps you've guessed, is genes. Your mother's nose wasn't copied onto your face, but half her genes were copied into your genome, and half your father's genes as well. Because genes are copied, particular gene variants can increase or decrease in frequency in the gene pool. And that means that genes, rather than organisms or groups or species, are the immediate targets of natural selection.

Next question: Which genes does natural selection favor? If we take an organism-centered view, the answer is that selection favors genes that are good for the individual – not in terms of happiness, pleasure, or survival, but in terms of inclusive fitness. If we take a *gene*-centered view, on the other hand, the answer is that natural selection favors genes that are good for *themselves*. As Dawkins famously put it, selection favors "selfish genes" – genes that act as if their one goal in life is to increase their representation in the gene pool. More precisely, the metaphorical goal of the genes is to get themselves copied at a faster rate than rival versions of the same gene (rival *alleles*).[16] Often, genes achieve this by boosting the survival and reproductive success of their bearers – giving them sharper teeth, for example, or more attractive tails. Sometimes, however, genes achieve their goal by boosting the survival and reproductive success of *other* organisms: organisms that have a better-than-average chance of possessing copies of the very same genes. That's what's happening, for instance, when genes induce their owners to make milk for their young or to explode in defense of the nest. Notice that, in both these last cases, the genes involved are still behaving selfishly – they're acting to increase their own representation in the gene pool. But the *organisms* are *not* behaving selfishly; they're paying a price to help other organisms. This illustrates an important point, and one that's often overlooked, namely that selfish genes can sometimes give rise to altruistic individuals.

From what I've said so far, you might assume that the gene's-eye view and inclusive fitness theory are simply two different ways of construing

the same basic facts about the world: one in terms of organisms maximizing their inclusive fitness, the other in terms of genes maximizing
their own fitness. As Dawkins and others have argued, however, the
gene-centered view is actually the more accurate way of conceptualizing
things. Not only does it explain everything that inclusive fitness explains
(genes helping their owners; genes helping their owners' kin), it also
explains certain things that inclusive fitness cannot. This includes *segregation distorter genes*. In sexually reproducing animals, genes normally
have a 50:50 chance of getting copied into each sperm or egg, and therefore a 50:50 chance of ending up in each offspring their owner produces.
Segregation distorters are genes that cheat the system, giving themselves
a better than 50:50 chance of making the final cut. By biasing the coin
flip, these outlaw genes can get themselves selected over competing alleles
without having any discernible effect on the organism's inclusive fitness,
or even while lowering it.[17] This is inexplicable on the assumption that
selection favors genes that maximize the organism's inclusive fitness. But
it makes perfect sense on the assumption that selection favors genes that
maximize *their own* fitness. Thus, the gene's-eye view goes beyond inclusive fitness theory, explaining things the latter cannot. This suggests that
it paints a more complete and more accurate picture of the nature of the
evolutionary process.

It's difficult to overstate just how radical a departure this picture is
from the traditional organism-centered view. For one thing, it rewrites our
understanding of adaptations. From an organism-centered perspective,
adaptations are designed to enhance the organism's inclusive fitness. From
a *gene*-centered perspective, on the other hand, adaptations are designed
to propagate the organism's genes. To be more precise, adaptations are
designed to propagate the genes giving rise to them. Spikes and shells are
designed to propagate the genes giving rise to spikes and shells; peacocks'
tails are designed to propagate the genes giving rise to peacocks' tails;
and nipples are designed to propagate the genes giving rise to nipples.
Of course, for any particular individual, a useful adaptation benefits *all*
the genes in its genome, not just those that helped build the adaptation.
But across vast numbers of individuals and over vast periods of time,
the genes responsible for the adaptation come to share genomes with
essentially every other gene in the gene pool. The only genes they benefit
consistently and reliably are themselves. Thus, the ultimate effect of any
adaptation is to help to propagate only those genes that give rise to it.

As well as rewriting our understanding of adaptations, the gene's-eye
view rewrites our understanding of the place of the organism within the

evolutionary drama. In the traditional view, individuals replicate themselves, and they use genes to do it. In the gene's-eye view, *genes* replicate themselves, and they use *organisms* to do it. As Dawkins put the point, "A monkey is a machine that preserves genes up trees; a fish is a machine that preserves genes in the water."[18] Monkeys, fish, and all organisms are temporary life-support systems for their potentially immortal genes – the means that genes have "invented" to preserve and propagate themselves. Thus, evolution is not about organisms getting better at copying themselves. Evolution is about genes getting better at copying *them*selves. As a side effect, organisms get better at *copying genes*. And with that, we arrive at our final hypothesis.

> **Hypothesis #6:** Evolution is about the survival of the fittest genes. Genes are selected if they get themselves copied faster than rival alleles. Adaptations are designed to pass on the genes giving rise to them. And organisms are not survival machines, baby-making machines, grandchild-making machines, or even inclusive fitness machines. Organisms – from worms to groundhogs to humans – are *gene machines*: biomachines designed to propagate their hereditary material.

To avoid any misunderstandings, I don't mean that propagating your hereditary material is what you *should* do necessarily; you might have other ideas about what to do with your brief time on this planet. But just as a toaster used as a doorstop is still a machine designed to toast bread, you – whatever you choose to do with your life – are still a machine designed to propagate your genes. All of us are. It's what the priests, the sages and philosophers searched for in vain: the ultimate explanation for our existence.

For the Good of the Group?

There's one last complication to consider. The complication is called *group selection*. This is the idea that traits can sometimes be selected because they're good for the group, rather than good for the individuals possessing them. Group selection has a long history in evolutionary biology. In his 1871 book *The Descent of Man*, Darwin argued that, over the course of human evolution, groups of sympathetic, loyal, and brave individuals did better than groups of uncaring, disloyal cowards, and thus that the former virtues were selected over the latter vices. They were selected, according to Darwin, not because the *individuals* possessing them were more successful, but because their groups were. Nearly a century later, the zoologist V. C. Wynne-Edwards added flesh to the bones of this idea, arguing that animals routinely put aside their selfish interests for the good of the group, and that group-level selection accounts for this propensity.[19]

Group selection seems plausible enough at a casual glance. Among the experts, however, the entire subject is mired in controversy. Ever since George C. Williams launched a full-frontal assault on the concept in the mid-1960s, most biologists have flat-out denied that group selection could work.[20] Their main argument has been that organisms that sacrificed their own fitness for the good of the group would leave fewer descendants than those that put themselves first. The classic example concerns the cute little rodents called lemmings. Every now and then, lemming populations get out of hand. Left unchecked, the lemmings would burn through their resources, destroy their habitat, and ultimately wipe themselves out. It's widely believed that, in order to avoid this dire outcome, surplus lemmings magnanimously throw themselves off any nearby cliff, killing themselves for the good of the group. But could such a tendency evolve? The answer is no; the suicidal lemming is a zoological urban legend – and it's easy to see why. Imagine we start with a group of non-suicidal lemmings. One day, a mutant is born with a built-in tendency to throw itself off a cliff as soon as the population starts getting too large. Would this tendency be selected? Clearly not. By cutting short its life, the death-loving lemming would leave fewer offspring than its less civic-minded counterparts. Its self-sacrificial inclination, as admirable as it might be, would be flushed straight out of the gene pool. And even if such an inclination did somehow gain a foothold – if, for instance, our alien scientist genetically engineered a population of suicidal lemmings – the tendency wouldn't last long. Sooner or later, a selfish mutant would be born, or a selfish migrant show up, that was unwilling to throw itself off the cliff and instead stood aside and let other lemmings make that sacrifice. All else being equal, this free rider would have more offspring than the rest, and eventually the free riders would crowd out the altruists and take over the population. Selfishness would beat group-directed altruism.

Based on arguments like this, group selection was largely expunged from evolutionary biology during the second half of the twentieth century, and consigned to the dog house of bad ideas. Since the 1990s, however, biologists have started looking at the issue again, and some think that group selection might be a viable proposition after all.[21] The name most closely associated with group selection today is David Sloan Wilson. According to Wilson, we need to distinguish between two modes of selection: selection operating between individuals within a group, and selection operating between groups within a larger population. Within groups, selfish individuals do better than altruistic ones, just like the free-rider lemmings. At the same time, though, *groups* of altruists tend to do better than groups of self-interested individuals; they work together better and burn up fewer resources fighting among themselves. Thus, within-group selection favors selfishness, whereas between-group selection favors altruism. More precisely (and less happily), between-group selection favors within-group altruism coupled with antagonism toward other groups. Which selection pressure will win this Darwinian tug of war: within- or between-group selection? It all depends on the relative strengths of the countervailing selection pressures. Often, individual selection triumphs. In certain circumstances, however, group selection can get the upper hand. This, in Wilson's view, is exactly what's happened in our species.

People usually assume that group selection is inconsistent with the selfish gene approach, perhaps because Richard Dawkins (who came up with the phrase "selfish gene") is also one of the leading critics of group selection theory. But the assumption is mistaken. If group selection really happens, it involves the selection of genes that, through their effects on their owners, benefit the group as a whole. If these genes are selected over competing alleles then, by definition, they are selfish genes – genes that promote their own continuance in the gene pool. The only difference is that they do this by furthering the interests of the group, rather than the interests of their owners. As such, group selection is, in principle, consistent with the gene machine approach.

But does it actually happen? This is a thorny question, which we'll come back to in Chapter 5.[22] For now, let's just say that, if the group selectionists turn out to be right, group selection becomes another tool that we can use to tackle the alien's questions. We might argue, for instance, that groups of altruistic, religious, artistic, and musical individuals did better than groups of selfish, atheistic, aesthetically challenged, or tone-deaf ones, and that that's why the former traits prevailed. I'm not at all persuaded that this is the case; indeed, as I'll explain in Chapter 5, I think it's probably not. But to cover our bases, here's an optional, modified version of Hypothesis #6, which leaves the door open to group selection.

Hypothesis #6.2: Genes are selected to the extent that they propagate themselves in the gene pool. Often, they do this by helping their owners to survive and reproduce, or by helping their owners' *kin* to survive and reproduce. Sometimes, however, they may do so by helping their owners' *groups* to do better than other groups, even at some cost to their owners. Either way, adaptations are designed to pass on the genes giving rise to them. And human beings, along with all other organisms, are gene machines.

So, there we have it. The gene machine approach provides, I hope, a useful tool for understanding modern evolutionary theory, or what's often called the *neo-Darwinian synthesis*. I freely admit that it bypasses a lot of complexity. Organisms are gene machines only to the extent that they're shaped by their genes, and only to the extent that the genes in question got there via natural selection rather than random genetic drift. Some genes rise to prominence not because they're useful, but because they happen to be linked to other genes that are. And the scope of selection is limited by developmental biases and constraints. Overall, though, the gene machine approach provides a valuable way to understand the big-picture implications of the neo-Darwinian view of evolution – the view that, despite occasional confident proclamations to the contrary, is still the reigning paradigm in evolutionary biology.[23] Importantly, however, the gene machine theory doesn't just apply to the body. It applies as well to the mind.

Evolving Minds

It takes ... a mind debauched by learning to carry the process of making the natural seem strange, so far as to ask for the *why* of any instinctive human act. To the metaphysician alone can such questions occur as: Why do we smile, when pleased, and not scowl? Why are we unable to talk to a crowd as we talk to a single friend? Why does a particular maiden turn our wits so upside down? The common man can only say, "*Of course* we smile, *of course* our heart palpitates at the sight of the crowd, *of course* we love the maiden, that beautiful soul clad in that perfect form, so palpably and flagrantly made for all eternity to be loved!"

And so, probably, does each animal feel about the particular things it tends to do in presence of particular objects ... To the lion it is the lioness which is made to be loved; to the bear, the she-bear.

—*William James (1890), p. 387*

We've finished our tour of the evolutionary biology department. It's time to take the knowledge we acquired there back with us to the psychology department, and see what we can do with it there. That, in a nutshell, is the mandate of evolutionary psychology: to take theories from evolutionary biology and use them to illuminate our understanding of the human mind and behavior. At a very general level, this is an easy enough assignment. Evolutionary theory tells us that organisms are machines designed to propagate their genes. This implies that, to the extent that the mind is shaped by natural selection, the mind must be a mechanism designed to propagate its owner's genes. This idea – that the mind is a gene-propagating mechanism – represents a rather spectacular break from traditional, everyday views of the mind. But it's also somewhat vague. To really get to grips with the evolutionary approach to psychology, we need to zoom in on the details.

Recall that natural selection creates an illusion of intelligent design: It creates adaptations that look as if they were designed to perform a particular function, despite having no designer. This means that whenever we come across an adaptation, we can perform what's called a *functional analysis*. We can ask: "What is the purpose of this adaptation? What's it *for*?" Here's how we might apply a functional analysis to some of the anatomical features discussed already.

Question: What is the function of spikes and shells?
Answer: to help protect the organism from harm.

Question: What is the function of the sex organs?
Answer: to enable reproduction.

Question: What is the function of female nipples?
Answer: to provide sustenance for newborn offspring.

You get the idea. What evolutionary psychologists do is take this explanatory framework and apply it to the human mind. For example…

Question: What is the function of hunger?
Answer: Hunger motivates us to obtain the nutrients we need to build and run the body. It has the same function as the warning light that comes on in a car when the car gets low in fuel.

Question: What is the function of disgust?
Answer: Disgust motivates us to avoid infectious substances and toxins. In effect, the disgust system is border control for the body – or, as the psychologist David Pizarro put it, a built-in poison detector.

Question: What is the function of fear?

Answer: Fear motivates us to escape or avoid danger and harm: to run away from the lion or steer clear of the edge of the cliff.[24]

Question: What is the function of pain?

Answer: Pain motivates us to protect ourselves from tissue damage: to get the mousetrap off your finger, for instance, or to pull your fingers out of the fire before it turns them into ash. Pain also motivates us to protect an injury while it heals.

Question: What is the function of lust?

Answer: Lust motivates us to engage in certain activities which, until the advent of reliable contraception, resulted in the production of babies.

Question: What is the function of parental love?

Answer: Parental love motivates us to look after our babies and children, so that one day, they can start the whole cycle again.

Notice that not only does a functional analysis work for both physical and psychological traits, but the same selection pressures that shaped the former can also shape the latter. Roughly speaking, fear and pain were shaped by the same selection pressure that gave rise to spikes and shells; lust was shaped by the same selection pressure that gave rise to sexual organs; and parental affection was shaped by the same selection pressure that gave rise to the mammalian nipple. As we'll see later, humans may also possess psychological equivalents of the peacock's tail and the carpenter ant's propensity to blow itself up – in other words, psychological adaptations designed to attract mates and to facilitate the care of relatives other than offspring. In short, body and mind were fashioned by the same Darwinian forces.

Something else to notice is that, just as the eye is exquisitely designed to enable vision, so too our psychological adaptations are exquisitely designed to execute *their* evolved functions. Consider disgust. People everywhere find certain items disgusting. These include bodily secretions (vomit, diarrhea), rotting materials (decomposing carcasses, spoiled food), and certain animals (such as rats, flies, and maggots). Importantly, the list of universal disgust elicitors compiled by psychologists is virtually identical to the list of disease vectors independently compiled by epidemiologists – exactly as we'd expect if the disgust system were an adaptation designed to nudge us away from harmful microbes.[25]

As well as our built-in aversions, humans have the capacity to learn new ones. Everyone's had the experience of consuming a novel food or beverage, getting violently ill soon afterward, and then discovering, the next time they encounter the food or beverage, that they simply can't stand the thought of it. They've developed what psychologists call a *conditioned taste aversion*. The capacity to develop such aversions is common among animals that eat a wide range of foods, and its adaptive logic is clear: The food might have caused the illness, so best not to eat it again. Admittedly, the food and the illness are often unrelated, and thus the food may be perfectly safe. But although this sounds like a bug in the system, it's actually probably a design feature. The fitness costs of dying from eating poisoned food are much greater than the costs of shunning a safe meal and going hungry once in a while. As such, the system is set up to err on the side of caution. In this way, and in many others, the disgust system is well-engineered to carry out its evolved function.

So too is the fear system.[26] As with disgust, humans come factory-equipped with a proneness to certain fears, coupled with a capacity to learn new ones. Among the fears we seem innately disposed to acquire is the fear of snakes. Snakes have been persecuting primates for as long as primates have walked the earth (or, more often, "swung the trees"). It's no surprise, therefore, that we develop a fear of snakes much more readily than we develop a fear of, say, bagpipes.[27] And when I say "we," I don't just mean human beings. Many primates are snake-phobic; it's a common feature of the primate psyche.[28]

Another species-typical human fear is the fear of heights. Not all animals are burdened with this fear; if you've ever seen a spider wandering around the outside of a skyscraper, you know that spiders are not the slightest bit afraid of heights. They don't need to be; they're so small they can fall safely from any height. When we fall, in contrast, we go splat. Thus, a wariness of heights makes sense for us, and for other large animals – and that's presumably why we've got it. We can learn to be *less* scared, but most of us start out with a basic level of apprehension.

Of course, our proneness to such fears wouldn't be remotely useful if fear didn't help us to deal with the things that we're afraid of. Unsurprisingly, then, this ancient emotion has a range of features that fit it to that task. For one thing, fear involves physiological arousal, which readies the body for action (the famous fight-or-flight response). In addition, fear involves the adaptive allocation of attention. When we're running away from a lion, we tend not to get distracted and forget what we were doing. One of the only times that people truly live in the moment is when their lives are in danger. Once again, this makes good adaptive sense; it is, after all, one of the only times when we truly need to.

One more point: As soon as we start applying functional explanations to the mind, it's important to draw a clear and careful distinction between *proximate explanations* and *ultimate explanations*. A proximate explanation is one that focuses on the immediate causes of behavior. An example would be "People have sex because they enjoy it." An ultimate explanation is one that focuses on the evolutionary function of behavior: the effects for which the behavior was selected. An example would be "People have sex because sex results in the production of offspring." These explanations – enjoyment and reproduction – are not inconsistent with one another. Indeed, the latter explains the former; we have sex because we enjoy it, but we evolved to enjoy it because sex results in the production of offspring. The reason the proximate/ultimate distinction is important is that people make a pastime of mistaking ultimate evolutionary explanations for everyday psychological ones. Critics of evolutionary psychology are among the worst offenders. A fairly standard criticism of the field goes like this: "Evolutionary psychologists claim that we have sex in order to have children, but that's just not true – most of the time we have sex just for fun. In fact, often the *last* thing we want when we have sex is children!"[29] From what I've said already, it should be clear that this is simply a misunderstanding. When evolutionary psychologists argue that sex is about making babies, they're talking about the evolutionary function of the behavior, not what people want. The critics have failed to distinguish the ultimate from the proximate, the evolutionary mode of explanation from the psychological.

The generalized form of this error is the idea that, according to evolutionary psychologists, people have an innate motivation to pass on their genes, and that we're all constantly scheming about how we might achieve this. As the Harvard psychologist Steven Pinker points out, though, "If that's how the mind worked, men would line up outside sperm banks and women would pay to have their eggs harvested and given away to infertile couples."[30] Rather than having a very general motivation to propagate our genes, humans have a portfolio of more specific motivations – motivations to eat and drink, to run away from predators, to have sex and care for our young. Collectively, these lead us to act *as if* we're trying to propagate our genes, but without any strategizing on our part and without us having gene propagation as an actual, literal goal. To be more precise, our basic drives and motivations led *our ancestors* to act in ways that typically propagated *their* genes *in the environment in which our species evolved*. These motivations may or may not accomplish this goal in our current environment. I'll say more about this soon. At this stage, the thing to remember is that, although we're gene machines, we don't have a built-in motivation to pass on our genes, at either a conscious or an unconscious level.

Leaving the Comfort Zone

So far, we've looked at evolutionary explanations for just a few components of the human mind: hunger and disgust, fear and pain, lust and parental love. These examples are, I hope, uncontroversial. Evolutionary psychology, however, *is* controversial, and that's because evolutionary psychologists take this mode of explanation a whole lot further, applying it to things that many would rather chalk up to social forces and learning. This includes the fact that, on average, men are more interested than women in uncommitted sex with many partners, and that women are choosier about who they sleep with. It includes the fact that people prefer sexual partners who have healthy skin and symmetrical faces, and that men tend to put more weight than women on good looks and youthfulness in a mate. It includes the fact that people fall in love and get jealous if the person they're in love with gets involved with someone else. It includes the fact that people generally love their own children more than the next-door neighbor's children. And it includes the fact that most people are repulsed by the idea of mating with siblings or other close kin. According to evolutionary psychologists, these tendencies are not merely inventions of culture. They're part and parcel of human nature.

To some, this might sound like common sense. But in the halls of academia, at least until recently, the idea has been considered preposterous and perhaps even dangerous. For much of the twentieth century, psychologists and other social scientists attributed anything more complex than basic sensory capacities, unlearned reflexes, and a handful of simple emotions to learning, socialization, and culture. If men are more interested than women in no-strings-attached sex, it's because society encourages men to sow their wild oats but represses women's sexuality. If people find some physical attributes more sexually enticing than others, it's because they've been exposed to too many underwear ads or too many underweight models. If people are possessive of their lovers, it's because they've been socialized in a capitalist society to treat people like property, rather than to share and share alike. If people would rather care for their own children than care for all children indiscriminately, it's because they've internalized bourgeois family values. And if people are averse to mating with close kin, it's because the surrounding culture has somehow imparted this aversion. Where else could it come from? Every educated person knew that there were cultures out there somewhere in which everything was different and better... although most people could never name any. The tacit assumption was that human beings are "blank slates": instinct-free animals capable of learning essentially anything. As the anthropologist Ashley Montagu put it, "Man is man because he has no instincts, because everything he is and has become he has learned, acquired from his culture, from the man-made part of the environment, from other human beings."[31]

From this perspective, which the evolutionary psychologists John Tooby and Leda Cosmides call the *Standard Social Science Model* (or *SSSM*), biologists can contribute precisely nothing to the understanding of the human mind or behavior. And, to paraphrase Wittgenstein, whereof they cannot speak, thereof they must remain silent.[32]

For many decades, most academics acceded to this unwritten rule. But then sociobiology and evolutionary psychology arrived on the scene, like skunks at a party, and started challenging the old order. Certainly, to some degree, humans really are blank slates: We rely more on learning than does any other creature. Is it realistic, though, to think that everything beyond fixed reflexes and basic drives is purely a product of learning? Could we *just as easily* socialize women to be more interested in casual sex and pornography than men, and men to be more broody? Could people *just as easily* learn a preference for unhealthy skin as healthy skin, asymmetrical faces as symmetrical ones, or elderly mates as youthful ones? Could we just as easily learn to be overjoyed about our partners cheating on us as to have our hearts torn into a thousand pieces? Could we just as easily establish societies where children are cared for communally as ones where people care preferentially for their own offspring? Could we just as easily learn to fall in love with siblings as with equally attractive non-relatives? If your answer to these questions is no, then you agree with the evolutionary psychologists: These things are not shaped exclusively by learning and culture. No doubt, learning and culture influence them to a greater or lesser extent. But some things come more naturally to us than others. In that sense, there's such a thing as human nature.

The above argument is really just an appeal to intuition. Before we can take it seriously, we need to do a number of things. The first is to say *why* these traits might have evolved. Why do we have the nature we do, rather than the nature of a lion or a dung beetle? It's easy enough to understand the evolutionary functions of fear, lust, and disgust. But what are the functions of mate preferences, sexual possessiveness, and an aversion to incest? This is where the gene machine theory really starts earning its keep. If you want to understand a tool or a machine, it helps to understand what it's designed to do. The same applies to the mind. Pinker made this point well in his book *How the Mind Works*:

> In rummaging through an antique store, we may find a contraption that is inscrutable until we figure out what it was designed to do. When we realize that it is an olive-pitter, we suddenly understand that the metal ring is designed to hold the olive, and the lever lowers an X-shaped blade through one end, pushing the pit out through the other end. The shapes and arrangements of the springs, hinges, blades, levers, and rings all make sense in a satisfying rush of insight. We even understand why canned olives have an X-shaped incision at one end.[33]

In the same way, when we understand what the *mind* is designed to do – namely to pass on the genes giving rise to it – various aspects of mind and behavior suddenly make sense in an equally satisfying rush of insight. Why are more men than women interested in having multiple sex partners? Maybe it's because the best strategy for passing on one's genes differs for men and women: A man who has, say, five sexual partners in a year could potentially have five bouncing babies; a *woman* who has five sexual partners in a year would have no more babies than she would if she'd only had one partner.[34] Why are we attracted to people with healthy skin, symmetrical faces, and youthful appearance? Maybe it's because these traits go hand-in-hand with health and fertility, and therefore people with these preferences leave more descendants than those with a fetish for infertile, unhealthy mates. Why are people possessive of their mates and lovers? Maybe it's because jealousy motivates people to prevent their mates from straying, thereby increasing the chances of having and raising children with them – and for men, because it also decreases the chances of ending up unknowingly raising someone else's child.[35] Why do people care more for their own children than the next-door neighbors' children? Maybe it's because our own children carry more of our genes than the next-door neighbors' do (unless, of course, there's been some hanky-panky... and the gene machine theory might shed some light on that, too).[36] Why are we repulsed by the prospect of mating with close kin? Maybe it's because the offspring of incestuous liaisons are more likely to have genetic defects, and thus that any genes that discourage incestuous mating have a very good chance of persisting.[37]

We see, then, that there are plausible evolutionary explanations for traits that were formerly explained solely in sociocultural terms. The next question is how we can choose between the evolutionary explanations and the traditional sociocultural ones. By way of an answer, consider a simple fact about the human condition, confirmed every day in family homes around the world: Parents can bribe their children with candy but not with broccoli. Why is this? Is it because children have been socialized to prefer candy, but could just as easily have been socialized to prefer broccoli? Presumably not. If anything, parents try to socialize their children to prefer the broccoli, only to discover that their progeny are much less malleable than they'd hoped. Furthermore, our sweet tooth isn't just an idiosyncratic preference of Western cultures. People everywhere love sugar-rich food: fruit, honey, honey ants, pop tarts. And it's not just people! *Most* primates have a sweet tooth. Most cats, on the other hand, don't. Lions couldn't bribe their cubs with broccoli *or* with candy. That's because lion nature differs from human nature. Different animals evolve different natures to fit them to their different ways of life.

Again, I hope this example is uncontroversial. But similar arguments can be mounted for other, more contested claims about human nature. First, just as children's sweet tooth survives even in the face of parental efforts to get them hooked on broccoli, so too sex differences, mate preferences, jealousy, nepotism, and incest aversion all survive even in the face of strong social pressure to eradicate them. In the last century or so, for instance, people experimented with gender-neutral parenting, free love communes, and communal child-rearing arrangements, but none of these experiments had more than limited success, and most were dismal failures. In contrast, things that are obviously not evolutionary products – fashions in hair and clothing, for example – change like the wind.

Second, contrary to rumors spread by social scientists, the phenomena under discussion transcend cultural boundaries. Despite strenuous efforts spanning more than a century, anthropologists have yet to unearth a culture where the women go to war while the men take care of the kids, where people are just as attracted to the elderly as to those in the prime of life, where people are indifferent to their spouse's "extracurricular" sexual activities, where there's no special bond between parents and their biological offspring, or where brothers and sisters routinely marry one another and have successful marriages. If you've heard otherwise, I'm afraid you've been misled – or so I'll argue in later chapters.

Third, and to my mind most persuasively, many of the traits we're discussing are found in other animals – animals that resemble us in evolutionarily relevant ways. So, for example, in mammalian species where males are larger than females, males also tend to be more interested in sexual variety for its own sake, more prone to physical violence, and less involved in childcare. In species where individuals choose their mates, they usually show a proclivity for sexual partners displaying indicators of health, vigor, and good genes: healthy red faces or feathers, symmetrical tails, signs of fertility. In species where individuals routinely interact with both kin and non-kin, kin favoritism is ubiquitous – indeed, it's even found in plants and slime molds. In pair-bonding species, couples commonly exhibit analogues to human jealousy: Among gibbons, for instance, males chase away rival males and females chase away rival females. And in a wide variety of species, individuals avoid mating with close kin. When we find these tendencies in other animals, we don't hesitate to explain them in terms of natural selection. When we then find exactly the same tendencies in human beings, the most parsimonious and plausible explanation is that they're products of selection for us as well.

Of course, we should always be open to the possibility that, in our species uniquely, we have the same traits and tendencies but for an entirely different reason: learning, socialization, or culture. We should always insist, however, on strong evidence before accepting this biologically improbable scenario. Until that time, natural selection is the most reasonable explanation for the traits we share with other animals.[38]

The Mating Mind

We've seen that natural selection can build psychological equivalents of shells and spikes, sexual organs and nipples. But can it build psychological equivalents of the peacock's tail? In other words, can it build psychological traits designed to attract sexual partners? According to the evolutionary psychologist Geoffrey Miller, it can. In Miller's view, many of the traits that most strikingly distinguish our species from other animals are not survival tools but mating ornaments, comparable to the peacock's tail, the mandrill's multi-colored face, and the bright red rump of the baboon. The clearest-cut examples are traits like art, music, and humor, which Miller argues evolved purely as mating displays and have no other evolutionary function. But a number of other traits, including language, intelligence, and morality, are multi-functional adaptations: They're part-survival tool but also part-mating display. According to Miller, human beings are exceptional in the extent to which our minds have been fashioned by sexual selection, just as peacocks are exceptional in the extent to which their tails have. "Sexual selection," he writes, "transformed a small, efficient ape-style brain into a huge, energy hungry handicap spewing out luxury behaviors like conversation, music, and art."[39] From this perspective, our minds are less survival devices than they are sexually selected entertainment systems. Many of the traits that we view as our noblest and most elevated turn out to have a much more down-to-earth purpose: They're designed, at least in part, to charm prospective mates into bed.

This is a novel hypothesis, to say the least. Until recently, says Miller, evolutionary theorists tied themselves in knots trying to find survival benefits or group benefits in phenomena such as art, music, and humor. As a result, they either came up with far-fetched adaptationist hypotheses, or concluded that the traits in question were not adaptations at all, but rather social constructs. In Miller's estimation, this was one of the biggest mistakes of twentieth-century evolutionary biology. It stemmed from the assumption that adaptations must be efficient and sensible survival tools, when in fact they're often fun and frivolous sexual displays. If Miller is right, then the early evolutionary theorists were like peacocks who set about trying to find the survival advantages conferred by their unwieldy tails, and then – when they couldn't find any – concluded that their tails must be social constructs. This is not a trivial mistake.

Miller's approach is controversial, but it's worth taking seriously. As the peacock's tail reminds us, mate preferences are not merely products of evolution; they're *causes* of evolution as well. And if mate preferences can shape our bodies, there's every reason to think that they can also shape our minds – after all, psychological traits, like physical ones, are almost always partially heritable.[40] Here's an example of how it could work. Men and women prefer mates who are intelligent and funny. In our ancestral past, these preferences may have created a selection pressure for greater intelligence and a better sense of humor, in exactly the same way that the peahen's mate preferences created a selection pressure for larger, glitzier tails. Over many generations, our preferences may have driven the evolution of our big brains and our funny bones. Simple! Of course, this then leaves us with a new question: Why would people have these particular preferences in the first place? One possibility is that the preferences are arbitrary: It could have been anything; it just happened to be intelligence and humor. But Miller favors another hypothesis. In his view, intelligence and humor – along with art and music and various other cultural displays – are Cuban cigars: hard-to-fake indicators of fitness. Just as it's hard for peacocks to grow an attractive tail, it's hard for human beings to grow a brain capable of these frivolous feats. Individuals who are up to the task must have especially good genes. That's why a sexual preference for these traits is adaptive. It gives your offspring a better genetic start in life.

Many psychologists – including even many evolutionary psychologists – have misunderstood Miller's theory, assuming the idea is that men attempt to charm women into bed, and that women simply give the green or red light. But that's not the claim. The claim is that behaviors such as art, humor, and creative displays of intelligence evolved in a context of *mutual mate choice*: Ancestral men and women both attempted to charm one another with these talents (albeit men more fervently), and both were choosy about their long-term mates (albeit women somewhat more so than men). As we'll see in Chapter 3, this is because both sexes, not just females, typically helped to care for their young.[41]

Having come this far, people often have a number of misgivings about the evolutionary psychologists' project. One common misgiving runs roughly along these lines:

> Evolutionary psychologists argue that everything people do is aimed at spreading their genes. But if that's the case, why do we use birth control? Why we do we spend so much time watching porn? Why do we adopt unrelated children? Why do we lust after food that makes us fat and wrecks our health? Why do we eat and smoke and work ourselves into an early grave? Why do we take mind-altering, body-punishing drugs? And why is mental illness so common? Doesn't all of this go against evolutionary psychology?

The answer is no, and the reason is simple: *Evolutionary psychology does not claim that everything people do is aimed at spreading their genes.* The idea that it does is a common misunderstanding of the field. It's an understandable misunderstanding, given the idea that organisms are gene machines. But it's a misunderstanding nonetheless. We need to draw a sharp distinction between the claim that organisms are designed to pass on their genes (which is the gene machine view) and the claim that *everything an organism does* is designed to pass on its genes. The latter would imply that every behavior is adaptive. However, not only do evolutionary psychologists deny this, but they have a range of tools at their disposal for explaining our maladaptive behavior. Arguably, the most important is the concept of *evolutionary mismatch*. That's what we'll look at next. (For a discussion of other common misgivings about evolutionary psychology, see Appendix A.)

Exiles from Eden

> Although most psychologists were faintly aware that hominids lived for millions of years as hunter-gatherers or foragers, they did not realize that this had theoretical implications for their work.
>
> —*John Tooby and Leda Cosmides (1992), pp. 96–97*

Imagine you're a hedgehog. You're out foraging one night when you come to a road. You start making your way across the road when, all of a sudden, two white eyes appear on the horizon, burning bright as suns. The eyes belong to a noisy metal monster, and the monster is heading straight for you. What should you do? One thing you probably *shouldn't* do is stop where you are and roll up into a spiky ball. That might be a good idea if a predator were poking around, trying to work out whether you'd make a good meal. But it's not such a good idea when two tons of metal are about to turn you into a pancake. Your best hope in that situation would be to keep on moving, as fast as you can, and try to get off the road before the metal monster gets there. As a headline from the spoof news site, *The Daily Mash*, put it, "Rolling into a ball not as good as running like f*ck, hedgehogs told." Sadly, though, that's not what you're going to do. You're going to roll into a ball and get squished.

Does this tragic tale undermine an evolutionary account of hedgehog behavior? On the contrary, an evolutionary account makes good sense of the story. It's safe to assume that, during the period when the ball-rolling defense evolved, it was usually useful. Most of the time, it probably still is. But hedgehogs spent the bulk of their evolutionary past in a world without roads, highways, and humans piloting fast-moving metal shells. Natural selection doesn't have foresight, and thus it couldn't equip hedgehogs in advance with a road-specific defense mechanism. And there hasn't been nearly enough time to evolve such a mechanism since humans started littering the landscape with roads and cars. Thus, every time a hedgehog bumbles onto a road, its protective response is out of sync with its environment. This is an example of a phenomenon known as *genome lag*, *evolutionary disequilibrium*, or – my preferred term – *evolutionary mismatch*. The mismatch in question is between the hedgehog's present environment and the environment in which it evolved.

In many ways, humans are in the same boat as the hedgehog: We're not adapted to the strange new world we've created for ourselves – this world of straight lines, right angles, and strict schedules; of cars, shaved faces, and designer jeans; of mirrors, cameras, and cities as large as ant colonies. But what world *are* we adapted to? To use the lingo, what is our *environment of evolutionary adaptedness* or *EEA*? This turns out to be a tricky question, and there are several answers floating around in the evolutionary psychology literature. By far the best known, though, is the idea that humans are adapted primarily for life as hunter-gatherers living in the savannah and woodlands of Pleistocene Africa. That's the world in which we spent most of our evolutionary history, and therefore that's the world we're designed to inhabit – the human equivalent of the hedgehog's road-free forest. As we'll soon see, this claim needs some pretty substantial tweaking. But it's a reasonable place to start.

First things first: If humans are designed primarily for life as hunter-gatherers, how did our hunter-gatherer ancestors live? The naysayers tell us we could never know, but we've actually got a fairly clear picture. This comes from extensive research in paleoanthropology (the study of ancient "bones and stones"), plus two centuries' worth of ethnographies of the world's remaining hunter-gatherers. The first lesson we can take from this research is that our ancestors lived in many different ways, just as humans do today, and thus that we need to be wary about drawing any hard-and-fast conclusions. At the same time, though, certain generalizations are clearly warranted. For most of our evolutionary history, our ancestors lived in relatively small, relatively egalitarian groups, with a mix of relatives and non-relatives. They were nomadic and had little in the way of possessions. They made tools, used fire, and cooked their food. Both sexes had "jobs" outside the home: Men did most of the hunting; women did most of the gathering. Mothers provided the bulk of the hands-on childcare, but other adults, including grandmothers, fathers, and other relatives, often pitched in as well. Infant and child mortality were high, and people of all ages were vulnerable to predators and disease. Violent altercations were common, including periodic dust-ups between rival groups. People had spoken language but not written language. They didn't have agriculture or domesticated animals. And they didn't have birth control, baby formula, modern medicine, stored food, police, lawyers, or governments. In short, as Tooby and Cosmides famously observed, life for our ancestors was like a camping trip that lasted a lifetime.[42]

According to evolutionary psychologists, the hunter-gatherer camping trip isn't just ancient history; it's central to understanding the animal we are today. There are three main reasons why. First, we were hunting and gathering when we first evolved in Africa several hundred thousand years ago. Thus, our initial emergence as a species was shaped by the selection pressures associated with the hunter-gatherer lifestyle. Second, our species evolved from earlier human species, which had been hunting and gathering for millions of years before we appeared on the scene. Thus, hunting and gathering represents an extremely deep trend in our lineage. And third, after taking the hunter-gatherer baton from our premodern ancestors, humans continued to live as hunter-gatherers in Africa for most of the rest of our history. It's only in the last 70,000 years or so that a handful of modern humans ventured out of Africa and began spreading around the globe. And it's only in the last *10,000* years that some people gave up the hunter-gatherer lifestyle, and adopted agriculture instead. As evolutionary psychologists like to say, 10,000 years is a mere blink of an eye in evolutionary terms. Moreover, agriculture wasn't common until at least 5,000 years ago, so that's half a blink. We've done some evolving since then, but not a lot. So, what do you get when you take a hunter-gatherer and transplant it into the modern world? According to evolutionary psychologists, what you get, more or less, is you and me.

This places us in an awkward predicament. Biologically, we're largely the same animal that roamed the Pleistocene savannah: a pack-hunting African ape. Culturally, though, we're unrecognizable. If our alien scientist looked at two layers of human relics, one from 10,000 years ago and one from today, it would probably assume it was looking at the handiwork of two different species. But we're the same species, and although our surroundings have changed, our core nature is largely the same. As a result, we're mismatched with our current environment. As Robert Wright observed in his book *The Moral Animal*, "We live in cities and suburbs and watch TV and drink beer, all the while being pushed and pulled by feelings designed to propagate our genes in a small hunter-gatherer population." Or as S. Boyd Eaton put it, modern humans are "Stone Agers in the fast lane."[43]

Critics of evolutionary psychology like to point out that human evolution didn't just come to a standstill when agriculture took off or when humans migrated out of Africa. But no evolutionary psychologist is claiming (or should be claiming) that it did. *Homo sapiens* are not in Hardy-Weinberg equilibrium. (Google it if you're interested; it's an important concept in population genetics.) The claim is not that there's been *no* evolutionary change since the dawn of agriculture or the African exodus. The claim is that there hasn't been *much time* for such change. Most significantly, there hasn't been enough time to evolve any complex new adaptations. Just as we haven't evolved any new bodily organs in the last 10,000 years (a spare pair of hands, say), we haven't evolved any new psychological adaptations, either: a new emotion, for example, or a new cognitive faculty. It's very unlikely that we've evolved any psychological adaptations specifically for agricultural life. And it's *extremely* unlikely that we've evolved any psychological adaptations specifically for life in industrial or post-industrial societies: adaptations that fit us to city living, for instance, or capitalism, science, or tech.[44]

But although evolutionary psychologists don't deny that we've done some evolving since we gave up our hunting and gathering ways, it's fair to say that they've sometimes underestimated quite how much we've done. Since the field first coalesced in the late 1980s, biologists have come to realize that evolutionary change can sometimes happen extremely quickly. Rapid evolution has been documented in various species, our own included. In fact, one of the biggest discoveries in the study of human evolution in the last few decades is that not only has our species continued to evolve over the last 40,000 years, but the pace of our evolution has picked up.[45] Human evolution has been particularly rapid since we took up agriculture. Gregory Cochran and Henry Harpending describe our evolutionary trajectory since that time as an ever-accelerating, 10,000-year explosion.[46] Among other things, there's been strong selection on genes related to disease resistance, diet, and nervous system functioning.

What caused the explosive increase in the pace of our evolution? The main culprit seems to be culture. Until recently, prominent evolutionary biologists such as Stephen Jay Gould argued that, as human culture became sufficiently sophisticated, human beings freed themselves from the dominion of natural selection. We developed clothes and shelters to escape the elements, weapons to guard against predators, and food extraction techniques to stave off famine. As a consequence, according to Gould, evolution in our species ground to a halt.[47] More recent scholarship suggests, however, that this has things exactly back to front. As Cochran and Harpending argue, species only stop evolving if their environment is static. But humans' cultural capacity meant that our environment was in a constant state of flux. As such, humans were continuously exposed to new and powerful selection pressures. In addition, as we mastered agriculture, our populations grew much larger. More people meant more mutations, which meant more opportunities for useful new gene variants to filter into the gene pool. For both reasons – new selection pressures and new mutations – culture did not call an end to human evolution. It did the opposite; it put it into overdrive. The upshot is that, although we probably don't have any complex new mental adaptations, we may differ in important ways from our Pleistocene ancestors.

Our growing knowledge of rapid recent evolution has overturned many previous claims about our species. Consider the idea that human beings are African apes. Some of us are, certainly: Those who live in Africa or whose recent ancestors did. And all of us are to some degree: We can survive in colder regions, but only if we cover ourselves in artificial fur and build protective bubbles around ourselves (you're probably inside such a bubble right now). Nevertheless, some human beings are partially adapted, biologically, to non-African environments. Europeans and East Asians have light skin so they can soak up enough UV radiation from the sun to synthesize vitamin D – a crucial adaptation for agriculturalists in regions of the world with little sunlight.[48] They, to some degree, are European or East Asian apes. The Inuit are stocky and short-limbed because this body shape better conserves heat – a crucial adaptation to their icy Arctic home. They, to some degree, are Arctic apes. The idea that all human beings are African apes is a good first approximation, but that's all it is. The same applies to the idea that humans are hunter-gatherers decked out in modern garb. Again, some of us are, but some populations are partially adapted to agriculture and dairy farming: They've evolved, for example, to digest milk even after the age of weaning, or to digest starchy foods more efficiently. These populations aren't simply mismatched hunter-gatherers. To some extent, they're agricultural animals. I'll say more about this in Chapter 6.

Where does this leave us? To begin with, it's safe to say that the original conception of the EEA focused too strongly on one place (the savannah), one time (the Pleistocene), and one way of making a living (hunting and gathering). As I mentioned earlier, though, the original conception is not the only one on offer. There's another important conception, which solves many of the problems of the first. It comes from two of the founders of evolutionary psychology, John Tooby and Leda Cosmides. In their view, the EEA "is not a place or time. It is the statistical composite of selection pressures that caused the design of an adaptation."[49] Taking this approach, there's no single EEA for our species; instead, every adaptation has its own EEA, and there's no reason to assume that the EEA for any given adaptation will fit neatly within the borders of the Pleistocene savannah. The EEA for the visual system, for instance – a world of objects emitting or reflecting light – long predates the Pleistocene and still exists today. In contrast, the EEA for light skin in high-latitude populations came after the African exodus, and the EEA for lifelong milk drinking came after the relevant populations took up dairy farming. For all these examples, the selection pressures shaping the adaptations were not limited to the Pleistocene, the savannah, or the hunter-gatherer lifestyle. The idea that every adaptation has its own EEA captures this effortlessly; the original conception struggles.

That said, it is worth stressing that the original conception wasn't entirely off the mark. Humans spent the better part of their evolutionary history living as hunter-gatherers, and therefore the selection pressures associated with this chapter in our history are likely to loom large for our species. As important as they are, though, they're not the whole story.

There's still some debate about where the new discoveries leave the concept of mismatch. Some commentators, including the evolutionary biologist Marlene Zuk, argue that recent, rapid evolution is a major stumbling block for mismatch explanations of human behavior.[50] But this overstates the case. As rapid as evolution can sometimes be, unambiguous examples of mismatch are easy to find. To take just one example, descendants of the Dutch have lived in sunny South Africa for many centuries, and yet still have pale skin. In other words, their skin tone is mismatched with the UV levels in their environment, even after all this time. Mismatch is common in our species, and there's no reason to think that this applies to the body but not the mind. Indeed, as we'll see in the next section, plenty of psychological adaptations are out of whack with the modern environment.

With all this in mind, let's return to the questions that launched this section. First, does the fact that people smoke cigarettes falsify evolutionary psychology? No! On the contrary, evolutionary psychology provides a cogent explanation for the fact that people smoke *even though* it's bad for them: Cigarettes are an evolutionary novelty and there hasn't been time to evolve an aversion to them. Second, does the fact that people use contraception falsify evolutionary psychology? No again! As Pinker points out, had premodern environments "contained trees bearing birth-control pills, we might have evolved to find them as terrifying as a venomous spider."[51] They didn't, though, and therefore we don't fear them at all. The same basic argument applies to pornography, recreational drugs, and the institutionalized adoption of non-relatives. Natural selection couldn't anticipate these novel stimuli and circumstances, and there hasn't been time for selection to carve out new adaptations to deal with them. Thus, all these bewildering behaviors have the same root cause: Modern humans are a fish out of water. We're living anachronisms. And that's why a lot of what we do makes about as much adaptive sense as a hedgehog rolling into a ball in the face of oncoming traffic.

A Fish Out of Water

Evolutionary mismatch provides a fairly straightforward explanation for some of the crazy behavior that so puzzled the alien scientist. This includes the fact that so much of the food we pour down our gullets is so bad for us. Our diet is obesogenic: It makes us fat. It's also carcinogenic, diabetogenic, and cardiovascular-disease-ogenic. Why are we so deeply attracted to foods that, in a very real way, are unfit for human consumption?

The answer is that our appetites evolved in a food landscape quite unlike the one we inhabit today. As Michael Power and Jay Schulkin put it, "We evolved on the savannahs of Africa. We now live in Candyland."[52] In the pre-Candyland environment, we didn't have access to copious supplies of sugary, starchy, or salty foods. Our best bet was to eat as much of these things as possible on the rare occasions we could. To that end, natural selection equipped us with powerful appetites for sugar, starchy carbohydrates, and salt. But now that we have easy access to unhealthy doses of these once-rare foodstuffs, this is no longer a recipe for success. Our preferences are mismatched with our environment. The problem is compounded by the fact that much of our food is now unnaturally appetizing. A significant portion of our diets consists of *supernormal stimuli*: human-made products that activate our senses far more powerfully than any natural stimulus ever could. Sweets and candy, for instance, are human-made super-fruit; they push the same buttons that fruit does, but they push them much harder. Funnily enough, most of the actual fruits we find on our supermarket shelves are supernormal stimuli too. Through selective breeding, strawberries, mangoes, and most fruits are larger, sweeter, and juicier than anything found in nature. It's almost impossible these days to avoid unnaturally tasty food.

Of the various health hazards associated with our modern diet, obesity is the most conspicuous indicator that we didn't evolve in the kind of food environment we now occupy. For most of the history of life on Earth, the biggest danger for most animals was eating too little. Today, for the first time, the biggest danger for many human beings is eating too much. The obesity epidemic has spread to every continent, despite strong social pressure against it. The total weight of all the adult humans on the planet is 287 million tons. According to one estimate, around fifteen million of those tons come from the excess weight that people are carting around with them. Needless to say, the problem is worse in some places than in others. A ton of human beings in North America, for example, is around twelve people, whereas a ton of human beings in Asia is around seventeen.[53] And in some parts of the world, the obesity epidemic has even spread to our pets. In the United States, believe it or not, there's now an Association for Pet Obesity Prevention.

Obesity no doubt has a number of causes. But mismatch is clearly one of them. In a world flooded with pizza, chocolate, and fizzy drinks, and with billboards and banner ads constantly reminding us of pizza, chocolate, and fizzy drinks, it's all too easy to pile on the pounds. Natural selection hasn't had time to furnish us with defenses against these seductive poisons. There have, however, been various cultural responses to the problem – everything from fad diets to stomach stapling to parental warnings such as "Don't eat too much of that chocolate cake; it's not good for you and it'll make you fat." Doctors tell us the solution is to take lots of exercise and switch to foods that don't taste very good. Unsurprisingly, most people find this advice difficult to follow. As such, another cultural response to the obesity epidemic is the Fat Acceptance Movement. Like obesity itself, this is ultimately a product of evolutionary mismatch.

Mismatch explains another tendency that puzzled our alien friend: Why aren't our fears properly apportioned to the risks in our environment? Why, when we decorate our houses for Halloween, do we opt for plastic snakes and spiders, rather than plastic cigarettes or condoms – things that are now much bigger threats to life, limb, and reproductive success? And why, when we try to teach our children to fear roads and electrical outlets, do they stubbornly insist on fearing snakes and monsters instead? To be clear, the mystery is not that we fear certain things more than we need to. A little excess fear is exactly what evolutionary principles predict. As Randy Nesse and George Williams put it, "the cost of getting killed even once is enormously higher than the cost of responding to a hundred false alarms."[54] That's why most animals are neurotic: why they're more anxious and easily spooked than a rational, Spock-like assessment of the evidence would warrant. Thus, the mystery isn't that we worry too much; the mystery is that the rank order of our worries is jumbled up. People are more likely to develop phobias of snakes and large predators than they are of things in their environment that are much more likely to kill or harm them, such as handguns, speeding cars, and rising sea levels.

The explanation should be clear. It's presumably no coincidence that the things we fear out of all proportion to the risks – snakes, large carnivores, and the like – were threats to survival and reproductive success throughout most of our evolutionary history, whereas the things that we don't fear nearly enough – cars and global warming – appeared on the scene more recently. Evolutionary psychologists have argued persuasively that humans have an evolved wariness of recurring ancestral threats: something akin to an ancestral species memory.[55] The strongest case is for snakes, but other plausible candidates include spiders, heights, darkness, angry faces, and strangers. Of course, we don't *only* fear ancestral threats. But we do seem to fear them more readily than novel ones. We need strong cultural interventions to make us suitably afraid of Big Macs, sunbathing, and climate change. Not so snakes or spiders. Often we need to learn *not* to be so afraid of these things.

Our outsized fear of ancestral threats, and our excessive consumption of junk food, are the best-known examples of evolutionary mismatch. But they're really just the tip of the mismatch iceberg. One less-discussed example is found in the classroom. Play is common among mammals and other intelligent animals, and we've got a fairly good idea what it's for. In a nutshell, play is practice for the kinds of tasks that the young animal will engage in as an adult.[56] This is also more or less the function of human schools: They prepare children for the roles they'll play later in life. Why, then, don't children enjoy school more? Why do they enjoy doing homework about as much as adults enjoy doing taxes? Why would they rather be playing? Indeed, why isn't schoolwork *viewed* as play?

You know the answer. We didn't evolve in the kind of world we now inhabit, and thus the skills we learn in the classroom – math, writing, and the rest – are not the kind of skills we evolved to master.[57] Here, as elsewhere, culture goes against the grain of human nature. How do we know? Because children don't like school!

Another area where the mismatch concept is doing good work is in the field of *evolutionary medicine*.[58] Practitioners in this field use the tools of evolutionary theory, mismatch among them, to shine a light on matters of health and illness. One of the field's success stories is its explanation for breast cancer. For most of our history, women spent the bulk of their reproductive years either pregnant or breastfeeding. Women don't menstruate when they're pregnant, and they tend not to menstruate when they're breastfeeding either, at least not in hunter-gatherer conditions. The net result is that, until recently, most women had only a hundred or so menstrual cycles in their lifetimes. Things are very different today.

Women hit puberty earlier, have fewer pregnancies, and spend a smaller fraction of their lives nursing their young. They therefore have many more menstrual cycles than their ancestors: as many as four hundred.[59] This is more than their reproductive systems are designed for, and it exposes them to unnaturally high levels of ovarian hormones and unnaturally frequent hormonal fluctuations. This in turn increases women's risk of breast cancer – as well as conditions such as anemia and endometriosis. These maladies were almost unheard of until recently. Like obesity, they're essentially diseases of modernity, or what are sometimes referred to as *mismatch diseases*. As Daniel Lieberman notes in his book *The Story of the Human Body*, other possible mismatch diseases include acne, allergies, asthma, cavities, flat feet, heart disease, hemorrhoids, high blood pressure, impacted wisdom teeth, lower back pain, osteoporosis, short-sightedness, and type-2 diabetes.[60]

The list of mismatch diseases also includes some psychological maladies. In many ways, the world today is a primate paradise. Compared to any other period in human history, we've got lower infant mortality rates, longer lifespans, less violence, greater wealth, and more opportunities to pursue the goals and the lifestyles that suit us.[61] We should be over the moon... but we're not. Most of us are *reasonably* happy, sure. But we're hardly ecstatic, and some of us are simply miserable. As Geoffrey Miller observes, the world has never been better, and yet many people have to take special medications to avoid suicidal despair.[62] Now obviously, life has never been a picnic. However, some aspects of the modern world may be misaligned with human nature in ways that produce novel psychological problems – problems that, like breast cancer and endometriosis, are largely diseases of modernity.

One likely candidate is postpartum depression. This life-sapping malady afflicts a significant minority of women in modern industrial societies, but is rare or even absent among hunter-gatherers. What are they doing right that we're not? According to Jennifer Hahn-Holbrook and Martie Haselton, the list is long. Women in hunter-gatherer societies are more likely than modern women to live near kin, and thus to have plenty of social support. They space out their births more than we do, which means that they never have more than one fully dependent youngster at a time. They always breastfeed their young, and breastfeeding is associated with the release of hormones that help insulate people from stress. Their diets are richer than ours in omega-3 fatty acids, and less rich in sugar and starch. And they're far more physically active than us, and always get plenty of sun. All these things make hunter-gatherer mothers much less prone to postpartum depression than women in modern societies. Indeed, modern societies look almost as if they were designed to maximize new mothers' chances of getting depressed.[63]

Another possible disease of modernity is ADHD: attention-deficit hyperactivity disorder. In an essay entitled *Orangutans on Ritalin*, the developmental psychologist Gabrielle Principe noted that:

> No animal other than us modern humans – our hunter-gatherer ancestors included – suffers ADHD. But plenty of today's elementary school children, who spend eight hours a day jammed inside a classroom, do. The American Psychiatric Association considers it a mental disorder. But it is also exactly what you'd expect if you put any juvenile (insert your choice of species here) behind a desk, made it do seatwork, told it to concentrate, and didn't let it out to play.[64]

Psychiatric diagnoses like ADHD take a problematic behavior and locate its cause in the person, rather than in the environment. For some diagnoses, such as dementia and schizophrenia, this is entirely appropriate: Though we can often change sufferers' environments in constructive ways, the core of the problem lies within their skulls. When it comes to ADHD, however, things are much less clear. The conventional explanation is that ADHD is a problem with the brain mechanisms controlling attention. But many children with ADHD can concentrate for hours on the things that interest them. This suggests that the problem is not their attentional capacity, but rather that they're not interested in the things the world insists they should be – the things they're doing at school.[65] They're bored! If this is right, then what's the best way to understand ADHD? Is it that afflicted children have malfunctioning brains? Or is it that we've cooped them up like battery chickens, when these little primates are designed to be running around, talking, and arguing? Young boys, in particular, find it difficult to sit still and concentrate for long periods of time, not because they're defective but because they're male primates.[66] To be sure, for many children, school is fine. However, for a significant minority, it's a round hole to their square peg. And many children diagnosed with ADHD would fit in perfectly well in an ancestral-type environment, where they could run and play all they wanted. This suggests that, at least in some cases, ADHD is more a mismatch problem than a brain problem. Of course, this doesn't mean that it's *not* a problem. We can't just disestablish schools or let kids run wild. But we should always bear in mind the words of the American feminist Gloria Steinem, who asked, "If the shoe doesn't fit, must we change the foot?"

A Spandrel in the Works

What would an alien scientist make of cows? What would it think these creatures were designed to do? The alien's first impression might be that cows are machines designed to suck up grass and convert it into two things: milk and methane. The grass goes in one end, the milk and methane come out the other. Being more familiar with cows, we know that the milk and the methane don't have equal standing. Cows are machines designed ultimately to turn grass into more cows, and milk plays an important part in that process: Mothers turn the grass into milk, and then their calves turn the milk into more cow flesh (theirs). Methane, on the other hand, is just a by-product of digestion. Cows don't produce it for some adaptive purpose. It's simply a side effect.

There's an important lesson here. So far, in trying to explain the human mind and behavior, we've talked a lot about adaptations. But not everything is an adaptation. Many traits – from the redness of blood to the shape of the chin – are accidental *by-products* of adaptations. They're what the paleontologist Stephen Jay Gould and geneticist Richard Lewontin nicknamed "spandrels": an architectural term, which refers to the triangular spaces that appear when a curved arch is enclosed within a rectangular frame (look, in your mind's eye, at the upper corners of the resulting structure).[67] These spaces, said Gould and Lewontin, have no purpose; they're just what happens when you put an arch and a frame together. Many traits found in human beings and other animals are the same. They weren't specifically favored by natural selection, but came along for the ride with traits that were.

Here's an example. Children everywhere suck their thumbs to comfort themselves. Similarly, baby elephants sometimes suck their trunks, presumably for the same reason. Thumb-sucking and trunk-sucking are not adaptations in and of themselves. It's hard to imagine how either activity could enhance the survival or reproductive success of the sucker or the sucker's kin. More than likely, they're by-products of something else. That something isn't hard to find; it's the instinctive sucking behavior found in all mammals. This appears on the first day of life, and has a clear evolutionary rationale: obtaining milk from the mother. To reward this fitness-enhancing behavior, natural selection has made sucking a pleasurable activity for infants. In effect, thumb-sucking babies (and trunk-sucking elephants) have found a way to cheat the system – a way to siphon off some of that pleasure without actually doing the job the pleasure was designed to reward. Thus, thumb-sucking isn't something that natural selection specifically built into our behavioral repertoire. It's just something that happens once in a while when you have a creature with a thumb, an instinctive tendency to suck, and the ability to learn new tricks.

Not all by-products fit this mold. A very different type of by-product is the male nipple. Female nipples have an obvious evolutionary function: feeding one's offspring when they're not sucking their thumbs or trunks. But with a few possible exceptions, such as the male Dayak fruit bat, male mammals do not nurse their young. Why, then, do they bother with nipples? It turns out that male nipples are a by-product of the developmental process that gives rise to males and females in the first place. Mammalian embryos begin life in a gender-neutral state, as little blobs of protoplasm with the potential to develop into either males or females. At around six weeks, they begin diverging from this initial state and develop the standard male or female forms, depending on their genetic sex.[68] Because mammalian embryos can go either way, they need all the relevant equipment to build a baby of either sex. And that's the reason – the only reason – that males have nipples. Female nipples are an adaptation, male nipples a by-product of that adaptation.

A similar story has been floated for the female orgasm. Like female nipples, the male orgasm has an obvious evolutionary rationale. You can be virtually certain that on the night you were conceived, your father had an orgasm. But did your mother? Not necessarily. (Sorry to put these ideas in your head!) Does female orgasm have an evolutionary function? Ironically, scientists have yet to come to a satisfying conclusion about this matter. Many think it does, but some are skeptical. In the latter camp is one of the pioneers of evolutionary psychology, Donald Symons. Symons argues that, just as male nipples have no function but are by-products of female nipples, so too female orgasm has no function but is a by-product of male orgasm.[69] Women, and at least some other female primates, have the machinery that makes orgasm possible, but only because that machinery is adaptive for males. Like I say, there are also plausible adaptationist hypotheses for female orgasm. Some argue that it boosts the odds of conception, others that it helps to guide female mate choice.[70] But if the by-product theory is correct, then female orgasms are a little like handstands: We weren't specifically designed for them, but given the basic setup of the human body, they're something that we're capable of. In effect, female orgasm is a cultural invention, building on a fortunate quirk of women's reproductive physiology.

(Incidentally, some people are offended by the by-product explanation for female orgasm. In their view, to say that female orgasm is merely a by-product, rather than an adaptation in its own right, is to devalue it.[71] The first thing to say about this is that, even if it were true, it would be irrelevant to the question of whether the by-product explanation is accurate. Female orgasm is either a by-product or it's not, regardless of our preferences. The second thing to say, though, is that, as it happens, there's no good reason to think that the by-product explanation devalues female orgasm. The only reason we'd think it did is if we tacitly assumed that if something is an adaptation, it's valuable, whereas if it's not, it's not. But why assume that? Many adaptations are things that no moral person would value, and many things that we do value are not adaptations. As such, people who object to the by-product hypothesis are doing something strange: They're saying that it's hugely important to them that female orgasm turns out to be an adaptation, like vomiting, cobra venom, and the scorpion's stinger, rather than a "mere" cultural invention, like science, medicine, and the Universal Declaration of Human Rights. This is not a position we ought to take seriously. We value orgasms because they're pleasurable and emotionally satisfying, not because they came about via one evolutionary process rather than another. For more on this general issue, see Appendix A.)

By-product explanations may help to unravel some of the mysteries raised by the alien's-eye view of *Homo sapiens*. Why, for instance, do so many people devote so much time to pornography? No one would argue that it's because those of our ancestors with a taste for the X-rated were more likely to survive and reproduce, or to help their kin to survive and reproduce. Clearly, it's not an adaptation. Instead, the penchant for porn is a by-product of other psychological attributes that themselves are adaptations. At the top of the list is the tendency to become aroused in response to visual sexual stimuli. As we'll see in the next chapter, this tendency is notably stronger in men than in women, which immediately starts to explain a rather obvious sex difference in the consumption of pornographic materials.[72] For now, however, let's just say that the porn habit is not the product of a porn-specific adaptation. It's a spandrel.

It's also not our only evolutionarily puzzling online behavior. Another is our addiction to photos and videos of cute baby animals. The Internet is saturated with kittens, puppies, and baby orangutans in wheelbarrows. Why do so many people take such delight in staring at infant members of other species? It's not as if, say, porcupines enjoy staring at baby chickens. As with porn, our love of these nonhuman animals is probably not an adaptation. More than likely, it's spillover from psychological mechanisms designed for more human-centered purposes. There's a certain cluster of traits that people everywhere find irresistibly cute.[73] This includes big round eyes in the center of the face, a small nose, and plump, stubby limbs. Our affection for creatures with these features presumably evolved to motivate us to care for our own infants and toddlers. But the same features are found in many other infant mammals, and even in the adult members of some nonhuman species. As a result, we often feel affectionate and protective toward these individuals as well – not because it's adaptive, but just because adaptations aren't perfect. By the way, as you might already have noticed, the spillover hypothesis doesn't just explain our fondness for cute animal videos. It also hints at an explanation for a much older and more pervasive phenomenon: our habit of keeping pets.

There's another way to cash out the by-product hypothesis – one that may prove useful in solving the alien's dilemmas. This is Steven Pinker's *strawberry cheesecake hypothesis*.[74] Cheesecake is one of the world's most popular desserts. Unless we're watching our weight, most of us would choose it over fruit. This is curious, though, because we evolved to eat fruit but we didn't evolve to eat cheesecake. The explanation, of course, is that, like much of our food today, cheesecake is a supernormal stimulus: an artificial concoction that presses our evolved buttons more strongly than any natural substance, and that therefore packs more of a punch. According to Pinker, art, music, and fiction are directly analogous to strawberry cheesecake. They're not adaptations; they're technologies we've developed to artificially stimulate our brains in ways we find enjoyable. They're hacks! Art, for example, draws much of its power from species-typical aesthetic preferences: preferences for symmetry, certain colors, and particular types of scenery. These preferences plausibly evolved to help us choose suitable mates, suitable food, and suitable habitats. Successful art – art that goes viral – satisfies these preferences outside their natural domain, giving us a little jolt of pleasure in the process. Without the evolved preferences, humans wouldn't have invented art. In that sense, art is a by-product.

Other cultural phenomena may be by-products in just the same sense. Morality may be a by-product of our capacity for empathy, our sense of fairness, and emotions such as guilt, shame, anger, and disgust. Religion may be a by-product of our tendency to explain puzzling events in terms of the actions of human-like agents, our fear of death and disaster, and our ability to ask questions such as "How did the world begin?" and "Why is there something rather than nothing?" Music may be a by-product of the brain mechanisms involved in analyzing the emotional overtones of people's speech. And science may be a by-product of our intelligence, our curiosity, and our capacity to share ideas and build on what others have achieved before us. As we'll see later, cultural evolution has a crucial part to play in explaining how these by-products took the particular forms they did. Still, if the by-product explanations for art, morality, religion, music, and science are accurate, the implication is that some of the most important phenomena of human life are the evolutionary equivalents of sucking one's thumb (or one's trunk, if one is an elephant).[75]

Evolutionary theory solves one of the great mysteries facing the human mind: Where did the mind contemplating these mysteries come from? The short answer is that the mind is a product of natural selection, designed to propagate the genes giving rise to it. Evolutionary theory doesn't answer the deeper question of why matter has the capacity to become conscious when it's organized and functioning in a particular way: why the pulsing gray porridge of our brains gives rise to the Technicolor of conscious experience. Nonetheless, given that it apparently does have this capacity, the theory explains how it is that some of the matter on at least one little planet is in this peculiar state. Thus, a curious fact about natural selection is that, even though it's a completely mindless process – a process without foresight or understanding – it has given rise to creatures that have minds, that have foresight, and that even have some basic understanding of the universe of which they're a part.

3

The SeXX/XY Animal*

An Academic Culture War

"Everyone knows that men and women are different... except social scientists."

I first heard this wisecrack as a graduate student in psychology, and it instantly rang true. In everyday life, most people recognize that the sexes differ. We see it at school; we see it at work; we see it in our kids and in ourselves. To start with, we know that men and women have different bodies and reproductive equipment, that men are generally larger and stronger, and that women generally live longer. But we also know that the differences are not just physical. We know that men watch more sports and more porn, whereas women watch more rom-coms and read more. We know that men are more inclined toward violence and more likely to end up in prison, whereas women are more likely to take sensible precautions. We know that men are more interested in things and machines, whereas women are more interested in people. And we know that men are more likely to go into "nerdy" professions such as math or engineering, whereas women are more likely to go into the caring professions and to spend more time looking after children.[1]

No one denies, of course, that there are short men and female sports fanatics, or that there are violent women and hands-on dads. And everyone knows that there's lots of overlap between the sexes – that if we lined everyone up by height, for example, there'd be more women than men at the short end of the line and more men than women at the

---

* Thanks to Darwin Stewart-Williams and Jane Stewart-Williams for suggesting the title of this chapter.

tall end, but that there'd be plenty of both in between.[2] Still, the idea that there are *average* differences between the sexes seems blindingly obvious to most people. Perhaps equally obvious is the idea that these differences are not merely cultural conventions. They're partly in the nature of the beast.

To our alien scientist, these observations might appear uncontroversial – banal, even. However, since the latter decades of the twentieth century, to voice such thoughts on a university campus has been to wander into a political minefield. What was common sense to most people throughout most of human history transforms into heresy the moment one crosses the threshold of the ivory tower. Anyone who broaches the topic can expect to be met with one of three responses.

The first is flat denial: The difference you think exists, doesn't; you've simply picked up an unfounded stereotype like a child picks up an infection. This argument is usually deployed only in response to claims about *psychological* sex differences; even the most dedicated sex-difference deniers have a hard time denying, for instance, that men are taller than women. But for any sex difference that people can't see with their own two eyes, there are those who will claim it's a myth.

A second response is to concede that the difference exists, but to argue that it's purely a product of discrimination. Deep down, men and women have the same desires and drives; women are just prevented from acting on theirs. There are just as many women as men who want to be CEOs and mathematicians, but most are stopped in their tracks by sexist expectations or unconscious bias in the workplace. A more extreme version of this story is the idea that we live in a patriarchal society: one in which men have privilege and power whereas women are second-class citizens, and in which women are channeled into preordained roles, even against their deepest wishes.

The third response is to concede that, sure, men and women do sometimes have different desires and drives, but to argue that this is entirely due to the fact that they're socialized in different ways. From day one, society rubs people's faces in their gender. The first thing we say when a baby enters the world is "It's a boy!" or "It's a girl!" We then color-code this new person and set about teaching them what it means to belong to their gender tribe. We give boys toy guns and Lego, girls frilly dresses and dolls. We teach boys to be tough and strong, girls to be kind and nurturing. And we push boys to be engineers or doctors, girls to be teachers or nurses. It would be naive to think that none of this has any effect. If it didn't, why bother socializing children at all?

In short, the traditional view in the social sciences is that stereotypes of the sexes definitely aren't true – but that if they are, it's all down to social forces: bias, gender labeling, gender-typed toys, cultural norms, media role models, and other pernicious influences. Most important, the differences are not *innate*. The idea that they might be isn't just factually inaccurate; it's dangerous. The danger is that claims about innateness will stall or reverse the progress of women in society.[3] Indeed, some social scientists insinuate that that's the hidden agenda lurking behind such claims. The sexologist John Money, for instance, once stated that "Nature is a political strategy of those committed to maintaining the status quo of sex differences." In a similar vein, the sociologist Jessie Bernard dismissed studies of sex differences as "battle weapons against women."[4]

One might wonder what evidence Money and Bernard had to back up these rather serious accusations. To be fair, though, it is easy to see why people are sensitive about this issue. We don't have to look far into the past to find scientists issuing extremely sexist, poorly evidenced proclamations regarding innate sex differences. To give a particularly jarring example, in 1879, the social psychologist Gustave Le Bon wrote that:

> In the most intelligent races, as among the Parisians, there are a large number of women whose brains are closer in size to those of gorillas than to the most developed male brain ... Without doubt there exist some distinguished women, very superior to the average man, but they are as exceptional as the birth of any monstrosity, as, for example, of a gorilla with two heads.[5]

With skeletons like this in the closet, it's little wonder that people are wary of the whole notion of innate sex differences. The idea that there are no differences, or that any differences come from nurture alone, seems safer and more optimistic.

How *plausible* is it, though? Can we really believe that the unlettered masses have been hallucinating sex differences for all these years? Surely not. As anyone but a social scientist can plainly see, there are genuine differences between men and women.[6] And is it really so crazy to think that at least some of these differences go gene-deep, rather than just culture-deep? Is anyone who thinks that there might be a genetic contribution really just trying to keep women down, or having the wool pulled over their eyes by the people who are?

Funnily enough, at around the same time that the Nurture Only theory of sex differences was really starting to dig in its heels in the social sciences, a very different theory was taking shape elsewhere in the academy: in the evolutionary biology departments. Whereas social scientists focused exclusively on sex differences in humans, evolutionary biologists focused on sexual dimorphism in nonhuman animals. (Sexual dimorphism is a fancy name for sex differences: *di* means two; *morph* means form.) And whereas social scientists explained sex differences almost entirely in terms of learning and culture, evolutionary biologists focused on the role of natural selection. In their view, male and female animals differ from one another because the things that enhance male fitness don't always enhance female fitness, and vice versa. Wherever this is the case, selection can favor different traits in females than the ones it favors in males. Initially, this explanatory framework was extended only to nonhuman animals. It wasn't long, however, before a handful of mavericks began to wonder whether an evolutionary explanation might apply to us as well. Perhaps, they suggested, the sex differences we see in our species aren't solely products of learning and culture after all. Perhaps they're built into us at the ground level, just like they are in many other animals.[7]

Given the political climate of the time, it's little surprise that, as these ideas began to gain traction, the reaction was often less than warm. Many old-guard social scientists responded to the new ideas not with arguments and evidence, but with anger, indignation, and personal attacks. In their view, the people promoting the biological theories – the sociobiologists and later evolutionary psychologists – weren't just doing neutral science. They were concocting post hoc explanations for tired, sexist stereotypes: stereotypes that put men on a pedestal and women in a bad light. Worse than that, they were giving a pseudoscientific justification for a regressive political agenda – one that aimed to push women back into the kitchen and out of the public sphere.[8] (For a general discussion of criticisms of this kind, see Appendix A.)

With the benefit of hindsight, it's fair to say that the reaction to the evolutionary approach was a vast *over*reaction, and a rather unfair one at that. The sociobiologists and evolutionary psychologists weren't arguing that distinguished women are as rare as two-headed gorillas, or anything remotely like it. Indeed, if the new theories painted an unflattering picture of either sex, it wasn't women. As we'll soon see, evolutionary psychologists argue that men are naturally more violent than women, more prone to infidelity, and more prone to taking stupid, life-threatening risks. If these findings are battle weapons in a war against women, then evolutionary psychologists have chosen a strange way to wage that war. (Of course, it's rather strange that they're waging it at all, given that many evolutionary psychologists – including many of the field's founders – are women themselves.) But in any case, what matters in the end is not whether the claims of evolutionary psychology are pleasant or unpleasant, welcome or despised. What matters is whether they're true. With that in mind, let's wander into the minefield. Where do our sex differences come from?

The Evolution of Sex Differences

In Chapter 1, I made a comment that might get me in hot water with my psychologist colleagues. I suggested that psychology has made a lot less progress than it could have because too many psychologists know too little about other animals. This applies as much to sex differences as to any other area. As a corrective measure, I suggest that we begin our exploration of human sex differences by taking a counterintuitive step. I suggest that we *forget all about* human beings and fix our attention instead on our nonhuman cousins. What kind of sex differences do we find in the rest of the animal world? The answer is that species differ enormously, and that for any sex difference you care to name, there are always a dozen exceptions – and then a dozen more. Still, you can't have an exception without first having a general rule, and the general rules are usually the best place to start in our quest to make sense of the world. Without further ado, then, here are ten of the most common sex differences found in the animal kingdom.[9]

1. In many species, males and females differ in size. In most, the females are larger. Among spiders, for instance, the males are often microscopic compared to the giant females. Among large vertebrates, on the other hand, the size difference is often reversed, and males are the larger sex. Thus, male gorillas are twice the size of females, and bull elephant seals are three or four times the size.

2. Males in many species have a stronger sex drive than females, and a more powerful appetite for new and numerous sexual partners. This reveals itself in many unexpected ways. Male fur seals have been observed trying to mate with king penguins; male snow monkeys have been observed trying to mate with deer; and male jewel beetles have been observed trying to mate with beer bottles that vaguely resemble female jewel beetles. Even more jaw-dropping, a male mallard duck was once observed raping a fellow male who'd recently flown into a window, dying on impact.[10] All these examples show that the male sexual response can be rather indiscriminate, to put it politely.

3. The flipside of this coin is that females are often much choosier than males about their sexual partners. A common sight in nature is a hapless male trying desperately to impress a female, and the female turning away, completely uninterested. Another common but less amusing sight is a male violently trying to force a female to mate with him, and the female trying desperately to escape. Both cases illustrate females' greater sexual choosiness.

4. Males are often more ornamented than females. The peacock's tail is the standard example, but it's only one of many. From birds to reptiles to insects, males in many species are adorned with head crests, throat sacks, or bright, patterned feathers, or have a repertoire of songs, dances, and other party tricks designed to impress the females.

5. In many species, males "pay" for sex. For example, among black-tipped hanging flies, an amorous male will present a female with a nice, juicy insect, and then copulate with her while she distractedly enjoys her meal. If the female finishes before he does, she simply walks away – game over. If, on the other hand, *he* finishes first, he snatches the insect away and tries to woo another female with the leftovers.[11] In other species, the "payment" is less direct. Thus, in many birds, fish, and frogs, males maintain territories where the females lay their eggs and rear their young. In return, the males get to sire their offspring.

6. Males are generally more aggressive than females, and spend more of their spare time beating each other up. The prototypical examples, seen in a thousand nature documentaries, are male deer locking horns and bull elephant seals locked chest-to-chest in combat. But again, we find a thousand more examples in every corner of every continent. To begin with, male kangaroos routinely fight for females. The fights look a lot like boxing matches except that, unlike human boxers, the kangaroos periodically rear back on their tails and kick their opponents with both feet. Similarly, male ring-tailed lemurs engage in what are called "stink fights": They impregnate their tails with their scent and then waft it at their opponents. If that fails to sort things out in a gentlemanly fashion, they then start violently jousting with each other. In these and other cases, the males are either fighting directly for females, or fighting for the resources, territories, or status that are needed to attract the females' interest.

7. Males often come equipped with a frightening arsenal of built-in weapons. These include everything from antlers and spikes to supersized fangs. Males also often have built-in defenses, such as tough skin and reinforced skulls. Females, in contrast, are normally less well armed and less well armored.

8. In many species, females grow up faster than males, a phenomenon known as *sexual bimaturism*. In gorillas and elephant seals, for instance, females reach reproductive maturity several years before males. Likewise, among satin bowerbirds, females reach reproductive maturity at around two years of age, whereas males *don't* reach reproductive maturity till around seven.

9. Females tend to live longer than males.[12] Dawson's bees are a memorable example. By the time the females of this Australian bee species lay and provision their eggs, all the males are dead – killed in a vicious struggle to fertilize as many females as possible. The end result is a society consisting solely of females.

10. Last but not least, in species where parents care for their young, females usually do most of the caring. Among tigers, for example, females are the primary caregivers, while the males are deadbeat dads. This is a fairly standard arrangement among mammals, and among parental animals in general.[13]

So, there we have it: ten of the most common sex differences found in the animal world. Now the hard part. Where do these differences come from? Darwin saw this as one of the central mysteries in biology and, in typical Darwin fashion, took some large steps toward solving it. As described in Chapter 2, his sexual selection theory explains how ornaments such as the peacock's tail evolve through female mate choice, and how armaments such as the deer's antlers evolve through male–male competition. Darwin never explained, however, why it's usually the males that sport the ornaments and armaments, and usually the females that are ultra-finicky about their mates. The answer to these questions would have to wait until the second half of the twentieth century, and the work of the evolutionary biologist Robert Trivers, a colorful character who's sometimes described as the Albert Einstein of sociobiology. Among Trivers' seminal contributions was his *parental investment theory*.[14] To unpack the logic of the theory, let's start with a quiz.

Question 1: What's the largest number of children that any man has ever had?

Take your best guess before reading on.

(This is a filler sentence so that you don't inadvertently read on before guessing.)

The answer is... have you taken a guess yet? The answer is *888*. The proud father was a Moroccan emperor named Ismail the Bloodthirsty, who reigned from 1672 to 1727. Needless to say, there wasn't just one *Mrs.* Ismail the Bloodthirsty. Ismail had a harem of hundreds of fertile young women. Some scholars argue that not all of those bloodthirsty babies could have been Ismail's, and that some of his wives or concubines must have got up to some mischief. But mathematical simulations suggest that it's perfectly possible; all he'd have to do would be to have sex once or twice a day, every day, for roughly thirty years.[15] Either way, it's easy to see that a man could potentially sire a large army of offspring, just as long as he could find a sufficient number of collaborators.

Time for our next question.

Question 2: What's the largest number of children that any woman has ever had?

Again, take your best guess before reading on.

(This is another filler sentence. What's your guess?)

The answer isn't twenty. It isn't forty. The answer is *sixty-nine*. The mother was a nineteenth-century Russian peasant called Valentina Vassilyev. Vassilyev started her reproductive career early, finished late, and had lots of multiple births: twins, triplets, and quadruplets. At the risk of stating the obvious, both she and Ismail the Bloodthirsty are about as far removed from normal as it's possible to get; it's rare to meet someone who has sixty-eight siblings, let alone 887. Nonetheless, comparing their records illustrates an important point, namely that the maximum number of offspring a male can sire is much higher than the maximum number a female can. If all the stars were perfectly aligned, a man could have dozens or even hundreds of offspring. A woman, in contrast, would struggle to have twenty, even if she had a harem of hundreds of fertile young men and sex twice a day for thirty years. And we all know this; that's why, although you were probably surprised by Ismail's record, you were even more surprised by Valentina Vassilyev's.

The sexual asymmetry in maximum offspring number is not unique to humans. In many species – the vast majority, in fact – the ceiling number of offspring for males is higher than that for females. The most important reason for this is parental investment.[16] That was Trivers' great insight. In most species, females invest more time, energy, and resources into their young than do males. Among mammals, for instance, it's the females that get pregnant, the females that nurse the young, and the females that handle most of the childcare. Just as an author who specializes in epic novels will publish fewer works than one who churns out short stories, high-investing females will have fewer offspring than successful low-investing males. The implications of this simple fact are far reaching.

To start exploring them, let's imagine an extreme case: a hypothetical species in which males invest essentially nothing into their offspring, whereas females invest a great deal.[17] I hasten to add that humans are not like this, and that the sex differences in our species are a lot more modest. Still, our hypothetical species will bring out the basic logic of Trivers' theory and provide a firm foundation for understanding the complexities of the human case. The first thing to notice is that because males in our hypothetical species invest next-to-nothing in their young, they can potentially produce a huge number of rug rats; all they need to do is mate with a huge number of females. That, of course, is easier said than done. What's clear, though, is that any traits that increase a male's chances of doing it would have a very good chance of being selected.

What sort of traits would fit the bill? A good start would be a strong, undiscriminating sex drive and a desire to mate with as many females as possible. Males possessing such drives and desires would be like guided missiles, seeking out every sexual opportunity available. Animals don't use contraceptives, and thus these guided-missile males would leave more offspring than males with low libidos, males with impossibly high standards, or males that only had eyes for one female. Their offspring would then tend to inherit their fathers' sexual proclivities. The net effect is that these male proclivities would be more common in the next generation than they had been in the last. Over many generations, the sex-crazed males would crowd out the more sex-sane ones until one day, eventually, the crazies would be the norm.

The equation is very different for the females in our hypothetical species. Because the females invest a huge amount into their offspring, they can't drastically increase the number of offspring they have simply by mating with lots of males. What they can do, though, is make sure they mate with the *best* males: the healthiest, the ones with the best genes, the ones best able to lavish resources on them. Females with a penchant for such males, and an aversion to lesser specimens, would have healthier, genetically superior offspring – offspring that would have a better-than-average chance of surviving to breeding age. These offspring would inherit their mother's choosiness, and female choosiness would become more and more common with each new generation. Thus, in the extreme case (and let me remind you that humans are nowhere near this extreme), the males would evolve to prioritize quantity of mates, whereas the females would evolve to prioritize quality.

And it doesn't stop there. As soon as the females evolved their exacting standards, those standards themselves would become a new selection pressure acting on the males. Males who measured up would have more offspring, and therefore the traits the females preferred would become more and more common among the male of the species. If the females preferred males with big noses, for instance, big-nosed males would have more offspring, and the average male nose would get progressively bigger: a multi-generational Pinocchio effect. Similarly, if females insisted that males provide a "nuptial" gift in exchange for sex, or that males possess territories in which the females can rear their young, males would evolve to do those things. The general rule is that, whatever the females want, the males evolve to provide it.

That, at any rate, is one path to reproductive success for males. Another would be to focus less on persuading females to mate with them than on eliminating the competition: other males. In that context, males with a hair-trigger temper and a willingness to take on rivals would have an advantage over peacenik males. Of course, unless *everyone* was a peacenik, they'd need something to back up the threat. A bigger body would be a step in the right direction. After all, when it comes to fighting, size matters; that's why there are weight classes in boxing. And if a male's big body came equipped with big, Arnold Schwarzenegger muscles, or with horns or spikes or fangs, that too would help secure victory. Any genes contributing to the development of these traits would stand a good chance of getting copied into lots of new bodies. Over time, those genes – and the traits they helped to build – would become more and more entrenched in the species. Males, in other words, would evolve into bigger, badder fighting machines.

This evolutionary waltz toward bigness and badness would have a number of downstream effects for our hypothetical species. First, the males would take longer than the females to reach full-blown adulthood. This is because it takes longer to grow a larger body, and longer to develop the skills necessary for holding one's own in a competitive social environment. Second, the females would typically live longer. There are several reasons for this. One is that, because males engage in risky conflict with rivals, males are more likely to wear themselves out and get themselves killed. Another is that, because females are the primary caregivers, female fitness is dependent not merely on producing babies but on sticking around to look after them. For low-investing males, in contrast, sticking around makes less difference to their lifetime reproductive success. Risking everything for a chance to win the reproductive bonanza might be worth the gamble.[18]

Our hypothetical species tells us something important, namely that most of the recurring sex differences we see in the animal world trace back to another, more fundamental difference: the difference in the maximum number of offspring that males vs. females can produce. So far, however, we've looked at just one pattern of sex differences: what I and my colleague Andrew Thomas call the males-compete/females-choose (or MCFC) pattern.[19] This is a common pattern among the animals, but it's not the only one. Importantly, though, the same variable that explains the MCFC pattern – maximum offspring number – also explains the others.

To begin with, maximum offspring number explains what initially looks like an exception to Trivers' rule: the *sex-role-reversed species*.[20] These are species in which the usual MCFC pattern is turned on its head. The best examples are jacanas: tropical wading birds sometimes known as Jesus birds for their apparent ability to walk on water, and sometimes known as lily trotters for their actual ability to walk on lily pads and other floating vegetation.[21] In many jacana species, the females are larger than the males, spend more time fighting each other, and – if they're sufficiently dominant – put together harems of males. The males, on the other hand, are smaller than the females, less aggressive, and less fixated on accumulating mates. Jacana role reversal is so complete that scientists initially mistook the females for males and the males for females. Another example of a sex-role-reversed species is the Gulf pipefish: a small fish that looks like a seahorse straightened out with hair straighteners, and that is in fact a close cousin of the seahorse. Like jacanas, Gulf pipefish buck the MCFC trend. Female pipefish are ornamented with silver stripes, whereas male pipefish are ultra-choosy about their mates, preferring older, larger females.[22]

At first glance, the sex-role-reversed species appear to falsify Trivers' theory. On closer inspection, however, they provide some of the strongest support there is that Trivers hit the nail on the head. That's because, in all these gender-bending species, the *males* invest more into offspring than the females. Among jacanas, for instance, the males man the nests, look after the eggs, and care for the chicks. Similarly, among Gulf pipefish, the males incubate the eggs in a specialized brood pouch. As a result, the *females* in these species can potentially produce more offspring. Like an anti-gravity machine, this flips all the normal selective forces in the opposite direction, producing male-like females and female-like males – just as parental investment theory would predict.

Up until now, we've treated sexual dimorphism as if it were an all-or-nothing affair. Needless to say, it's not; dimorphism comes in degrees. At one end of the spectrum, some spiders are so dimorphic that scientists can't tell whether a given male and a given female belong to the same species unless they catch them mating. At the other end, male and female love birds are so similar it virtually takes a blood test to tell them apart. (Somehow, though, they seem to manage it; if they didn't, they'd have a tough time reproducing.) Once again, however, sex differences in maximum offspring number explain the cross-species variation. Consider two of our ape cousins: the gorillas and the gibbons. Gorillas are *polygynous*; successful males maintain a harem of females. As a result, the maximum number of offspring that a male can produce is much higher than that of a female, and the males fight to be among the few with a harem. Gorillas are therefore highly sexually dimorphic: The males dwarf the females. Gibbons, in contrast, are *socially monogamous*: They live in isolated family groups with one adult male and one adult female, and the males have little scope to spread their seed. As a result, the maximum number of offspring that a male can produce isn't much higher than that of a female, and gibbons are *sexually monomorphic*: Males and females are almost identical in size.

Sexual monomorphism is rare in mammals, but it's common in birds. In roughly 90 percent of birds, males and females form pair bonds and work together to care for their young.[23] Although males can sometimes boost their reproductive output by cheating on their mates, pair-bonding and biparental care dramatically reduce the gap between males' maximum offspring number and females'. Thus, birds are among the most sexually monomorphic groups of animals. (Peacocks, our go-to example of sexual dimorphism, are actually highly atypical of our feathered friends.) Pair-bonding birds aren't just similar in size and parental tendencies; they're also often similar in their approach to mating. In most species, as we've seen, there's a clear division of labor in the mating game: Males compete for mates and females choose from the males on offer. In many pair-bonding birds, on the other hand, *both* sexes compete for mates and *both* are choosy about their sexual partners. In technical terms, these species have a system of *mutual mate choice*.[24]

At the start of this section, I asked you to forget about human beings. It's now time to recover from our self-imposed amnesia and face our species head on. How do we fit into this scheme?

Are People Peacocks or Penguins?

Are human beings highly dimorphic creatures like peacocks and gorillas – species in which some males mate with lots of females and others mate with none? Or are we more like gibbons and love birds – species in which males and females form pair bonds and the sexes are reasonably similar?

The short answer is: It's complicated. On the one hand, humans exhibit most of the sex differences found in our Top Ten list. First, men are larger than women: around 10 percent taller and 20 percent heavier. Men are also notably stronger than women, especially when it comes to upper-body strength.[25] Second, girls hit puberty earlier. For a period in childhood, girls and boys are comically mismatched; the girls suddenly tower over the boys. Girls' more rapid developmental trajectory may be part of the reason that they do better than boys at school; even when they're the same age, the girls are, in effect, older. Third, despite maturing faster, women typically outlive men.[26] This is the reason that, when you see a tour bus full of seniors, most of them are women. Women's greater longevity is found in nearly every nation, including most of those that still have high rates of maternal death during childbirth. From Jesus to Jim Morrison, men live faster and die younger than women.

In addition to these physical and developmental differences, humans exhibit most of the behavioral and psychological differences found in the Top Ten list. As we'll discuss soon, on average, men have a stronger desire for sex with multiple partners and a greater propensity for violence, whereas women are choosier about their mates and more concerned about a mate's resources. Furthermore, both sexes implicitly treat sex as a resource possessed by women and sought after by men.

These differences make good sense in the light of parental investment theory, because on average, women invest more into offspring than men. For starters, women's minimum contribution to the production of any child is a nine-month pregnancy, followed (at least until recently) by several years of breastfeeding. *Men's* minimum contribution is a little smooth talking, a few minutes of sex, and a dollop of protoplasm. On top of the sex difference in the minimum contribution, women also tend to do more of the childcare, not just in the West but in every culture for which we have data.[27] Men often do some of it, but rarely as much as women. Thus, physiologically and behaviorally, men invest less in their offspring. As a result, the maximum number of offspring that a man can have is greater than the maximum a woman can. It therefore stands to reason that humans check most of the boxes on our Top Ten list.

At first, it seems like an open-and-shut case: Humans are typical mammals with the typical pattern of mammalian sex differences. As we look more closely, however, things start to get more complicated. Humans diverge from the mammalian prototype in three important ways. First, although men and women are clearly not identical, the sex differences in our species are notably smaller than those in many others. For example, men are larger than women, but this difference pales into insignificance when humans are considered shoulder-to-shoulder with animals such as gorillas, orangutans, and elephant seals. The same holds as well for most psychological sex differences. Indeed, most psychological differences are smaller than the difference in height. The sex difference in interest in casual sex, for instance, which is one of the largest psychological sex differences in our species, is only around half the magnitude of the height difference. Thus, while men and women overlap in height, they overlap even more in their desire for casual sex. Generalizing the point, humans are dimorphic, but the level of dimorphism for most traits is relatively modest.[28]

Second, humans are not a species in which males alone compete for mates, or females alone are choosy about their mating partners. If we *were* such a species, men presumably wouldn't have any mate preferences and women would never have their hearts broken when the man they chose didn't choose them back. Similarly, if men weren't choosy about their mates – if they were interested solely in quantity rather than quality of mates – women would never have to worry about making a good impression on a date, would never fuss over their appearance before going to a bar or a nightclub, and would never pick apart the evening later, cringing at every awkward joke or misstep. Instead, they'd put on their most comfortable shoes and an old jumper, and spend the night evaluating men's appearance and courtship displays, never once expecting to be evaluated themselves. Does this sound like any dating scene on planet Earth? No. On this planet, both sexes compete to attract the other's eye and both are choosy about their mates.[29] Herbert Prochnow once described courtship as "the period of dating during which a girl decides whether she can do any better." But in our species, courtship serves this function for both sexes, especially in the context of serious, long-term relationships.

Third and finally, humans are not a species in which males are ornamented whereas females are drab. *Both* sexes have sexual ornaments: traits such as enlarged breasts in women and a V-shaped torso in men, which appear at puberty and which the other sex usually finds attractive.[30] It might be argued, in fact, that when it comes to physical appearance, women are more ornamented than men. As we'll see later, women's looks are more important in the mating game than men's – the exact opposite of what we find in peacocks and most other animals. And as we might expect in light of this asymmetry, women are also rated as more aesthetically appealing, by men and women alike.[31]

In a number of ways, then, humans are very different from most mammals. Indeed, in certain ways, we're more like the average bird than the average mammal. What's going on? To answer this, compare your own love life to that of a typical peacock. For the peacock, an entire relationship is a quickie: Male and female meet, mate, then go their separate ways, like two multi-colored ships that pass in the night. They don't fall in love, they don't settle down, and the duties of child-care fall entirely on the females' metaphorical shoulders. That's why peahens are choosy about their sexual partners whereas males are not. It's also why males alone have to woo their mates with their tails and extravagant displays.

Humans are a lot of things, but we're not peacocks. We're known to have the occasional quickie-length relationship, sure. But unlike peacocks, we're also known to fall in love. Right this moment, all around the globe, thousands of people are falling in love – some willingly; others against their better judgment. And when the twinkle in their eye turns into a babe in their arms, men often help to care for this tiny new individual. Certainly, men rarely log as many childcare hours as women, and some men are as parental as mud. But men in all known cultures log many more childcare hours than any peacock, and more than most other male mammals.[32] *All* male polar bears are deadbeat dads, for instance, but most male humans are not. In our species, the phrase "absentee father" is a term of abuse. In most mammals, it wouldn't be, any more than the phrase "non-lactating male" would. Male parental care just isn't on the menu.

Why is it on the menu for us? Ultimately, it traces back to the fact that making a new human being is an extremely expensive enterprise. Human offspring are utterly helpless at birth, and take a lot longer to grow up than the young of any other species. According to one estimate, it takes ten-to-thirteen million calories to ferry a human child from birth to nutritional independence in a traditional non-state society.[33] (My colleague and former PhD student, Andrew Thomas, tells me that this works out to between 18,000 and 23,000 Big Macs.) The costliness of our offspring is largely a consequence of our oversized brains. As a result of these ravenous organs, human children require more care than mothers alone can provide. Unlike most female apes, human mothers need help. As the sociobiologist Sarah Hrdy points out, this help can come from many sources, including grandparents (especially grandmothers), older siblings (especially sisters), aunts, grandaunts, maternal uncles, mother's friends, and miscellaneous hangers-on.[34] Often, though, it comes from the father.[35]

As men started helping to pick up the tab for their young, three things happened in short order. First, men evolved to be roughly as picky as women about their long-term mates – in other words, to be as picky as they could get away with, given their own desirability. This follows from the same logic as female pickiness: The more you invest, the wiser your investment needs to be. Second, women evolved to compete for the best investors. When males' only contribution to making offspring is sperm, females don't need to compete for male investment; not to put too fine a point on it, but there's more than enough sperm to go round. However, when males also invest time, energy, and resources into offspring, male investment becomes a scarce and valuable commodity, and females evolve to compete for it: They try to impress males and outshine other females. Third, because both sexes were choosy about their mates, both evolved their equivalents of the peacock's tail. Men's mate preferences may have helped shape women's breasts and hourglass figure, for instance, and women's mate preferences may have helped shape men's V-shaped torsos.[36] The fact that both sexes possess these secondary sexual characteristics, rather than only men, is one line of evidence that both sexes evolved to be choosy about their mates, rather than only women. If that *hadn't* happened – if men were not choosy – women would be as drab as peahens.[37]

In summary, pair-bonding and paternal care lowered the number of offspring that even the most successful men could sire (Ismail the Bloodthirsty notwithstanding). As a consequence, humans evolved a system of mutual mate choice, much like what we see in most birds. Humans *do* still exhibit most of the sex differences found in our Top Ten list. And it makes good sense that we do; after all, despite all the pair-bonding and paternal care, some men have always managed to have more offspring than any woman ever could.[38] However, men have had less scope to do this than males in many species, and therefore the sex differences found in our species are nowhere near as pronounced. Humans are inbetweeners, suspended on the tightrope between highly dimorphic animals like peacocks and highly monomorphic animals like gibbons.

That established, our next task is to survey some important human sex differences, and assess the arguments and evidence for an evolutionary account of their origin. Let's begin by looking in some detail at a topic that some find endlessly fascinating but others find simply abhorrent: the sex difference in casual sex.

Keeping It Casual

To get the party started, consider the following joke – a favorite of the evolutionary psychologist Donald Symons.

> An Irishman, an Italian, and an Iowan are arguing about which bar is the world's best.
> "The best bar in the world is Paddy's Pub in County Cork," says the Irishman. "After you've bought two drinks at Paddy's, the house stands you to a third."
> "That's a good bar," says the Italian, "but not as good as Antonio's in Old Napoli. At Antonio's, for every drink you buy the bartender buys you another."
> "Now, those sound like mighty fine bars," says the Iowan, "but the best bar in the world is Bob's Bar and Grill in Des Moines. When you go into Bob's you get three free drinks and then you get to go in the back room and get laid."
> The Irishman and the Italian are astonished to hear this, but they are forced to admit that Bob's Bar and Grill must indeed be the best bar in the world. Suddenly, however, the Italian gets suspicious.
> "Wait a minute," he says to the Iowan. "Did that actually happen to you personally?"
> "Well, no, not to me personally," admits the Iowan. "But it actually happened to my sister."

It's a straightforward joke, but a gender-neutral alien scientist would struggle to get it. We only do because we're familiar with certain stereotypes of the sexes: stereotypes about how much they value no-strings-attached sex, how easily they can get it if they want it, and the extent to which they perceive it as a favor granted vs. a favor received. The joke therefore raises two questions. First, is there any truth in the stereotypes? And second, if there is, why? Are the sex differences inscribed in our DNA, like male vs. female genitalia? Or are they arbitrary cultural habits, like wearing a dress vs. wearing a suit?

First things first: Is there really a difference? Casual observation certainly suggests as much. A few years ago, I read about a Chinese paranormalist who persuaded a young woman that her vagina was haunted, and that in order to exorcise the evil spirits that dwelt within, he would have to have sex with her. He even charged her $3,000 for the service. The next day, the woman went to the police and the "sexorcist" was arrested. The specifics of this story are surprising, but the underlying motives are not – and it would be much *more* surprising to read about a woman who tricked a man into sex by convincing him that his penis was haunted. Similarly, it would be surprising to hear parents warn their sons that "Women are only after one thing," or for male friends to advise each other "Just because she buys you dinner doesn't mean you owe her sex!" Unless you're an alien, such statements would sound decidedly odd.

So, casual observation suggests that there's a difference. The problem, of course, is that casual observation is notoriously unreliable; that's why people so often casually disagree about these sorts of matters. Ideally, our views will be shaped primarily by more rigorous, scientific evidence. The good news is that there's now a mountain of such evidence addressing this question, and addressing it from multiple angles. One angle is simply to ask people about their sexual proclivities – not face-to-face but in anonymous questionnaires. In survey after survey, the same picture emerges. First, on average, men have a stronger sex drive than women. As Mignon McLaughlin observed, "A nymphomaniac is a woman as obsessed with sex as the average man." Men have more sexual dreams than women, and spend more time daydreaming about sex. Men masturbate more; according to an old joke, 98 percent of men report in surveys that they masturbate and the other 2 percent lie. Men are more likely to regret passing up an opportunity for casual sex. And men are much more likely to view their virginity as a source of embarrassment and an albatross around their necks to be cast off as soon as humanly possible. The social category "reluctant virgin" is populated almost solely by men, as is the category "involuntary celibate" (or "incel").[39]

The survey data point to other interesting trends as well. On average, men are more willing to engage in sex in the absence of a close relationship. They're more willing to jump into bed with someone they barely know. They report wanting more sexual partners in the next month, the next year, the next decade, and over the course of the rest of their lifetimes. They report more sexual fantasies, and more fantasies involving romance-free sex and an ever-changing roster of partners. And they're more likely to cheat on their partners, even when the relationship is going well.[40] The studies showing these kinds of sex differences aren't just limited to Western nations; the same results have been found in every major world region, including Africa, the Americas, Asia, Europe, the Middle East, and the Pacific. The differences even show up in gender-egalitarian, sexually liberal nations such as Norway.[41] Certainly, some women enjoy casual sex and some men don't; we're talking about a difference in the average scores of men as a group vs. women as a group. But that average difference is one of the most solid findings in psychology.

Self-report studies are a good start, but they're never completely convincing. The comedian Lenny Bruce used to ask the men in his audience how many had had blowjobs. (This was back when blowjobs were still a novelty in polite society.) Most male hands went up. He then asked the women how many had *given* blowjobs. No hands went up. "*Someone's* lying," he concluded. And that's the problem with self-report studies: People often lie about sensitive issues – and even when they tell the truth as they see it, their perceptions may be muddled or misguided to a greater or lesser extent. Before we can rest our case, we need to corroborate the self-report findings with studies looking at people's actual behavior "in the wild."

Many such studies exist. These include, for a start, real-world experiments. In one of the most famous studies in all of psychology, researchers had a team of young men and women approach members of the other sex on a busy campus and say: "Hi. I know you don't know me, but I've been watching you for a while, and I find you very attractive."[42] It sounds a bit stalkerish, but this was in the days before mobile phones and nobody called the police. The research assistants then asked one of three questions: (1) "Would you go out with me tonight?"; (2) "Would you come over to my apartment tonight?"; or (3) "Would you go to bed with me tonight?" (To be clear, this wasn't a multiple-choice question; that would have made this already-weird situation even weirder. Each person in the study was asked only one of the three questions.)

The results managed to be at once surprising and exactly what you'd expect. For the first question ("Would you go out with me?"), there was no sex difference; around half the men and half the women said yes while the other half demurred. For the second question ("Would you come over to my apartment?"), a large sex difference opened up: 69 percent of men said yes, as opposed to just 6 percent of women. But for the last question ("Would you go to bed with me?"), the gender gap was huge: 75 percent of men said yes... as opposed to 0 percent of women.

Not only did more women than men turn down the kind offer of sex, among those who did, there was a striking sex difference in the manner of the refusal. Most of the men were apologetic, explaining that they were married or had a prior engagement, and in some cases asking if they could get a rain check. The women, in contrast, were not apologetic. Typical responses included "You've got to be kidding" and "What's wrong with you?" None of the men asked what was wrong with the woman offering him sex. Incidentally, history doesn't record how the participants in the study reacted after being told they were in an experiment. If the study's findings are anything to go by, though, there were a lot of disappointed guys at the end of the day, and a lot of relieved women.

The would-you-go-to-bed-with-me study was conducted in the United States in the late 1970s, at the height of the Sexual Revolution and before the AIDS crisis. Despite that, the sex difference was glaring. And it wasn't an isolated finding. Similar studies have been conducted in other countries and later decades, and they've all come to the same conclusion: Even in the most sexually liberated nations on Earth, men are more likely – far more likely – to say yes to a stranger who offers them sex.[43] This doesn't imply, of course, that no woman has any interest in casual sex. Some do. Some even have fantasies about sex with total strangers, even if relatively few ever act on them.[44] But their fantasies tend not to involve men walking up to them out of the blue, telling them they've been watching them, and then explicitly asking for sex. If the person doing the asking were a movie star, things might be different.[45] In the normal course of events, though, the sex difference is as big as an elephant.[46]

There are various other ways to measure the elephant. One is to look at the behavior of people who, for one reason or another, have relatively few constraints when it comes to getting the kinds of sexual relationships they want. Gay men and lesbians are one such group. In the relationship arena, gay men don't have to compromise with women, and lesbians don't have to compromise with men. As such, their sexual behavior gives us an unusually clear window on the sexual inclinations of men and women in general. And just as we'd predict based on the assumption of greater male interest in casual sex, gay men have more sexual partners than straight men, whereas lesbians have fewer partners than straight women. One study found, for instance, that in San Francisco in the 1970s (again, prior to the AIDS crisis), 75 percent of gay men reported having had more than a hundred sexual partners, and 28 percent more than a thousand. In contrast, only 2 percent of gay women had had more than a hundred partners, and none more than a thousand.[47] It's not that gay men have a stronger desire for casual sex than straights (or "breeders," as they're sometimes known in the gay community). As Donald Symons notes,

> heterosexual men would be as likely as homosexual men to have sex most often with strangers, to participate in anonymous orgies in public baths, and to stop off in public restrooms for five minutes of fellatio on the way home from work if women were interested in these activities.[48]

The problem, from the straight male's point of view, is that most women are not interested.

Gay men and lesbians aren't the only people who can generally get the kinds of sexual relationships they want. Extremely, unfeasibly attractive people can usually do so as well. These annoying individuals tend not to have to compromise with the other sex, even if they're straight, just because they're such prized commodities. And their romantic histories again back up the claim that men are keener on casual sex than women. Several studies indicate that ridiculously handsome men are more likely than run-of-the-mill men to rack up large numbers of sexual partners. Beautiful women, on the other hand, are not. This isn't because the latter have trouble attracting partners; it's because they more often opt for fewer partners and longer-lasting relationships.[49] We can all think of exceptions, no doubt. But the general rule is sound.

Of course, you don't have to be beautiful to get what you want if you've got other things going on. People with unrestricted political power – warlords, kings, and emperors, for example – can generally get their own way in the mating game even if they look like the back end of a bus. These individuals thus offer yet another window onto the average Joe or Josephine's true desires. And sure enough, when we look at the sexual antics of the preposterously powerful, we find the standard sex difference writ large. Ismail the Bloodthirsty turns out to be just one example of a much more general trend. According to the anthropologist Laura Betzig, in all the ancient civilizations of the world – including those of the Aztecs, the Babylonians, the Chinese, the Egyptians, the Incas, the peoples of the Indian subcontinent, and the Zulus – powerful men accumulated large harems of nubile young women. Equivalently powerful women, such as Cleopatra, did *not* accumulate large harems of nubile young men.[50] They could have but they didn't. Admittedly, Ismail and his ilk are far from typical human males. Still, they exhibit in an extreme form a sex diffe-rence found to a more moderate degree among the general rank-and-file.[51] Notice that Betzig's work has another implication as well, namely that the sex difference in the desire for multiple partners isn't limited to modern times or to Western populations.

A final, very different way to test for the sex difference is to look at consumer preferences, and in particular the kinds of entertainment people favor. To start with, as every non-alien knows, men consume more porn than women – a lot more. Porn is particularly popular with young men, middle-aged men, and older men. Some women like porn too, but far fewer than men, and those who do like it tend not to like it quite so often. Like men's sexual fantasies, pornography involves numerous, nameless partners, and sex stripped of love and commitment. Bruce Ellis and Donald Symons dubbed the fantasy world depicted in porn *pornotopia*, observing that,

> The most striking feature of male-oriented pornography is that sex is sheer lust and physical gratification, devoid of encumbering relationships, emotional elaboration, complicated plot lines, flirtation, courtship, and extended fore-play; in pornotopia, women, like men, are easily aroused and willing.[52]

Whereas men are the main consumers of porn, women are the main consumers of romance novels and rom-coms.[53] Not all women enjoy "romance porn," as it's sometimes called, but more women do than men. Romance novels serve up a fantasy world very different than the world of pornotopia. They usually involve sex, but they focus a lot more on mate selection and on building an emotional bond. In a typical romance novel, woman meets man; man is a bit of a jerk; woman reluctantly finds herself attracted to man; woman eventually tames man and the couple live happily ever after (or happily for the foreseeable future). The gulf between romanceotopia and pornotopia tells us a lot about the sexual psychologies of men and women. It also tells us why the sexes don't always see eye to eye. Dispensers of marital advice often recommend that men and women explore their mutual sexual fantasies. As the comedian Bill Maher once noted, however, "There are no such things as mutual fantasies! Yours bore us; ours offend you."

OK, fine; this is overstating the difference. We shouldn't read *too* much into men's porn addiction. Porn, after all, is not meant to satisfy every male desire; it's purely an aid to masturbation. As such, the fact that porn doesn't feature a lot of flirtation, courtship, or romance doesn't imply that men have no interest in these things in other contexts – that is, when they're not masturbating. Plenty of men interrupt their regular diet of pornography to watch the occasional rom-com, even when they're not forced to do so by their partners. And plenty of women who love a good rom-com also enjoy a sporadic encounter with porn. Nonetheless, the porn gap and the rom-com gap are not trivial – and like all the evidence we've surveyed, they point to a very real difference in the sexual natures of women and men.

This difference has a powerful impact on the dynamics of heterosexual relationships. Because men often have a stronger desire for sex, and especially sex of the no-strings-attached variety, intercourse is often treated as a resource that women possess and men pursue.[54] Even when the woman enjoys the sex as much as the man, it's still tacitly seen as a favor that she does for him, rather than the other way round (unless, that is, he's a movie star). This view of sex as a female service reveals itself in a number of ways. As evolutionary psychologist David Buss points out, women can normally land a more attractive partner if they offer sex without commitment, whereas men can land a more attractive partner if they offer commitment as well.[55] Men are more likely than women to give potential sexual partners gifts such as flowers and extravagant meals.[56] Men are virtually the sole consumers of prostitutes, and prostitutes are overwhelmingly women; as the sociologist Pierre van den Berghe observed, "The male prostitute, unless he caters to homosexuals, is an economic redundancy, constantly undercut by eager amateur competition."[57] And female porn stars are paid more than the males – a genuine gender pay gap.[58] The radical feminist Andrea Dworkin summed it up well:

> A man wants what a woman has – sex. He can steal it (rape), persuade her to give it away (seduction), rent it (prostitution), lease it over the long term (marriage in the United States) or own it outright (marriage in most societies).[59]

The conclusion is hard to duck. Men, on average, are more interested than women in casual sex and sexual novelty. The only question now is why.

Explaining the Casual Sex Gap

> Given our explicitness on this issue, when a critic describes the theory as proposing that "men are promiscuous, women are monogamous," one can only wonder about the person's scholarship, training, or eyesight.
>
> —*David Buss (2003), p. 225*

"Why can't a woman be more like a man?" asked Henry Higgins in *My Fair Lady*. Applied to the topic of casual sex, this question underpins one of the most heated debates in psychology. Why are men more interested than women in uncommitted sex and sexual variety? On one side of the debate are those who argue that it's all down to social pressure, socialization, and culture. On the other side are those who argue that, even if these factors nudge things around a little, or even a lot, the ultimate roots of the difference lie deep in our evolutionary past.

We've already seen the basics of the evolutionary explanation. A good way to make the logic of the explanation crystal clear is to imagine that you live in a premodern society – the kind of society in which most of our evolution took place – and that your one and only goal in life is to pass as many of your genes as possible into the next generation. What would be the best way to achieve this goal? The answer, it turns out, depends a lot on whether you're a man or a woman. If you're a man, your thinking might go like this:

> OK, the best way for me to spread my genes would be to have as many children as I can. Each child will need lots of parental care. So, one option would be to settle down with a fertile female, get her pregnant, and help raise a bunch of kids. On the other hand, if I'm particularly attractive to women, or for some other reason have lots of sexual opportunities, I could try my hand at a different approach: I could mate with as many women as possible. If I were to mate with, say, five women in a year, I could potentially have five kids. I probably wouldn't be able to help care for them all, so some might not make it to adulthood. But some surely would, and more than likely I'd end up with more kids if I took this approach than if I limited myself to just one woman. And of course, I could always do a bit of both: some fathering and some philandering. At the very least, I shouldn't be too picky about my casual sexual partners. After all, if I have a fling with a suboptimal partner, it's no big deal – it costs me little and I'm back on the market again almost immediately. Good times!

Now, with the exception of the deranged fertility doctor who surreptitiously impregnated his clients with his own sperm instead of the husbands' (true story), no one actually tries to figure out how they can pass on as many of their genes as possible. It's easy, though, to imagine genes that push around men's preferences, motivations, and emotions in such a way that they cause men to act *as if* they were trying to do exactly that: genes that increase or decrease sex drive, for instance, or that increase or decrease the tendency to bond. Any such genes would stand a good chance of being selected.

That's the male half of the equation. Now, what if you were a *woman* who wanted to pass on as many genes as possible? Well in that case, your thinking might go something like this:

> Just like a man, the best way for me to spread my genes would be to have as many kids as possible. But I could never have as many as the most successful men. Whereas a man who mates with five women in a year can potentially have five children, if *I* mate with five *men* in a year, I'll only have as many children as I would if I'd mated with one. Sure, there may be some benefits to having multiple partners. But nine times out of ten I'd be better off if I held out for a super-fit guy who'd give me super-fit kids, or a good provider who'd help me look after the kids – or if possible, a guy who'd do both. At the very least, I should keep well away from any man who clearly doesn't measure up. After all, whereas a man who hooks up with a suboptimal partner can be back

in the game straight away, if *I* hook up with a suboptimal partner, it could tie up my reproductive resources for at least nine months – and if I decide to keep the kid, for several years after that.

Again, no sane person would ever think like this, but again, they don't need to for the theory to work. Notice that I've tried to word the above in such a way as to avoid a common confusion: the idea that, according to evolutionary psychologists, men are only interested in racking up notches on the bedpost, whereas women are only interested in snagging a life-long mate. This view is doubly wrong. First, to say that men are more interested than women in casual sex is not to say that men are any less interested in long-term, committed relationships. I've mentioned already that pair-bonding and biparental care are a big part of the human repro-ductive repertoire, and we'll look at this in more depth in Chapter 4. The important point for now, though, is that both sexes are capable of falling in love and forming long-term relationships, and thus that long-term relationships must generally have been adaptive for both sexes: men as well as women. The desire for casual sex is just one component of men's mating psychology.[60]

Second, the fact that men are more interested in casual sex doesn't imply that women are *not* interested. Many are, and most evolutionary psychologists argue that casual sex was sometimes adaptive for women in our ancestral past. Casual mates may have provided meat or other resources, they may have helped out with the kids, or they may have had better genes than the guy who was willing to get serious.[61] But racking up sexual conquests didn't boost women's fitness in the immediate and powerful way that it did for men. We shouldn't be surprised, therefore, that on average, women today are *less* interested than men in racking up sexual conquests. Moreover, given that even a single roll in the hay could potentially have saddled a woman with a nine-month pregnancy and several years of childcare duties, we shouldn't be surprised that women have higher standards than men for casual sexual partners. Madonna once described herself as "selectively promiscuous." This is probably a good description of most women's approach to casual sex (the occasional drunken mistake notwithstanding).[62]

That, at any rate, is the evolutionary psychologists' story. There's a rival narrative, however, and that's the idea that the sex differences come not from evolution but from culture. Here's how this version of events might go. When it comes to casual sex, men and women are not playing on a level playing field. It's true, of course, that making a baby entails a greater biological expenditure for women than for men: Women get pregnant; women give birth; women nurse the young. However, unlike other animals, women *know* this. They know that whenever they have sex just for fun, they're playing Russian roulette with an unwanted pregnancy. Even if men and women had identical sexual desires, this would surely make casual sex less appetizing for women. Admittedly, pregnancy was a bigger worry before we had reliable contraception. But it's still going to weigh on women's minds more than it does on men's.

And pregnancy isn't the only concern. Men are larger and stronger than women, and more inclined to violence. This means that a woman places herself at a greater physical risk by going off with a man she doesn't know than a man does with a woman. She also places herself at a greater social risk. Even today, there's a sexual double standard in the West: Men who sleep around are viewed as heroes or lovable rogues, whereas women are viewed as sluts and "not marriage material." As Joan Rivers put it, "A man can sleep around, no questions asked, but if a woman makes 19 or 20 mistakes she's a tramp." Again, this is surely going to have an impact on women's sexual behavior. It's not evolution; it's just basic human rationality. People weigh up the costs and benefits of casual sex, and act accordingly.[63]

Besides, even if men and women *do* have differing sexual appetites, why assume that these come from evolution? Our appetites are powerfully influenced by the culture around us; that's why people in some cultures think that eating insects is about the most disgusting thing a person could do, whereas people in others think that eating bacon or cheese deserves that honor. Western culture is constantly teaching men that casual sex is bacon but teaching women that it's a fried grasshopper (or vice versa if you prefer the grasshopper). Girls are taught to keep their legs closed; boys are encouraged to sow their wild oats. Girls are taught that sex without love is a meaningless experience, boys that, as meaningless experiences go, it's a pretty damn good one. The sexual double standard might initially have appeared because parents worried that, if their daughter got pregnant outside of wedlock, she'd be left holding the baby – or that *they* would. But however it got off the ground, it soon became an unreasoned social norm: a moral belief that people soaked up from the culture around them and accepted independently of any consideration of the costs or benefits of casual sex.

Certainly, times have changed. The arrival of the pill in the 1960s greatly reduced the risks of casual sex for women, and as women started moving into the workplace, they no longer needed to barter their virginity and sexual favors for a slice of a man's paycheck. And just as we'd expect, premarital and recreational sex have become more common and more acceptable for women since that time. But culture changes only slowly, and the earlier attitudes linger on to some degree. In as much as this drives the sex difference in casual sex, it's not evolution *or* rational thinking. It's just a cultural habit.

So, we've got two possible explanations for the sex difference: the evolutionary explanation and the sociocultural alternative. It's time to start sorting the fact from the fiction, the wheat from the chaff, the men from the boys, and the women from the girls. Let's begin with the question of cultural change. It's clearly true that, since women got the pill and started earning their own money, casual sex has become more common and more acceptable for women. To a lot of people, this is proof positive that the casual sex gap couldn't be due to evolution. But the conclusion doesn't follow. The change shows that the gap couldn't be due *only* to evolution; it doesn't show that it's due only to culture. And just to be clear about this, no evolutionary psychologist denies that the pill and other environmental factors affect people's willingness to engage in casual sex. Of course they do! The only reason they wouldn't is if people's behavior were completely unaffected by the probable costs and benefits of different courses of action in their immediate environment. But no evolutionary psychologist would make such a ridiculous claim. The claim they do make is that the costs and benefits *aren't the whole story*. Environmental variables don't operate on a blank slate; they operate on minds that are somewhat differentiated by sex. To understand men and women's behavior, you have to take into account not only the environment but also the evolved sex differences.

With this in mind, men and women's sexual behavior since the 1960s actually fits better with the evolutionary psychologists' narrative than with its sociocultural rival. The pill removes one of the main risks of casual sex for women: the risk of getting pregnant and having to raise an unplanned child without the help of the father. Despite this, more than a half century after it first appeared, women are still less willing than men to engage in casual sex. This doesn't *prove* that the Nurture Only theory is false; it's no doubt possible to come up with a non-evolutionary explanation for the tenacity of the sex difference. But the tenacity of the sex difference isn't what we would have *predicted* on the basis of the Nurture Only theory. That should make us wary of any post hoc explanation for the persistence of the casual sex gender gap.

Other evidence presents a similar challenge to the Nurture Only view. Consider, for instance, the sexual behavior of lesbians. Unlike straight women, lesbians don't have to worry about getting pregnant *or* about being alone with a more physically powerful partner. If these worries were all that kept women from jumping into bed with every good-looking stranger, then lesbians would have more sexual partners than their straight female counterparts. As we saw earlier, however, they don't; they have *fewer* partners. This removes yet another block from the Jenga tower of the Nurture Only theory. It suggests that the casual sex gender gap couldn't just be a product of cost–benefit analyses related to pregnancy risk or personal safety.

It doesn't yet entitle us, though, to conclude that the gender gap is innate. After all, pill-takers and lesbians might simply be following the casual-sex gender norms of their society, or the dictates of their gender-role socialization as children. These are not unreasonable hypotheses; both, however, face formidable – and I would argue fatal – challenges. For one thing, the social pressures don't all point in the direction presupposed by the Nurture Only theory. Research has generally failed to demonstrate a consistent or pervasive sexual double standard in the modern Western world. Although the *belief* in the sexual double standard is still widespread, the double standard itself is much less so. Indeed, some people these days hold a reverse double standard, such that they judge *men* who sleep around more harshly than they judge women.[64] Certainly, "slut" is an insulting term. But so is "prude," and women who *don't* engage in casual sex are sometimes called that – by men who want to sleep with them, for a start. And while "slut" is usually reserved for women, there are plenty of pejorative terms for men who sleep around: womanizer, dirty old man, letch, sleaze, "worthless, lazy, good-for-nothing, womanizing asshole" – and so on.[65] At the very least, society gives us mixed messages. And yet the sex difference in casual sex persists.

More than that, the difference persists even when society pushes against it. As Donald Symons points out, men's stronger interest in casual sex and sexual novelty has survived society's best efforts to eradicate it. It has survived the efforts of parents, partners, and moralists to inculcate men with a healthy respect for monogamy. It has survived Christian moral teachings and threats of eternal damnation. It has survived cultural and legal institutions that endorse and incentivize lifelong monogamous marriage. It has survived worries that one might lose one's marriage, one's children, or even one's livelihood over an adulterous affair that won't stay hidden. And it has survived pop psychological attempts to stigmatize men's desire for casual sex by blaming it on psychosocial immaturity, psychological maladjustment, repressed homosexuality, low self-esteem, fear of commitment, a Peter Pan syndrome, misogyny, male entitlement, toxic masculinity, and rape culture. Meanwhile, women's greater reticence about casual sex has survived the efforts of some feminists and other thought leaders to persuade women to cast off the shackles of patriarchy and match men in the casual sex arena. This is all rather awkward for the Nurture Only theory. It suggests that, rather than being a *product* of culture, the sex difference in attitudes to casual sex often emerges *in spite* of culture.

Arguably, though, the most persuasive argument against the Nurture Only view is that sex differences in sexual inclinations and choosiness can be found in many individuals who have no gender norms, no socialization, and little in the way of culture: that rather sizeable group, so often overlooked by psychologists, known as *other animals*. The differences aren't found in all other species, but they are found in many, including most birds, mammals, and reptiles.[66] And when we find the differences in other animals, evolution is the only reasonable explanation. Why should humans be different? It's logically possible, of course, that the differences are products of evolution in squirrels, turkeys, and frogs, but of learning and culture in *Homo sapiens*. But it hardly seems likely. In other species, the differences appear when the ceiling number of offspring for males is higher than that for females. Humans meet this condition, and our species presumably evolved from earlier species that displayed the normal sex differences. As such, what the Nurture Only theory asks us to believe is that, in our lineage and ours alone, natural selection eliminated the normal sex differences, despite the fact that the selection pressure that initially created them was still operative.[67] Why would it do that? It's particularly perplexing given that, when we look around the world, we *still find* the sex differences that selection supposedly eliminated. Thus, the Nurture Only theory asks us to believe not only that selection eliminated the differences for reasons unknown, but that learning and culture then coincidentally reproduced exactly the same differences in every culture on record. This is not a compelling thesis. Cultural forces clearly influence people's willingness to engage in casual sex, and to some extent their desire to do so as well. But the idea that culture creates these sex differences out of nothing not only clashes with the available evidence, it clashes with everything we know about how evolution works.

Looking Good

Few tasks are as important in the life of a female animal, whether spider or spider monkey, as selecting a mate. And although no two females are exactly alike, a general trend stands out from the noise like coal in the snow: Females in a wide range of species show a distinct preference for males that can give their offspring a good start in life, either directly (by providing them with protection or resources) or indirectly (by providing them with good genes). Male animals, in contrast, are usually not so picky. The most that can be said about males in most species is that they prefer to mate with adult females of the same species – and even that's not an absolute requirement, as the monkeys, seals, and jewel beetles discussed at the start of the chapter make all-too uncomfortably clear. In short, in a great diversity of species, females are choosy about their mates but males basically aren't.

As we've seen already, and contrary to the view of some, humans are *not* that kind of species. Certainly, for brief or low-commitment relationships, men are more willing than women to relax their standards, if they have to. But for long-term, committed relationships, the sex difference in choosiness largely evaporates.[68] This fact is even reflected in everyday folk wisdom; the common stereotype is that men will *sleep* with anything that moves, not that they'll marry or have children with anything that moves.[69] Both sexes are choosy about long-term relationships: the relationships that dominate the romantic lives of the majority of women and men.

But are they choosy about the same things? Or do men and women look for different qualities in a mate? As is often the case, the answer is: It depends. In many ways, the sexes want the same things. Both want someone who loves them, who's good looking, and who isn't stupid, cruel, or unstable.[70] At the same time, though, men and women differ in the strength of their desire for certain traits. Two such differences have grabbed the lion's share of attention in evolutionary psychology. The first is that, on average, men place more importance than women on a mate's looks and youthful appearance. The second is that – again, on average – women place more importance on a mate's wealth and social status. As Warren Farrell put it, men see women as sex objects, whereas women see men as success objects. According to evolutionary psychologists, these tendencies have their origin in natural selection.

(Note that I'll be focusing here on the mate preferences of heterosexual men and women. This isn't because the preferences of non-heterosexuals don't matter, but because heterosexual preferences are the ones most plausibly crafted and honed by natural selection. They are, after all, the preferences that most reliably resulted in offspring.)

Let's start with the fact that men typically place more weight than women on a mate's good looks and youthfulness. The evidence for this sex difference comes from several sources. These include, to begin with, self-report surveys. The standard protocol here is to give people a list of traits and then ask them to rate how important each trait is to them when it comes to choosing a mate. In the vast majority of studies using this method, men rate good looks and youthfulness as more important than women.[71] This is true even among Western university students, who are usually less approving than the general populace of such allegedly sexist and old-fashioned preferences. Are people lying or confused about what they really want? It seems not. The preferences that people express in surveys are also evident in real-world behavior. Analyses of dating ads show that younger women get more responses than older ones, while men's age has less impact on their popularity.[72] When older men remarry, they tend to marry younger women; when older women remarry, they're not especially likely to marry younger men.[73] Men who are best positioned to get what they want on the mating market – wealthy men, for instance, or men who purchase mail-order brides – usually opt for younger women.[74] And sex workers charge more when they're younger, the exact opposite of what we find in almost every other profession.[75]

On top of all that, men tend to be more interested than women in what psychologists call *visual sexual stimuli*.[76] This is a big part of the reason that men consume more porn. It's not that women don't like sexual entertainment; they just generally prefer it in written rather than visual form.[77] As a result, most visual pornography is aimed at men. Pornographers have *tried* making porn for women, only to discover that there's not much of a market for it. Apparent exceptions, such as *Playgirl* magazine, turn out to be red herrings: They're purchased overwhelmingly by gay men, not by straight women. The fact that women are less interested in visual sexual stimuli is unlikely to shake up most people's view of the world. After all, if women *weren't* less interested, we'd live in a very different world from the one we now inhabit – a world where women respond to builder's bum in the same way that men respond to cleavage, where exhibitionists are the stars of women's sexual fantasies, and where every flasher lurking in the park is a sexual success story. I'm inclined to doubt that anything short of brain surgery could bring about a world like that.

Seen through a Darwinian lens, men's stronger interest in a partner's looks is initially quite mysterious. In most species, it's the females that care more about looks. Among peacocks, for instance, peahens will only mate with the males with the sexiest tails; males, in contrast, are less shallow and will mate with any female who'll have them. If peacocks had pornography, it would be the females, not the males, that would spend their time staring at images of the other sex, and the males that would complain about being treated as sex objects. And what's true of peacocks is true as well of most animals. In our species, however, it's the other way around. Why?

To solve this riddle, we first need to understand why anyone finds anyone else attractive ever. We'll discuss this in more detail in the next chapter, but the basic idea is as follows. Humans evolved to find certain traits attractive because, throughout our evolutionary history, those traits were statistical indicators that the people possessing them would make suitable mates. The traits in question include indicators that a potential mate is healthy, indicators that a potential mate is fertile, and indicators that a potential mate has good genes. If your one goal in life were to have as many healthy offspring as possible, then your best bet would be to have those offspring with people possessing these traits (assuming that they'd have you). Of course, in real life, people rarely think about health, fertility, or genes when choosing a mate. But that's because natural selection has done the thinking for us. It's equipped us with an automatic tendency to go weak at the knees for anyone possessing traits correlated with these evolutionarily significant variables. Such traits include everything from facial and bodily symmetry to lustrous hair to healthy, youthful skin.

That, in a nutshell, is why people care about their partner's looks. Why, though, do men care more? Simple: The traits that people see as good looking are often linked to youthfulness, and women's fertility is more tightly coupled to youthfulness than men's. Consider the world records for the oldest man and the oldest woman to have children. The oldest man was ninety-six. The oldest *woman* was sixty-six – three decades younger – and she was only able to have a child at that age through in vitro fertilization. Most women go through menopause between the ages of forty-five and fifty-five, after which their baby-making days are over. Men, on the other hand, can in principle keep making babies till the day they die.[78] Put simply, women have a narrower window of fertility than men. And that's why men have evolved to put more weight than women on a partner's youthfulness: Youthfulness is a more important indicator of fertility in women. Note that most animals don't have menopause, and thus that males in most species don't have a sexual preference for youthful females. Male chimps, in fact, prefer older ones.[79] The youthfulness preference isn't unique to humans, but it is rare, especially among the primates.

This is not to suggest, by the way, that women don't care about looks. Clearly they do; that's why good-looking men have more sexual partners than men less blessed in the looks department.[80] It's also not to suggest that a partner's age is never an issue for women. All else being equal, women find younger, fitter guys more physically attractive than older, less fit guys – guys whose main form of exercise is sucking in their guts when an attractive woman walks by. The point is that women are generally *less strict* about looks. They can more easily fall for a guy who's past his prime, if he ticks all the other boxes.

I've given you the evolutionary explanation for the good-looks sex difference; now let's consider some of the criticisms raised against it. The critics have two main lines of attack. The first is to argue that the sex difference doesn't exist; the second is to argue that, even if it does, it comes from culture rather than evolution. Tackling the first point first, many critics observe that, although evolutionary psychologists *claim* that men prefer youthful women, various lines of evidence contradict the assertion. For example, one of the most popular categories in online porn is "MILF," which features older women in the starring role.[81] And even in mainstream media, plenty of older women are sex symbols, many of whom date younger men (aka toy boys). This, the critics argue, falsifies the claim that men have an inbuilt, insurmountable preference for youthful women.

The critics have their facts straight, but their conclusion is wrong. First, while it's true that MILFs are popular in porn, younger women are *more* popular – precisely the pattern that evolutionary psychology predicts and explains. Furthermore, MILFs themselves are usually quite young; most are in their twenties or thirties. Porn featuring women outside that age range is not nearly as popular.[82] Second, it's true that older women can sometimes be sex symbols. However, the fact that atypically attractive women are still attractive in their forties or fifties doesn't undermine the claim that, in general, women are *more* attractive in their reproductive prime (as are men). If this seems unfair, it's probably because... it *is* unfair! But that, unfortunately, doesn't mean it's not true.

There's another problem with the men-like-older-women argument. From an evolutionary perspective, it's not actually age per se that matters in choosing a mate; it's outwardly visible traits that, in our ancestral past, correlated with youthfulness and fertility. If a woman stopped aging at twenty-two, men would still find her attractive when she turned 101. And in certain ways, women these days *do* stop aging, or at least age at a slower pace. The modern world is a lot less physically harsh than the environment our ancestors had to negotiate. Moreover, we've now got numerous ways to fake the signs of youthfulness, among them cosmetics and cosmetic surgery. Bearing that in mind, imagine that you took a photo of a postmenopausal woman and airbrushed it so it looked like she was twenty-five. You then showed the doctored photo to a bunch of men, all of whom agreed that the woman in the photo was very attractive. Would this disprove the evolutionary psychologists' claim that men tend to prefer younger, premenopausal women? Of course not; that would be a silly argument. The original argument – that modern men sometimes find older women attractive and that *this* disproves the evolutionary psychologists' claim – isn't quite so silly. But it is somewhere on the same continuum of silliness. Women in the developed world are, in effect, airbrushed. Because of our relatively cushy lives and our easy access to cosmetics, some postmenopausal women today look as youthful as many premenopausal women in our ancestral past. Thus, the fact that older women are sometimes seen as attractive doesn't undercut the evolutionary psychologists' story. On the contrary, the fact that younger women are still usually seen as *more* attractive, despite all the airbrushing, is testament to the power of the evolved preference for youthfulness.

It doesn't look like we're going to be able to wish away the sex difference. The next move for the critics would be to argue that the difference comes solely from culture. Culture, the critics might argue, *teaches* us that women's looks are more important than men's: Men's bodies are for action; women's are for looking nice. Culture also teaches us an arbitrary ideal of what "looking nice" means. It teaches us that youthfulness is central to attractiveness for women but much less so for men. Not only is this meme perpetuated by parents and peers, it's stamped into our heads, 24/7, by a mass media that bombards us with images of scantily clad young women, and by a beauty industry that hawks products designed to hide wrinkles, eliminate gray hair, and generally make women look younger. With all that going on, it's little wonder that we judge women on their looks more than we do men.

On first inspection, the Nurture Only view sounds plausible. The more we think about it, however, the more the aura of plausibility fades. First, if the preference for youthful women were purely a product of learning, the implication would be that people could just as easily learn to find elderly women more attractive than younger ones as they could the other way around, and that the beauty industry could just as easily brainwash women to buy wrinkle-enhancing cream as anti-wrinkle cream, gray hair dye as blonde or red. This seems a tad unlikely. To my mind, it's much more plausible that media and the beauty industry exploit pre-existing tastes than that they create new tastes from scratch. Second, we now have a truckload of evidence that the traits people consider attractive in women are closely linked to fertility.[83] Before this evidence came to light, the Nurture Only view committed us to just one claim: that the preference for youthfulness in women is an arbitrary cultural convention. Now, though, the Nurture Only view commits us to *two* claims: that the preference for youthfulness in women is an arbitrary cultural convention *and* that it's purely a coincidence that youthfulness in women is closely linked to fertility. The addition of this second claim renders the Nurture Only view much less plausible than it was – especially given the profound evolutionary significance of fertility in mate choice.

As well as these theoretical problems, several lines of evidence flatly contradict the Nurture Only view. First, the sex difference in the preference for good looks isn't just a local phenomenon, but appears to hold across cultures. In one of the first major studies in evolutionary psychology, one of the founders of the field, David Buss, surveyed more than 10,000 people from thirty-three nations, asking them what they valued in a long-term mate. As expected, in almost every one of Buss's thirty-seven samples, men placed more importance on looks than women. Furthermore, in every sample without exception, men reported desiring women a few years younger than themselves, whereas women reported desiring men a few years older (most of the participants were in their twenties).[84] Later research revealed that men's preference isn't for a younger woman per se; it's for a woman at or around the peak of fertility.[85] Thus, teenage boys tend to find women who are somewhat *older* than themselves more physically attractive, despite the fact that these women have approximately zero interest in most teenage boys, and despite the social norm that men should date younger women.[86]

One might object that, although the Buss study canvassed a much wider range of cultures than most research in psychology, all its samples were drawn from modern, industrialized nations. But the anthropological record extends the scope of our gaze to small-scale, traditional societies, and largely paints the same picture. For example, the anthropologist Napoleon Chagnon reported that, among the Yąnomamö of the Amazon basin, men speak of their lust for women who are *moko dude* – that is, who are sexually mature but haven't yet had a child. The literal translation of the phrase is *harvestable* or *perfectly ripe*.[87] A cherry-picked anecdote? It seems not. A systematic survey of traditional folktales from preindustrial nations, tribes, and bands found that, in every world region, men were depicted as caring more than women about a mate's appearance. The only plausible explanation for this cross-cultural convergence is that art was imitating life, rather than the other way around, and that the sex difference itself is found in all these world regions.[88] The apparent universality of the sex difference is a strong challenge to the idea that the difference is merely an arbitrary cultural invention, comparable to driving on the left-hand side of the road rather than the right. If it were that, then where are all the cultures where women care more about looks than men, or where men prefer older women?

There's one more argument against the Nurture Only hypothesis which I find particularly persuasive. The claim, as you recall, is that media and culture teach us that women's desirability is more dependent on their looks and youthfulness than men's. This leads to a number of testable predictions regarding the mate preferences of lesbians and gay men. First, if the claim were true, we'd expect that lesbians would put roughly as much importance on a mate's looks and youthfulness as do straight men. After all, lesbians are primarily attracted to other women, and were exposed to the same cultural propaganda as everyone else about what makes women attractive. Second, we'd expect that gay men would put *only as much* weight on a mate's looks and youthfulness as do straight women; after all, gay men are primarily attracted to other men, and were exposed to the same cultural propaganda as everyone else about what makes men attractive. Those are the predictions, but is that what happens? It's not. As Douglas Kenrick and colleagues have shown, gay men place just as much importance on looks as straight men, and lesbians place only as much as straight women.[89] This gives us a strong reason to doubt that sex differences in the preference for good looks and youthfulness come largely from media or any other social influence. Genes and prenatal hormones are much more likely culprits.

A Lust for Wealth and Status

Let's turn now to sex differences in the importance of a partner's wealth and status. I'm sure I don't need to remind you which direction these differences go in, but in case you've been asleep all your life, the finding is that, on average, women have a stronger preference for both of these traits than men. As with the sex difference in the importance of looks, the evidence for the status and wealth differences comes from a variety of sources. In self-report surveys, women consistently rate a mate's wealth and status as more important than do men.[90] And again, people's expressed preferences translate into actual behavior. In dating ads, men who mention that they have a good job get more responses than ne'er-do-wells.[91] Similarly, at the other end of the relationship lifecycle, women are more likely than men to dump a partner who loses his job, lacks ambition, or is lazy.[92] Power is part of the package as well. Henry Kissinger once described power as the ultimate aphrodisiac, but this really only applies to half the population. As US congresswoman Patricia Schroeder observed, among politicians, middle-aged men become magnets for sexual attention the moment they attain high office; middle-aged women, on the other hand, attract no more sexual attention in that role than they would if they were waitresses. Because women place more importance on wealth and status, and men more on looks and youthfulness, wealthy, high-status men often partner up with young, attractive women.[93] The reverse happens as well sometimes, but it's less common.

Of course, men and women start pairing up long before they're old enough to accumulate significant wealth or status. Even then, though, women find traits that *predict* wealth and status more alluring than do men: traits such as confidence, competence, and raw, unbridled ambition. Frank Harris was only exaggerating a little when he observed that, "A man without ambition is like a woman without beauty."[94]

Whereas men's stronger preference for good looks initially seems somewhat anomalous from a Darwinian perspective, sex differences in the importance of wealth and status map more easily onto what we see in other animals. Females in many species – from bullfrogs to widowbirds to black-tipped hanging flies – prefer males that can furnish them with food, resources, or living space. Evolutionary psychologists have come up with two main theories about why the same seems to be true of humans. The first ties the trend to pair-bonding and biparental care. Because men commonly invest in their young, women tend not to choose their mates solely on the basis of how healthy they are or how good their genes happen to be. They also take into account how willing and able men are to invest in them and their offspring. Ancestral women who preferred men who were more willing and more able had more surviving offspring, and thus the preference inevitably spread.

The second theory is that ancestral men who succeeded in monopolizing resources and clawing their way to the top of the social ladder tended to be genetically fitter men, and that that's why women evolved their sexual fetish for them. Wealth and status, on this view, are hard-to-fake signals of fitness, broadcasting the quality of their owner's genetic endowment to the world. Women evolved a proclivity for men displaying these traits, not because such men would invest in their kids, but just because they'd give those kids their superior genes. Note that the good genes explanation and the male investment explanation are not necessarily incompatible. It's possible that both factors contributed to the evolution of the wealth and status preference. Different selection pressures can simultaneously favor the same trait.

Consistent with an evolutionary explanation for the sex differences, it's not only Western women who place a premium on a mate's wealth and status. In his cross-cultural survey of human mate preferences, David Buss found that, in every one of his thirty-seven samples, women valued a mate's financial prospects to a greater extent than men – on average, twice as much. The difference appeared in monogamous societies and polygamous ones, in capitalist societies and communist ones, and in every racial group and religious group netted in the study.[95] Comparing his findings with earlier research, Buss and his team discovered that, in the United States, women's stronger preference for financial prospects persisted right through the twentieth century, even surviving the Sexual Revolution and the second wave of the feminist movement in the 1960s and 70s.[96] Meanwhile, anthropological research suggests that the sex difference isn't limited to modern, industrialized nations. Jonathan Gottschall and colleagues, in their analysis of the traditional folk tales of bands, tribes, and preindustrial societies, found that, in every world region without fail, women were depicted as caring more than men about a prospective mate's wealth and status – presumably because, in every world region without fail, women actually *do* care more than men about these traits in a mate.[97]

The evidence for the sex differences is strong. How strong, though, is the evolutionary explanation for them? No doubt, women do often desire a man with wealth. But is this because wealth is *sexy* in the way that a chiseled jaw or a V-shaped torso is? Or are women – some women – simply making utilitarian judgments about the obvious benefits of bagging a wealthy husband? The latter is certainly possible. When an attractive young woman marries an elderly millionaire, we don't think "She's sexually attracted to him because of his money." We think "She's *not* sexually attracted to him; she's using her looks and sexual favors to get access to his cash. Hopefully she at least likes the guy." And if it's all about utilitarian calculations, an evolutionary approach may be the wrong way to go. Evolutionary theory does a better job of explaining our raw desires than our cold, utilitarian calculations.

Of course, we'd then need to ask why it is that more women than men make this particular calculation. But the psychologists Alice Eagly and Wendy Wood think they have an answer to that – an answer that doesn't directly invoke evolution.[98] Until recently, women had little scope to work outside the home or to earn their own money. This meant that their only real career option was to find a husband who'd support them. In that situation, women clearly needed to put more weight than men on a mate's wealth and status. We don't need to posit a specific evolved preference to explain this; we only need to assume that women know what's good for them. To the extent that the trend persists today, even in the face of women's growing financial independence, it's only because old habits die hard. An important selling point of Eagly and Wood's theory is that it purports to explain the fact that the sex difference in the preference for wealth is found in every culture: According to the theory, this is a side effect of the fact that, in every culture, men control the wealth. And why do men control the wealth? Perhaps for no other reason than that men have bigger muscles than women, and thus that only men stand a chance of winning any altercation over resources. If this is right, then we can explain all the facts without having to invoke any evolved psychological sex differences.

To test their theory, Eagly and Wood re-analyzed Buss's cross-cultural data. As expected, they found some evidence that the sex difference in the importance of financial prospects in a mate shrank in cultures in which men and women were more equal and thus women less dependent on men.[99] Admittedly, this was the case for only one of the four measures of gender equality they looked at. Nonetheless, if it turns out that the sex difference really does narrow as women gain financial independence, this would suggest that the difference is shaped, at least in part, by the unique and changeable economic arrangements of the societies in which people live.

The question is, though, whether these economic arrangements are the whole story. And the answer is that they're almost certainly not. First, the sex difference doesn't *disappear* in more gender-equal nations, or even shrivel to a trivial magnitude. It's still there, and it's still a moderately sized difference. Second, other research suggests that, in Western nations, women who are wealthy themselves place no less importance on a mate's resources; instead, they often place somewhat more. Meanwhile, poor men are no more interested in finding a wealthy mate than men who are better off.[100] None of this fits easily with the idea that people's valuation of a partner's wealth is shaped primarily by practical reasoning about how to obtain wealth otherwise unavailable. It is possible that, when economic concerns are removed from the picture, the remaining sex difference is somewhat smaller. Still, the residual difference is plausibly attributed to natural selection. As with earlier examples, it's hard to explain the pattern of sex differences and similarities in our species without considering both evolution and culture.

Mind over Matter

I think we're ready to call it. Sex differences in the desire for good looks, wealth, and status in a mate were crafted by natural selection. They're not unaffected by cultural forces; nothing is. But culture doesn't conjure these differences into existence from thin air. They're part of our nature, and probably have been for a long, long time. If that's true, the implications go much further than we might initially surmise. Earlier, I touched on the idea that mate preferences are not merely products of evolution; they're also *causes* of evolution. Peahens prefer males with dazzling tails, and as a result, males' tails evolve to be more and more dazzling with each passing generation. What happens, then, when men have a stronger preference than women for a good-looking mate? And what happens when women have a stronger preference for a mate possessing wealth and status? The answer is that men and women, in effect, selectively breed each other for the traits they most want in a partner.

This might explain a number of unique and otherwise inexplicable facts about our species – facts that would bamboozle our alien scientist. In many species, including many birds, lizards, and insects, the males are gaudy and ornamented, whereas the females are drab and sensibly "attired": They blend in with the background rather than standing out and risking grabbing the ravenous attention of a passing predator. In *our* species, if anything, it's the other way round. To see what I mean, look at any modern Western wedding; it's the woman who dresses in the most expensive, extravagant outfit, while the man dresses in a standard-issue penguin suit. The bride, in other words, is the peacock, the groom her drab sidekick. This is a puzzling reversal of the usual pattern in nature. Whereas humans see male peacocks as more aesthetically appealing than peahens, an aesthetically inclined peacock (or alien scientist) would probably see women as more aesthetically appealing than men. *We* certainly seem to; that's why beauty pageants are usually competitions among women.[101] The fact that men place more weight on a mate's appearance might explain where the sex difference in attractiveness originally came from. Just as humans selectively bred fruit to be sweeter and dogs to be friendlier, men selectively bred women to be the better-looking sex.

But women have done their fair share of selective breeding as well. Have you ever known a person who was obsessed with wealth and status? If so, the odds are fairly good that that person was a man. Some women are obsessed with these things, and many men are not. But *more* men than women are obsessed, and the average level of interest in wealth and status is higher among men.[102] Why might that be? Women's preference for wealthy, high-status mates is plausibly part of the answer. While men were busy breeding better-looking women, women were breeding men with a hunger for status and resources.[103] This in turn may be part of the reason that men are better represented in the upper echelons of society: better represented among CEOs and politicians, for instance, and better represented among the people who work eighty-hour weeks and rake in impossibly large salaries. No doubt, other factors help to shape the occupational landscape. But status motivation is quite possibly an important piece of the puzzle, and ancestral women's sexual choices may help to explain the sex difference in this domain.

A History of Violence

On the day I started work on this section, the morning news was dominated by a rather horrific story. A young bank teller had been shot dead during a botched robbery. The killer had fled in a stolen van, and was now being chased down the motorway by a convoy of police cars. The chase was beamed live from news helicopters to TVs and computer screens all around the world. As viewers watched in morbid fascination, the felon careened through traffic, running several cars off the road in the process. Eventually, the felon pulled off the motorway and attempted to escape into the hills on foot, the police in hot pursuit. After several tense minutes, the felon pulled a gun on the cops – this part wasn't televised, luckily – and was promptly killed in a hail of gunfire. It was later revealed that the felon was a career criminal with a history of violent crime stretching all the way back to high school.

Now tell me: Are you picturing a male or a female felon? If you look back at the last paragraph, you'll notice that I didn't actually specify the felon's sex. Nonetheless, I'd be willing to bet that you were picturing a man. Don't worry – you weren't being sexist; you were simply playing the odds. Most men are not especially violent, but most people who *are* especially violent are men. (To be fair, as the philosopher Christina Hoff Sommers points out, most people who protect us from the violent minority are men as well.[104] Men, we might say, are double-edged swords.) And rare though they might be, men such as our felon (who, as you've probably guessed, was just a figment of my imagination) are the extreme of a more general trend, namely that men are more violent than women, more in-your-face aggressive, and more prone to taking risks.[105]

The evidence for these differences is easy to find. Men get into more fist fights than women. They play more violent video games and watch more violent movies. They're more likely to be hospitalized for punching walls. They're more likely to fantasize about killing another person. They're more likely to *actually* kill another person. And they're more likely to kill themselves.[106] For low-level violence – a slap or a shove – the gender gap is surprisingly small.[107] But as we move up the scale, from minor to more extreme forms of violence, the gap grows larger with every step. By the time we arrive at the most extreme form of one-on-one violence – homicide – men are virtually the sole perpetrators. Globally, more than 90 percent of homicides are committed by men. Most victims are men as well: around 70 percent.[108] Interestingly, the figures for chimpanzees are nearly identical; one chimpicide study found that 92 percent of killers were males and 73 percent of victims.[109]

Why are men so much more prone to violence than most women? According to evolutionary psychologists, it's because violence generally paid off more handsomely for our male ancestors than it did for ancestral females. And *that's* because – you're going to get tired of me saying this – men can potentially produce more offspring. To spell out the evolutionary logic behind this claim, let's again take an extreme case: elephant seals.[110] Elephant seals are highly polygynous; during the breeding season, successful males maintain a large harem of females, while the majority of males are consigned to celibacy. If a male can win control of a harem, and if he can fight off any challengers, he can potentially end up with a huge number of offspring – many more than any female. Fighting is a risky game, of course, and he may end up with no harem and no offspring. He may even end up dead. However, if he *doesn't* fight, he'll *definitely* end up with no offspring. As such, fighters have more offspring overall, and the cycle of violence continues. And it's not just fighters vs. non-fighters; *better* fighters have more offspring than their less pugilistically gifted rivals. As a result, over the course of elephant seal evolution, males evolved, step by bloody step, to be larger, stronger, and more pugnacious than females. Male elephant seals today are veritable fighting machines.

Female elephant seals, on the other hand, are anything but. If a female dukes it out with other females to be the first to mate with the harem master, she's not going to increase the number of offspring she has. All the females will mate with him eventually, and there's little advantage to being first in the queue. Indeed, if you have to risk injury to get there, being first may be actively disadvantageous. For that reason, females don't rush headlong into battle like the males. Wherever possible, they take the path of least resistance.

Like I say, elephant seals are an extreme case and not all species have taken this road. First, in many species, males have evolved to show off to females, rather than to beat up other males – think peacock's tail rather than deer's antlers. Second, species vary in terms of how much the males compete at all, whether through showing off *or* through fighting. And third, in many species, both sexes compete for mates, not just the males. The best way to get a handle on the cross-species variation is to imagine that each species can be represented by a pair of faders, like the volume faders on your sound system: one for males; one for females. The settings of the faders represent the level of reproductive competition within each sex. At the lowest setting, all the males in the group (or all the females) have exactly the same number of offspring; at the highest, one male (or one female) has all the offspring and no one else has any. Thus, the fader settings span from zero reproductive competition at one extreme to maximum reproductive competition at the other. Now obviously there's no species on Earth for which the faders are set to either extreme for either sex. The general rule, though, is that, for each sex, the closer the fader is set to maximum reproductive competition, the stronger the selection pressure on members of that sex to be among the lucky few that win the reproductive lottery, and thus the more intensely members of that sex evolve to compete with each other for the prize.

For elephant seals, the male fader is set to high reproductive competition while the female fader is set to low. As a result, males fight incessantly during the mating season whereas females are relatively peaceful. In peacocks, it's the same, except that male–male competition is more about showing off than fighting. In other species, however, the faders are set very differently. In some monogamous species, the female fader *and* the male fader are set to low reproductive competition, and thus there's not much fighting or showing off in either sex. And in some sex-role-reversed species, the female fader is set to higher reproductive competition than the male, and thus there's more fighting and showing off among the females. But although every possible setting is represented somewhere, some are more common than others. Because in most species the ceiling number of offspring for females is lower than that for males, the female fader generally doesn't stray too far from reproductive equality. The male fader, in contrast, has a wider range of possible settings. That's why, when we do find a sex difference, it's usually the males that are more competitive. Still, the further from reproductive equality the female fader is set, the more that the females in that species will compete for the best mates.

How are the faders set for human beings? As a human being yourself, you won't be too shocked to hear that the settings are very different for us than for elephant seals. The reason should by now be familiar. People commonly form long-lasting, largely exclusive pair bonds, and men commonly help care for their young. This lowers the ceiling number of offspring for men, and drags the male fader closer to reproductive equality. Thus, although male–male conflict is reasonably common in our species – more so than the female–female equivalent – human males are peace-loving hippies compared to bull elephant seals. Moreover, because men often invest in their young, women who manage to attract a good long-term mate have a big Darwinian advantage over those who don't: They're likely to have more surviving offspring (or were, at any rate, for most of our evolutionary history).[111] This nudges the female fader *away* from reproductive equality. Consequently, women compete for mates in a way that would make little sense for the average female elephant seal. The overall effect of these sex-specific selection pressures is that sex differences in competitiveness, aggression, and violence are smaller in our species than they are in many others.[112] But "smaller" doesn't mean "non-existent," and given that men have somewhat greater reproductive variability than women, it's no surprise that men are typically more competitive, more aggressive, and more violent than women. On the contrary, it'd be surprising if they weren't.

Male–male violence hasn't just left its mark on men's minds; it has left its mark on their bodies as well. Men are larger and more muscular than women, and more heavily armored: They have thicker brow ridges, for instance, and more robust jaws.[113] This is presumably because our male ancestors spent a fair amount of their time beating each other up and scaring each other off, with those who did so more successfully winning more status, more resources, and ultimately more mates and offspring. The evolutionary psychologist David Puts went as far to argue that, whereas women's bodies are largely the product of men's mate choices, men's bodies are largely the product of male–male competition. In other words, the female body is a peacock's tail, the male body a big pair of antlers. Consistent with this analysis, many uniquely male traits, including bulky muscles, bushy beards, and deep, booming voices, are more intimidating to men than they are attractive to women.[114] This hints that intimidation may be the main reason these traits evolved.

A big plus of the evolutionary explanation for men and women's differing proneness to violence is that it places human beings within the same theoretical framework that explains comparable differences in other animals. As always, though, humans are no ordinary animal, and thus we need to consider alternative explanatory avenues. The obvious alternative is that the sex difference is learned. Boys, it is often argued, are rewarded for their aggressive outbursts, whereas girls are rewarded for being passive, quiet, and ladylike. Boys are allowed to get away with murder – well, not murder exactly, but with lesser forms of violence. Girls, in contrast, never get this latitude. We call feminine boys "sissies," and confident girls "bossy." And these double standards persist into adulthood: When men take charge or speak their minds, we see this as assertive; when women do the same, we see it instead as aggressive. According to sociocultural theorists, we don't need to invoke ancient selection pressures to explain the sex difference in aggression. A more plausible explanation is right in front of our eyes.

By this point, you won't be too surprised to learn that, in my humble opinion, this proposal is deeply flawed. An initial flaw is that it's not only in the Western world that we find the relevant sex differences. Male–male competition is ubiquitous. Among the Trobriand Islanders of the Pacific, for example, men compete with each other to grow the largest, longest yams. (Freud would have a field day with *that*.) In every culture on record, competitive sports are dominated by men.[115] And in every culture on record, men are more violent and aggressive than women, especially with other males.[116] If the sex difference in aggression is just an arbitrary product of culture, why does it rear its ugly head in every human group?

To be fair, some sociocultural theorists have attempted to answer this question. Alice Eagly and Wendy Wood argue that, although men clearly do engage in higher rates of violence, this isn't a result of evolved differences in men and women's minds. In their estimation, it's an indirect effect of certain evolved differences in men and women's *bodies*. The differences they have in mind are that men are larger, stronger, and faster than women, and that women get pregnant and produce milk for the young. Because of these non-negotiable physical differences, men in every culture are funneled into social roles involving aggression and physical strength, whereas women are funneled into roles involving childcare. Like the ex-soldier who keeps on polishing his boots every day for the rest of his life, the roles we play in the world have enduring effects on our behavior and personalities. Over time, men actually become more aggressive and women become more caring. In short, Eagly and Wood provide a non-evolutionary-psychological explanation for the cross-cultural trend: The psychological sex differences are found in every culture because the *physical* sex differences are found in every culture. The physical differences are direct products of evolution; the psychological differences are not. If we reengineered our social roles, the psychological differences would quickly fall away.

It's a clever argument and one worth taking seriously. On balance, though, I don't think it flies.[117] To begin with, the Eagly–Wood theory raises some awkward questions. Why *wouldn't* natural selection create psychological sex differences as well as physical ones? The mere existence of the physical differences tells us that human males have been subject to stronger selection for aggression and violence than females. Why would this selection pressure shape our muscles, our skeletons, and our overall body size, but draw the line at our brains? And why would natural selection give men the physical equipment needed for violence but not the psychological machinery to operate it? This would make about as much sense as giving us teeth and a digestive system, but not a desire to eat. Furthermore, we know that the same psychological sex differences found in humans are also found in other species – species that don't have culturally elaborated social roles. Why would differences that clearly have an evolutionary origin in other species have an entirely different genesis in our own?

A further hole in the Nurture Only plot is that, as with earlier examples, many of the social forces invoked by the sociocultural theorists have an unfortunate habit of not existing. Consider the claim that society encourages males to be aggressive. This is probably true in some ways; we do sometimes give boys the message that they ought to be tough and not cry. Overall, though, we spend a lot more time discouraging *male* aggression than female. Why? Because males are more aggressive! Or consider the claim that we tell girls to be quiet and passive. Again, we probably do this sometimes. More often, though, we tell *boys* to be quiet and passive. Why? Same reason: Boys are louder and more disruptive! In the first piece of research I did as a graduate student, I found that people judged an aggressive act performed by a man to be less acceptable than the same act performed by a woman. The reason, it turned out, was that people perceived the female act to be less harmful.[118] This perception apparently translates into real-world behavior. In childhood, boys are punished more often and more severely for aggression.[119] Similarly, in adulthood, male defendants get harsher sentences for the same crimes, even controlling for criminal history.[120] Males, it seems, are more aggressive *despite* culture, not because of it. And this isn't just the case in the West. In most cultures, boys are taught not to be aggressive, but in all cultures, boys and men are more aggressive anyway.[121] Some argue that, even if culture doesn't create the aggression sex difference *ex nihilo*, it does still amplify a relatively trivial inborn difference. Often, though, culture may do the reverse: By clamping down on male aggression, culture may make the sex difference in aggression *smaller* than it would otherwise have been.

Other evidence pushes us to the same conclusion. The evolutionary psychologist John Archer points out that if the gender gap in aggression were due solely to socialization, it would presumably be smallest in the very young, and then grow steadily as the years went by. After all, the longer we live, the more time the forces of socialization have to sink their claws into our minds and behavior, and thus the bigger any gender gap should be. That's the theory; the reality, however, is quite different. First, the sex difference in aggression appears very early in life – usually before children take their first bite of their first birthday cake. From the moment they can move around under their own their steam, boys engage in more rough-and-tumble play than girls.[122] The same sex difference is found in other juvenile primates, and appears to be related to prenatal testosterone.[123] In humans, the sex difference shows up long before kids understand that they're boys or girls, so it can't just be that they're conforming to social expectations about what boys and girls ought to do.[124] In any case, children are terrible at conforming to social expectations, as any parent who's tried to persuade their progeny to sit nicely and quietly in a restaurant will readily confirm. And not only does the sex difference in aggression emerge early, it remains static until puberty. *Absolute* levels of aggression trend downward for both sexes; however, the gap between the sexes barely budges. If socialization creates the sex difference, why doesn't continued socialization before puberty pry the sexes apart?[125]

Second, as with many sex differences, the sex difference in aggression suddenly swells at puberty, and is larger among adolescents and young adults than among any other group. Like bull elephants in musth, human males often go a little crazy at this stage in the lifecycle. Margo Wilson and Martin Daly dubbed this the *young male syndrome*.[126] Males in the grip of this syndrome are more likely than any other demographic to be imprisoned, to kill someone, or to be killed by someone else: most often another young male. The behavioral geneticist David Lykken summed up the situation well when he observed that, if we could cryogenically freeze all the males in this age bracket, we would instantly eliminate most of the crime and violence that plagues human societies.[127] How would socialization theories explain the violence gap that opens up between the sexes at puberty? Is there a sudden surge in gender socialization – a surge which, for some unknown reason, happens at exactly the same stage of life in every culture and in many dimorphic species? Is it just a coincidence that this alleged surge in socialization comes at the same time as the massive surge in circulating testosterone that accompanies puberty in males?

Third and finally, after the violence and mayhem of early adulthood, male aggression steadily nosedives through the remainder of the lifespan. The socialization hypothesis offers no particular reason to expect this. But the decline in violence coincides almost perfectly with the decline in testosterone found in men throughout the adult years, and mirrors the decline found in males of other species. Once again, this is much easier to explain in evolutionary than in sociocultural terms. Taken together, the evidence strongly suggests that the sex difference in aggression in our species is a product of natural selection.

Mothers and Others

There's one last sex difference to consider: the most important of all. This is the difference that, for our species and most others, underlies all the rest: the sex difference in parental investment. As we've seen, sex differences appear in a species when one sex can potentially have more offspring than the other, and the main reason this happens is that one sex invests less in offspring. Usually, the males invest less, and thus usually the males can have more offspring. All the standard-issue sex differences – the sex differences in size, in sexual behavior, in choosiness and aggression – flow from this one, primordial difference. Parental investment, in other words, is the lynchpin that holds together the entire theoretical edifice.[128]

Why, then, do females usually invest more than males into offspring? Why not the other way round? To crack this case, we first need to take a step back and ask a different question: What determines whether a given individual is a male or a female in the first place? In everyday life, we identify people's sex by their appearance, their voice, their hair, and their clothes. But that can't be what *defines* male and female, because other animals come in male and female varieties, and yet most lack sex-specific hairstyles or fashions. Many would guess at this point that males and females can be defined by their sexual equipment: penises vs. vaginas. But that can't be right either, because not all animals are so equipped. In around 97 percent of bird species, for instance, the males are penis-free. Furthermore, in some nonhuman species, females have penis-like organs called *ovipositors*, which they use to lay their eggs. And in others, females have *pseudo-penises*, which they use, among other things, to intimidate each other. We see this, most famously, in hyenas.

So, is there *anything* that universally differentiates males and females? There is: Males produce smaller sex cells than females, which for animals means sperm rather than eggs. There are almost never exceptionless generalizations in biology, but this one is one of the few. It's true 100 percent of the time… but only because it's true by definition. Biologists *define* males as the sex that produces the smaller sex cells and females as the sex that produces the larger ones. Penises, pseudo-penises, and hairdos don't come into it.

The sex difference in the size of the sex cells was the very first sex difference: the original sexual inequality. Before there was pregnancy, lactation, or driving the kids to school, the only investment that either sex made in their offspring was making sex cells. Because females made the larger ones, females were the higher investing sex. Later, when more elaborate forms of parental investment evolved, they more often evolved in females, just because females had already started down the high investment track.[129] One consequence of this is that, in a wide array of species, many of the differences between sperm and eggs are mirrored in male and female bodies and behavior. For example, just as sperm are tiny and scramble around looking for large, stationary eggs, so too most male spiders are tiny and scramble around looking for large, stationary females. Similarly, just as eggs are often designed to nourish the growing embryo with a built-in supply of nutrients, so too female mammals are designed to nourish their offspring during pregnancy and nursing. In these ways and others, males in many species act like sperm and females like eggs. (Sorry for stereotyping sperm and eggs.)

Certainly, this isn't always the case. Although males always invest less into individual sex cells than females, the sex difference in parental investment can be reduced, eliminated, or even reversed at the level of behavior. We see some signs of this in our own species; men, as we know, often help to care for their young. But we shouldn't overestimate how much. While to some extent, men act like eggs, they're rarely as egg-like as women. In every culture for which we have reliable data, women spend more time caring for children than men.[130] Even in societies famous for their high levels of paternal care, the differences are far from trivial. For example, among the Aka pygmies, men hold their babies for an average of fifty-seven minutes a day – much more than men in most traditional societies. But women hold their babies for 490 minutes.[131] Thus, the question we need to ask is this: Why, given that selection started closing the parental gap in our species, did it never close it completely?

Evolutionary psychologists point to two main factors.[132] The first is *paternity uncertainty*. This refers to a fundamental sexual asymmetry in species with internal fertilization, namely that it's easier for females to identify which offspring are theirs than it is for males. If a baby comes out of your body, that's a pretty good clue that it's yours. As David Buss likes to say, as far as we know, no woman in the history of the species has ever given birth and thought: "Wait a minute! How do I know that this baby is mine and not some other woman's?" In contrast, if a baby comes out of the body of a woman you slept with nine months ago, that's not nearly as reliable a clue. As an old saying has it, maternity is a matter of fact, paternity a matter of opinion. A man who invests in his wife or partner's offspring is *probably* investing in his own offspring, but he *could* be investing in the offspring of another man. As we'll see in the next chapter, this doesn't happen particularly often in our species, but it doesn't happen never, either. And to the extent that it does happen, it reduces the average fitness payoff for male parental care. The obvious consequence? Men evolved to be less parentally inclined than women (on average, of course; not always). Animal studies support the general thrust of this argument: In species where males have greater paternity certainty, males invest more in their young; in species where males have less, they invest less.[133] The lesson for our species is that paternity uncertainty may have been one factor keeping a lid on male parental care.

A second factor, which I suspect is more important, is *mating opportunity cost*. The concept of opportunity cost comes from economics. To understand it, consider an argument that people sometimes give for using quack remedies: "I might as well try this new ancient natural spiritual medicine; after all, if it works, it works – and if it doesn't, it won't do me any harm!" The problem with this argument is that, even if the medicine really isn't directly harmful, it may still be harmful in another way. Specifically, the time and money you devote to one putative remedy is time and money you can't devote to another – one that might actually work. That's the opportunity cost. Exporting the concept to biology, *mating* opportunity cost refers to the fact that time and energy devoted to caring for kids is time and energy that can't be devoted to chasing new mates. This is true for both sexes, but the opportunity cost is potentially greater for men because... well, you know the reason by now. The implication is that, even if paternity were 100 percent assured, it would still make Darwinian sense for men to devote less time than women to parenting, and more time to seeking out mates. And that, argue evolutionary psychologists, is a big part of the reason that men are less parental than women.

If you know anything about academia, you won't be at all surprised to hear that ideas such as these have not always been greeted with open arms. Many social scientists deny that women have a maternal instinct, or that women's parental inclinations are naturally any stronger than men's. Some even view the idea as scientific cover for a regressive attempt to drag women back into the Dark Ages and undo the progress of the women's movement.[134] Sure, say the social scientists, women in the past did most of the parenting. But that's not because they have an evolved hunger for this role, in the same way that people have an evolved hunger for certain types of food. No. It's because the maternal role is the only role that traditional societies afforded women. Even today, there's a lot of pressure on women to squeeze themselves into that role. Women get flak for spending more time at the office than with their kids, and get even more flak for electing not to have kids in the first place. If women have an instinctive, all-consuming desire to care for children, why would the social pressure be necessary?

Of course, any reasonable sociocultural theorist would admit that it's not *just* immediate social pressure that leverages women into the parental role. Women do often want kids, and they often want them more than men. But why pin the blame for this on evolution? According to sociocultural theorists, to do so is to ignore the fact that, in a thousand different ways, subtle and not-so-subtle, girls are taught that they should prioritize family over careers, whereas boys are taught the opposite. This message comes from parents, from the wider community, and from books and ads and TV shows that flood children's fragile minds with gender-typical role models: stay-at-home mums and bread-winning, bacon-bringing dads. True, children get a lot less of this kind of indoctrination than they did in the past. But the sex difference in parental involvement is also a lot smaller now than it used to be. And that's surely just what we'd expect if the difference comes from culture.

What should we make of this argument? The first thing to emphasize is that evolutionary psychologists don't deny that culture and social roles have an impact on our child-rearing arrangements. How else could we explain the fact that, in earlier generations, Western mothers were usually stay-at-home mums, whereas today they're usually not, or the fact that Western fathers now spend a lot more time with their kids? The claim from evolutionary psychology, as I've mentioned before, is not that social roles have no impact, but that social roles aren't the whole story. Furthermore, the other part of the story – the innate contribution – probably helps to shape the social roles to some extent. In principle, we can invent any social roles we want. In practice, however, roles that mesh well with our evolved nature are more likely to persist than those that clash with it, at least in the absence of draconian social interventions.[135]

Why, though, should we assume that there *is* an innate contribution? An initial answer might be that the sex difference in parental care is found in every culture, and that this would be rather strange if the difference were merely an arbitrary social convention. But the sociocultural theorists have a response to this argument: It *is* a social convention but just not an arbitrary one. We've touched on this already. According to Eagly and Wood, the reason that women in every culture are assigned the childcare role is that, in every culture, women are the ones who get pregnant and who have the equipment for nursing the young.[136] We might also add that, once the maternal role is entrenched in a society, it soon becomes self-sustaining. Girls see women playing that role, and naturally want to emulate them. And adults *expect* girls to take that role, so they socialize them into it. The upshot is this: Women do the bulk of the childcare, not because of any innate *psychological* differences, but as an indirect effect of certain innate *physical* differences – those related to pregnancy and lactation.

Once again, this is a perfectly reasonable hypothesis, but once again it runs up against some pretty hefty counterarguments. Why would natural selection give women the physiological apparatus for maternal care but not the psychological drives needed to operate it? Why would selection create sex differences everywhere but in the brain? And why would selection take this approach only for our species? As we've seen, in most parental species, females do most of the parenting.[137] For other animals, the social role explanation isn't remotely plausible, and no one would deny that the sex difference is innate. The question, then, is why would the *same* sex difference be a product of an entirely different cause in humans and humans alone?

The answer is that it probably wouldn't. Consistent with this assessment, various lines of evidence show that the sex difference in parental motivation is not due solely to culture. First, even when culture pushes in the opposite direction, women are still more parental than men. An example comes from the Israeli communes known as kibbutzim. The kibbutzim were founded on radical socialist principles, and the founders sought to eliminate supposedly bourgeois traditions such as women being primarily responsible for childcare, and parents caring exclusively for their own biological children. To that end, kibbutz children were housed together in large communal quarters, rather than shacking up with their parents. But although this arrangement looked fine on paper, in practice it rapidly disintegrated. Parents – mothers in particular – hated it, and before too long started insisting that their children live with them. Some of the kibbutz men resisted for a while, but eventually they capitulated. Thus, rather than being forced into the motherhood role by men, the women had to rebel against the men to take on that very role.[138]

An analogous phenomenon can be seen in the modern West. As noted, sociocultural theorists commonly argue that women are pressured to put kids above their careers. These days, however, the pressure often goes the other way. Girls are taught that the working world is their oyster, and women are encouraged to lean in and rise through the ranks. Meanwhile, the occupation "stay-at-home mum" has been tacitly blacklisted, especially among middle- and upper-middle-class women. And yet despite it all, women are still more likely than men to want to stay at home and care for the kids, or to want to work part-time so they can spend more time with the kids – even when they feel that they shouldn't want this and that they're letting down the sisterhood.[139] These women aren't being pressured to do these things, any more than the kibbutz women were. Often, they're going against the prevailing social pressures. Many social scientists "blame" the sex difference in parental care on Western norms and values. But we can turn this claim on its head. The sex difference appears in the West *despite* the fact that so many people prize gender equality and a fair division of labor – and despite the fact that bottle feeding and baby formula mean that men can now care for infants just as ably as women.[140]

The sociocultural theorist might argue at this point that the remaining sex difference reflects cultural inertia rather than evolution: Our parental traditions haven't yet caught up with our ideals and our technology. But other evidence suggests that it goes deeper than that. First, the sex difference in parental inclinations appears early in life. Just as boys engage in more rough-and-tumble play than girls, girls engage in more play-parenting – an interesting pair of findings given that, across species, play in the young functions as practice for the behavior that the adults naturally specialize in.[141] Certainly, the early onset of the differences isn't bullet-proof evidence of an innate contribution; in principle, the differences could still come exclusively from socialization. But it doesn't seem likely. Girls who were exposed to high levels of testosterone in the womb engage in less play-parenting than other girls, and more rough-and-tumble play. Do parents socialize their daughters to be parental only if the daughters were exposed to normal levels of prenatal hormones? Clearly not. Instead, girls exposed to high levels of male hormones develop male-typical patterns of behavior even when their parents encourage them to act in a more feminine way – in other words, despite socialization, not because of it, yet again.[142] The only realistic explanation is that prenatal hormones help shape people's parental inclinations, and that in the normal course of events, they help shape average differences between the sexes in this domain.

A further reason to doubt that the sex difference in play-parenting comes solely from socialization is that we see the same difference in other primates.[143] Among chimpanzees, for instance, juvenile females will often pick up logs or sticks, and start cradling and caring for them. Sometimes, they'll even lay them down to sleep in their nests. (Yes, chimps sleep in nests; all the nonhuman great apes do.) In effect, girl chimps play with dolls. Boy chimps sometimes do too, but it's rarer.[144] The idea that the human sex difference in play-parenting is purely a product of culture would be much more plausible if we didn't find exactly the same difference in closely related species for which it clearly *isn't* a product of culture. And if the human sex difference in play-parenting has an evolutionary origin, it would be odd indeed if the human sex difference in actual parenting did not.

Note, by the way, that to say that women are naturally more parental than men is *not* to say that every woman should have to be a mother, or that men should be barred from the role of primary caregiver. People should do what they want. The point is simply that, like it or not, there are average differences in what men and women want, and that these are not due purely to social influences. The differences are rooted firmly in our biology.

Enemies of the Truth

To wrap up the chapter, let's return to the question we began with: What would an alien scientist make of human sex differences? We're now in a position to see ourselves as the alien would. The first thing it would notice is that certain psychological sex differences are found in every culture. To be sure, the magnitude of the differences varies somewhat from place to place and from time to time.[145] But if these differences were merely social inventions, why aren't there cultures where women are more interested in casual sex than men, where women go to war while the men stay home, or where men do most of the childcare? If these things are just indirect effects of evolved *physical* differences, why do the psychological differences persist even when the culture pushes against them?

Next, the alien would notice that the standard psychological sex differences are associated with certain hormones. Testosterone, for example, is linked to a variety of traits found more commonly in males than females, including a high sex drive, aggression, competitiveness, dominance, impulsivity, risk-taking, and status seeking.[146] Is it just a coincidence that the hormone associated with these male-typical behaviors is the same hormone tasked with turning a gender-neutral embryo into a boy and ultimately a man?

Finally, the alien would notice that the same sex differences found in humans are also found in many nonhuman mammals. In species where males invest less time and energy into offspring, the males tend to be larger, less discriminating about their sexual partners, and more aggressive, whereas the females tend to be quicker to reach puberty, choosier about their mates, and longer lived. When we see these differences in other mammals, we're not the slightest bit tempted to ascribe them to social forces. No one would argue, for instance, that male kangaroos fight more than females because kangaroo parents give boy kangaroos guns and girl kangaroos dolls, and no one would argue that female gorillas are more parental as a result of patriarchal gorilla gender norms. Every reasonable person accepts that the differences have an evolutionary origin. Moreover, every reasonable person accepts that the *physical* differences in humans have an evolutionary origin. Is it reasonable, then, to argue that the *psychological* differences in our species, though comparable to those found in other animals with the same physical differences, are the result of an entirely different cause – learning or culture – which just happens to recapitulate exactly the same package of differences?

Running the data through its supercomputer-like brain, the alien would quickly conclude that the standard psychological sex differences found in human beings have their origin in natural selection. What if the alien were then informed that many people believe that humans are exempt from the sociobiological laws that govern the evolution of other animals, and that with our language and culture, we've transcended the evolutionary process? The alien would probably shrug it off, pointing out that this is the same species that, until recently, held itself to be the center and purpose of the entire universe. When it comes to questions like these, many people are flat-Earthers. They imagine a chasm between humans and other animals that Darwin showed doesn't exist. Even some evolutionary biologists respect the apartheid between humans and their nonhuman cousins, arguing that although sociobiological theories clearly apply to other species, humans are a different kettle of fish. Can we really believe, though, that the sociobiological theories apply right across the animal kingdom, from bacteria to bugs to bonobos, but don't apply to one upright, uptight primate? That is a radical position. Where's the radical evidence to back it up?

The case for evolved sex differences is one of the strongest in all of psychology. And yet many psychologists and social scientists dismiss or downplay the idea that evolution played any role at all. Worse than that, in some quarters the idea is considered so abhorrent that merely stating it is treated as something akin to blasphemy. People are forced to perform pointless intellectual handstands and engage in superhuman feats of Orwellian double think, all to avoid acknowledging common sense truths about our species. Those who don't play ball – those who argue that the sex differences are real and that they almost certainly have an evolutionary origin – run the risk of being labelled sexists or neurosexists or a dozen other slanders.[147] Why, in a community that supposedly values free inquiry and the pursuit of truth, do scholars have to risk censure and personal attacks merely to state the obvious?

These issues are not just academic. The fact that many sex differences have an evolutionary origin has implications for how society deals with these differences.[148] In the bad old days, people tried to "cure" left-handedness and homosexuality with all manner of crackpot interventions. In our own, more enlightened times, we see this as a cruel and pointless practice. But people in the future may see our current efforts to "cure" sex differences as equally cruel and equally pointless – as today's equivalent of forcing lefties to write with their right hands, or of forbidding a son to play with dolls for fear it will make him gay. As Christina Hoff Sommers and other "freedom feminists" have argued, if basic sex differences are as deep-seated as handedness and sexual orientation, we should think twice about interventions designed to force males and females into the same mold: interventions designed to coax boys and girls to play with the same toys, for example, or to achieve a 50:50 sex ratio in every desirable profession where men currently dominate. We should question the automatic assumption that differences between the sexes necessarily imply discrimination against women. And we should ask what right we have to override people's preferences regarding their own lives and careers in order to enact *our* preference for a gender-neutral world.[149]

Of course, if certain sex-typical tendencies cause harm, then we have every right and every reason to try to eradicate them. To the extent that men are naturally inclined toward violence, for instance, we should aim to dampen this male-typical behavior, thereby reducing the size of the sex difference in violence. However, the justification for this is that violence is harmful, *not* that sex differences are somehow inherently bad. And when it comes to other sex differences – differences that cause no one any harm – what's the problem? Why not just let people be themselves?

4

The Dating, Mating, Baby-Making Animal

If an alien anthropologist dropped on earth, listened to pop music for a day or two, and browsed through a random assortment of self-help books, movies, and novels, it would quickly come to the conclusion that humans are obsessed with love, sex, and intimate relationships.

—*Garth Fletcher* et al. *(2013), p. 43*

The Conveyer Belt of Human Life

Imagine, if you will, that our alien scientist arrived on planet Earth with a hangover, and that its normally magnificent and penetrating intellect was initially somewhat dulled. Imagine further that, in its compromised state, the blurry-eyed alien didn't realize that human beings are naturally occurring organisms, and instead mistook us for factory-built machines. What would the alien think these "machines" were designed to do? Its first guess, as the hangover slowly subsided, might be that human beings are machines designed to make new machines of the same general kind – new human beings, in other words – which in turn make more new human beings, and so on. The alien would probably wonder why on Earth anyone would design a machine to carry out such a strange and apparently pointless task. But a well-traveled alien would have seen plenty of strange stuff in its time, and as hard as it might be to fathom the designer's motives, the alien would find it even harder to shake the impression that humans are machines designed solely to replicate their kind. We've got all the anatomical equipment needed to accomplish this task: sex organs, wombs, nursing apparatus. And we've got all the drives and desires needed to operate the equipment: sexual desire, parental affection, and the like.

Admittedly, having children isn't everyone's explicit goal in life. But it is the goal implicit in the design of our bodies and our basic drives. It's also the goal implicit in the developmental program that turns a fertilized human egg into a functioning human adult. Childhood culminates in sexual maturity; sexual maturity is, in effect, the goal that the developmental process aims to achieve. And once we pass the age of bearing and rearing children, our bodies begin to break down and crumble, eventually ceasing to function altogether. Humans are not machines designed to last forever; they're machines designed to last long enough to reproduce.

Or so the hungover alien might surmise. *We* know, of course, that humans are not actually machines – or to be more precise, that we're machines designed by natural selection, rather than made in a factory. And we also know that natural selection designed us not just to reproduce but to accomplish a somewhat broader task: passing on our genes. Still, there is a sense in which the alcoholic alien would be right. Although human beings are designed to pass on their genes in a number of different ways, by far the most important is making babies. That's what this chapter is about: the conveyer belt of human life. We'll cover all the major milestones, including choosing a mate, falling in love, fighting off rivals, popping out kids, and finally helping to shepherd those kids to adulthood so they can start the whole process again. Fair warning: By the time we're done, you might be put off sex and relationships for life...

Survival of the Prettiest

In 1838, at the tender age of twenty-nine, Charles Darwin's thoughts began to turn to marriage. By that point in his life, he was a rather accomplished young man. He'd spent five years circumnavigating the globe as ship's naturalist on *HMS Beagle*, and he'd since established a solid reputation as a scientist and respectable member of British society. It was time, he decided, to think about tying the knot. Darwin being Darwin, he approached this assignment in a methodical and logical manner: He drew up a list of pros and cons. Here's how the great naturalist's thinking went.

Marry

Children – (if it Please God) – Constant companion, (& friend in old age) who will feel interested in one, – object to be beloved & played with. – – better than a dog anyhow. – Home, & someone to take care of house – Charms of music & female chit-chat. – These things good for one's health. – but terrible loss of time. –

My God, it is intolerable to think of spending one's whole life, like a neuter bee, working, working, & nothing after all. – No, no won't do. – Imagine living all one's day solitarily in smoky dirty London House. – Only picture to yourself a nice soft wife on a sofa with good fire, & books & music perhaps. – Compare this vision with the dingy reality of Grt. Marlbro' St.

Not Marry

Freedom to go where one liked – choice of Society & *little of it.* – Conversation of clever men at clubs – Not forced to visit relatives, & to bend in every trifle. – to have the expense & anxiety of children – perhaps quarrelling – Loss of time. – cannot read in the Evenings – fatness & idleness – Anxiety & responsibility – less money for books &c – if many children forced to gain one's bread. – (But then it is very bad for one's health to work too much)

Perhaps my wife won't like London; then the sentence is banishment & degradation into indolent, idle fool –

In the end, Darwin decided that the joys of a soft wife outweighed the risk of banishment, and he came down firmly on the side of marriage. He proposed to his cousin Emma Wedgwood in November 1838, and the pair were married in January of the next year. It's not known whether Darwin ever confided in his wife about the method via which he decided to propose to her – hopefully not. Either way, though, their marriage was long and happy, and no one doubts, I think, that he made the right decision. Her too.

This anecdote raises an interesting question. With all due respect to Darwin, it's fair to say that most people don't approach the mating game in quite the way that he did. And to the extent that this is true of people, it's doubly true of every other animal on the planet, for whom the careful weighing of pros and cons is simply not an option. It's also true, however, that most animals – human and nonhuman alike – don't just mate at random. From a Darwinian perspective, it would be bizarre if they did; mate choice is just too important. As the famously grumpy philosopher Arthur Schopenhauer put it:

> The ultimate aim of all love affairs … is actually more important than all other aims in man's life; and therefore it is quite worthy of the profound seriousness with which everyone pursues it. What is decided by it is nothing less than the *composition of the next generation.*[1]

But if we don't make pros-and-cons lists, and we don't just mate at random, how *do* we make this most momentous of decisions? Easy: For the most part, we do it in the same way that we choose our food. We act on built-in preferences – preferences put in place by natural selection.

We touched on the question of mate choice in Chapter 3; now let's expand on the logic of our evolved mate preferences. To clear the ground for this endeavor, we first need to dislodge two common but misleading intuitions. The first is the idea that beauty is an objective property of the universe, like size or shape, and that we view certain individuals as attractive for the simple reason that they *really are* attractive. This is a deeply ingrained intuition. We all see that special someone, with their perfect smile and luminous skin, as inherently, objectively attractive, and we find it hard to believe that we only see them this way because that's how we're wired up. Presumably, though, a philosophically inclined chimpanzee would have the same intuition in regard to *its* love interests. A male chimp, for instance, would see a fertile female, with her hairy body and sexual swelling, as just as inherently, objectively attractive as we see each other, and would find it just as hard to believe that he saw things this way only because that's how selection built him. And an asexual alien scientist would be bewildered by the chimps *and* by the humans, and indeed by all sexually reproducing species. To the alien, even the most attractive among us would look like "ugly giant bags of mostly water" (to borrow a line from *Star Trek*). Every species has its own tailor-made standards of beauty and attractiveness, and there's no reason to think that the beauty standards of one big-headed primate – aka us – happen to correspond to objective aesthetic truths. Indeed, there's no reason to think that there even *are* objective aesthetic truths.

So, that's the first intuition we need to clear from the path. The second is its mirror image: the idea that perceptions of beauty, rather than being engraved in the bedrock of human nature, are in fact highly variable products of culture. On this view, standards of attractiveness that people naively assume are universal are actually inventions of Western societies, of advertisers, or of patriarchal power structures trying to keep women under the thumb.[2] To back up such claims (or in some cases, conspiracy theories), people point out that there are conspicuous differences in what people find attractive across historical epochs and geographical regions. In her own day, Marilyn Monroe was considered one of the most attractive women in the world; these days, she'd be considered overweight. Likewise, in earlier centuries, artists such as Rubens painted nude portraits of women who, by today's standards, could have stood to lose a few pounds. And as large as these historical differences are, they fade into insignificance when compared to the differences found among cultures.

Darwin discussed several of these in his 1871 book, *The Descent of Man.* At one point, for example, he noted that...

> The wife of the chief of Latooka told Sir S. Baker that Lady Baker "would be much improved if she would extract her four front teeth from the lower jaw, and wear the long pointed polished crystal in her under lip."

Later, in the same work, Darwin wrote that...

> Mr. Winwood Reade [a British explorer] ... admits that negroes "do not like the colour of our skin; they look on blue eyes with aversion, and they think our noses too long and our lips too thin."[3]

Based on factoids such as these, Darwin concluded that "It is certainly not true that there is in the mind of man any universal standard of beauty with respect to the human body."[4] Over the course of the subsequent century, most social scientists reached the same verdict. In my opinion, however, Darwin and the social scientists got it wrong, plain and simple. I don't deny that beauty standards differ somewhat across culture and time. But this only rules out the view that beauty standards are *100 percent* biological, with zero contribution from culture – an extreme view that nobody holds. The differences don't prove the equal-but-opposite extreme view that beauty standards are 100 percent *cultural*, with zero contribution from biology. And contrary to the latter view, there are good reasons to believe that biology does make a contribution. More than that, there are reasons to believe that the biological contribution is substantial.

To start with, the usual examples of cultural and historical variation in beauty standards are considerably weaker than they first appear. At her peak, Marilyn Monroe was *not* overweight by today's standards, even if she was somewhat curvier than the average modern fashion model. And it's not as if anyone looks at old photos of Monroe and thinks "Gross! I can't believe anyone ever thought she was hot." If a time-traveling alien catapulted Monroe into the present day, I doubt she'd be stuck for a date. As for Rubens, many of the women in his paintings fell within a range that would still be considered attractive today. Some were larger, sure. But no one has yet made a convincing case – or even tried to – that these paintings represented the mainstream sexual preferences of their time, as opposed to just those of the painter. Even back then, the work may have been great art but niche pornography. Consistent with this possibility, a recent analysis showed that Rubens was indeed an outlier; most artists of the time painted thinner women.[5]

On top of all that, several lines of evidence suggest that cross-cultural differences in beauty standards are relatively superficial. In one set of studies, participants were shown photos of women, and asked to rate their attractiveness. The raters came from a variety of ethnic groups, as did the women in the photos. Despite the diversity, however, there was strong agreement about who was most attractive, who was least, and who fell somewhere in between.[6] The fact that people from very different cultural backgrounds converged so powerfully in their ratings suggests that there's a biological contribution. In fact, it suggests much more; it suggests that the biological contribution dwarfs the effects of culture. A crystal in the lower lip might be prized in some societies, but it doesn't turn an ugly duckling into a swan. And in case you're wondering whether exposure to Western media explains the cross-cultural convergence, the evidence suggests not. First, the researchers conducting the study measured people's media exposure and found that it made no difference to how they rated the photos. Second, other studies show that even newborn babies prefer to gaze at faces previously rated by adults as attractive.[7] It's unlikely that babies could have acquired this preference from ads, billboards, or Hollywood movies in the first few days of life.

For all our differences, then, the core of what we find attractive seems to be the same for people everywhere. Beauty is in the eye of the beholder, but the beholder's eye has been shaped by natural selection, and thus doesn't vary greatly from age to age or from culture to culture. The next question is: Why? Why would natural selection "care" what we find attractive? To answer this, put yourself again in the shoes of a hypothetical ancestral human whose one and only goal in life is to have as many surviving offspring as possible. Given that goal, how would you set about choosing a mate? Here's how your thinking might go:

> OK, to begin with, I should pursue an adult member of my own species and the other sex.[8] That's pretty important. But it's also setting the bar quite low. There's plenty more I could add to my wish list. First, I should pursue someone who's fertile: someone who's capable of getting pregnant (if I'm a man) or of getting me pregnant (if I'm a woman). Second, I should pursue someone who has a clean bill of health: someone with a strong constitution, who won't pass any diseases to me or our shared offspring, and who might be around for long enough to help care for those offspring. And third, I should pursue someone who has good genes: genes likely to build strong, healthy children who have relatively few mutations and are resistant to the local parasites. These traits are interconnected to some extent. Fertility is partly dependent on health, and health is partly a matter of good genes. But the overlap between the traits isn't complete, so I should keep an eye out for signs of all three.

Suffice it to say, psychologically healthy men and women don't actually think like this when sizing up potential mates. And until recently, it wouldn't have done them much good if they had. Premodern humans couldn't demand that prospective mates furnish them with a doctor's certificate to prove they were fertile or healthy, let alone a genetic test to prove they had good genes. So how *do* we assess the reproductive prospects of potential mates and lovers? Well, strictly speaking, we don't. Instead, we have an automatic, unlearned tendency to find certain physical attributes attractive: attributes that, in our evolutionary past, were reliably associated with fertility, health, and genetic quality. Humans evolved to see people possessing these attributes as "good looking," and those not possessing them as "not-so-good looking." Unbeknownst to us, when we evaluate people in these terms, we are, in effect, evaluating them as potential sperm or egg donors. Beauty, from a Darwinian vantage point, is a certificate of good health, and courtship a process of shopping around for the best genes for our future offspring. In a sense, mate choice is a form of eugenics, and always has been.

Let's get down to the nitty gritty. Which traits have humans evolved to find attractive? Evolutionary psychologists now have a long list, including everything from healthy skin and lustrous hair to deep voices in men and youthful breasts in women. One trait that's been explored in great detail, however, is symmetry. Human beings, and most of the animals we're familiar with, are *bilaterally symmetrical*. That means that the right half of our bodies is roughly the mirror image of the left. The qualifier "roughly" is important here. Organisms vary in how closely they approach the ideal of symmetry. If you took a mugshot of yourself and split it down the middle, and then made a new face from each half by completing it with its reflection, you'd get a reasonable sense of how symmetrical you are. If you were perfectly symmetrical, the two faces would be identical. No one's perfect, of course, so you'd actually end up with two somewhat different faces: more like brothers than clones. The important point, though, is that the *more* similar the two faces are, the more symmetrical you are.

The reason this is important is that there's now strong evidence that symmetry is a crucial ingredient in attractiveness.[9] Experiments using doctored photos show again and again that highly symmetrical people are rated as more attractive than their lopsided counterparts, even when everything else is held constant: skin tone, health, you name it. Meanwhile, studies looking at real-world behavior find that, on average, highly symmetrical men start having sex at a younger age, accumulate more sexual partners, and ultimately have more children.[10] And humans aren't the only species with a taste for symmetry. Scientists have found the symmetry preference in many nonhuman animals as well, including rhesus monkeys, barn swallows, zebra finches, and Iberian rock lizards.[11]

Why is symmetry such a big deal? The short answer is that it's harder to grow a symmetrical face and body than it is to grow an asymmetrical one. It's harder for the same reason that it's harder to *draw* a symmetrical face and body than an asymmetrical one: There are more ways to get it wrong than get it right. Only the fittest specimens – individuals with good genes and good health – can hit the bullseye and achieve a highly symmetrical form. The less fit are more easily knocked off course by parasites, toxins, and other environmental stressors. The upshot is that a symmetrical face and body serve the same role in our species as the brilliant plumage of the peacock serves in theirs: They're honest signals of fitness. In a sense, we're all walking billboards, advertising the quality of our bodies and the quality of our genes to anyone with eyes to see.

This isn't merely data-free speculation. Many studies suggest that not only are symmetrical individuals perceived as more attractive, they also tend to be healthier, more fertile, and longer lived.[12] Admittedly, the evidence isn't completely consistent on this point.[13] At this stage, the jury's still out. If I had to make a bet, though, my bet would be that the symmetry–health link is real. For one thing, the link just makes good sense: The theory behind it is strong. For another, the link between symmetry and health has been found in other animals.[14] Humans are plausibly just one instance of a much more general trend.

At the risk of laboring a point that should by now be familiar, the claim is not that people think "Wow – this individual is more symmetrical than the population average. I hereby decide to be attracted to him." People just like what they like. However, the *reason* they like what they like is that, throughout the course of our evolution, these things were reliable indicators of good genes, good health, and fertility. Individuals who preferred symmetrical mates tended to have healthier, more attractive offspring, who ultimately gave them more grandchildren. As a result, just as night follows day, the symmetry preference became more and more common over time, until eventually it was the norm.

Symmetry is an important determinant of attractiveness for men and women alike. Other traits, in contrast, are important for one sex but not the other. One trait linked strongly to *women's* attractiveness is facial femininity. Women's faces differ from men's in a number of characteristic ways: They have larger eyes, more arched eyebrows, smaller noses, higher cheek bones, fuller lips, narrower chins, and smoother, softer skin. The more of these features a person possesses – the larger the eyes, the higher the cheek bones – the more feminine that person's face. Most men find feminine faces attractive, and it makes good sense that they do. Facial femininity is associated with youthfulness: Younger women have more feminine faces than older ones, and as we saw in Chapter 3, males in our species find traits linked to youthfulness particularly enticing. Furthermore, even among younger women, those with more feminine faces tend to be more fertile.[15] The preference for facial femininity is therefore functional rather than frivolous. It fixes men's roaming eyes on those women most likely to bear them children.

Another trait that men find appealing is the classic "hourglass" body shape: large bust, thin waist, relatively large hips. One way that scientists have examined this preference – or at any rate, the bottom half of it – is by looking at a metric called the *waist-to-hip ratio* or *WHR*. To work out your own WHR, take your waist size and divide it by your hip size. If your waist and your hips were exactly the same size, you'd have a WHR of 1, and you'd look like a tube. Most people's waist is smaller than their hips, however, and thus most people have a WHR of less than 1. In childhood, boys and girls have similar WHRs – somewhere in the region of .85 to .95. But as soon as puberty kicks in, girls start laying down fat on their hips, thighs, and buttocks. And as their hips expand, their WHRs drop. Boys' WHRs, on the other hand, stay approximately the same as they morph into men.

Like many traits that appear at puberty and that are found in only one sex, a low WHR is seen as attractive in women. The evolutionary psychologist Devendra Singh argued that, although there are historical and cross-cultural differences in the preferred *weight* for women, the preferred WHR is always roughly the same: around .7.[16] Again, this makes good sense. Not only does a low WHR distinguish women from men, it's a strong indicator of youthfulness and fertility. Older women – women whose reproductive days are behind them – have higher, more male-typical WHRs. In addition, among women of reproductive age, those with higher WHRs tend to be less healthy and less fertile: They have a harder time getting pregnant. Finally, women who are pregnant already, and thus not currently fertile, have higher WHRs as well. In short, if men were looking for a reliable cue that a prospective mate is the right sex, the right age, and capable of getting pregnant, they could hardly do better than WHR. It's little surprise, then, that men in many different cultures and countries – Cameroon, India, Indonesia, New Zealand, Papua New

Guinea, Samoa, the United States, perhaps even Ice Age Europe – do indeed find women with a lower-than-average WHR above average in attractiveness.[17]

We've looked at several traits that men find attractive in women. What about the other way round? Your first thought might be that, if men like feminine traits in women, women presumably like masculine traits in men. But here the waters get murky. On the one hand, a female preference for masculinity would seem to make some sense. Many masculine traits – a prominent brow, a square chin, a hulking upper body and deep, booming voice – develop under the influence of testosterone. For reasons that are not yet fully understood, testosterone-dependent traits are difficult to grow successfully. As such, only the fittest men get to be the proud owners of an ultramasculine face and body. That means that an ultramasculine face and body could potentially function as a costly display of fitness for men.[18] It would therefore stand to reason that women would have a taste for these masculine traits. Sure enough, many studies find that, just as straight men prefer women with an hourglass figure, straight women prefer men with a V-shaped torso, an athletic physique, and a voice like a low-pitched rumble. Men possessing these attributes tend to lose their virginity earlier, attract more sexual partners, and have more affairs than other men. Conversely, men with a pear-shaped body and a high-pitched, Mickey Mouse voice tend to attract less sexual interest and accumulate less sexual experience.[19]

So far so good. Like I say, though, the picture quickly starts getting more complicated. For many testosterone-dependent traits, women's preferences are all over the map. Beards are a good example. Some women like them, but others don't, and the popularity of beards rises and falls over time in a way that the popularity of women's breasts and youthful appearance does not. Furthermore, plenty of women actively dislike ultramasculine traits. Many studies suggest, for instance, that many women prefer slightly *feminized* male faces, especially for long-term relationships as opposed to flings.[20] When it comes to femininity in women, more is always better; when it comes to masculinity in men, you can easily have too much of a good thing. What explains this pattern?

Two possibilities spring to mind. The first is that the aversion to ultramasculinity is a side effect of the evolution of pair-bonding and biparental care. As I've mentioned already, and as I'll discuss in greater depth later in the chapter, pair-bonding isn't the only mating pattern that comes naturally to our species, and biparental care isn't the only childcare arrangement. Nonetheless, these things do seem to be central elements of the human reproductive repertoire. And in the context of pair bonds and parenting, masculinity isn't always a blessing. Ultramasculine men *can* be good partners, husbands, and dads, but they're always a bit of a wild card. Some are more interested in striving for status and seeking out new mates than in settling down and caring for kids, and some may be prone to violence. Perhaps then, over the course of our evolution, selection scaled back the strength of women's preference for masculinity.

A second possibility, which I touched on in Chapter 3, is that male-typical traits such as beards and deep voices may be more about intimidating other men than they are about attracting women.[21] In other words, these traits may be deers' antlers rather than peacocks' tails. To the extent that this is the case, the fact that women don't always find them attractive is beside the point: That's not why they evolved. Note that, as with earlier examples, these two explanations – pair-bonding and male–male competition – are not mutually exclusive. Both may help to explain women's mixed reactions to ultramasculine men.

Of course, some might ask whether an evolutionary explanation is needed in the first place – not just for the masculinity preference but for any human mate preferences. Couldn't it all come from learning and culture? As always, this suggestion does have an initial ring of truth: Humans learn all sorts of things, so why not this? As soon as we pop the red pill of the evolutionary psychological perspective, however, the Nurture Only view becomes a lot more difficult to swallow. Suddenly, we have to accept not only that our beauty standards are learned, but also that it's *purely a coincidence* that they map onto evolutionarily relevant variables such as health and fertility. Suddenly we have to explain why these supposedly arbitrary standards are found in every culture where scientists have looked for them. Suddenly we have to figure out why principles that apply right across the animal kingdom – principles such as that animals evolve a preference for healthy, fertile mates – don't apply to our species. And suddenly we have to explain why, if these principles *don't* apply to us, it looks exactly as if they do. An alien scientist wouldn't take the Nurture Only theory of human mate preferences seriously for a moment. Is there any good reason that we should?

Kissing Cousins

The singer Elton John once said in an interview that "There is nothing wrong with going to bed with someone of your own sex. People should be very free with sex." John then thought for a moment, and added "They should draw the line at goats."[22] Few would disagree with this last point, and in fact most people would draw the line long before goats. Of all the thousands upon thousands of objects in our environment – people, pets, pot plants, pens – most of us are willing to mate with only a tiny fraction. In that sense, humans are extremely picky about their sexual partners, and so are most animals. As we've already started to see, at least some of this pickiness comes to us courtesy of natural selection. But natural selection hasn't just equipped us with sexual *preferences*. It's also equipped us with a number of sexual aversions. In this section, we'll discuss one of the most important. By way of an introduction, let's conduct a little thought experiment – and let me apologize in advance for what I'm about to ask you to do. Please imagine each of the following situations…

- First: Having sex with your lover.
- Second: Your mother having sex with your brother.
- Third: Tongue-kissing your sibling.[23]

I know, I know – it's not a pleasant experience. People's reactions to these scenarios are as predictable as they are intense. And yet to an alien anthropologist, armed only with a social psychology textbook, the reactions would be entirely unexpected. According to standard treatments of the topic, interpersonal attraction is determined by variables such as proximity, familiarity, and similarity: We hook up with and fall in love with people who live near to us, who we see a lot, and who are similar to ourselves. Now even on the face of it, this is a paper-thin explanation. *Most* people in our social circles meet all three conditions, and yet we fall in love with barely any of them.[24] But there's another, more serious problem with the standard social psychological explanation. If variables such as proximity, familiarity, and similarity were as important as the textbooks tell us they are, people would be falling in love with close relatives left, right, and center. It stands to reason: We grow up with close relatives, we spend a lot of time with them, and they're generally more similar to us than anyone else in the world. But are close relatives the people we find most attractive? Clearly not. When it comes to sexual attraction, kinship is the ultimate deal breaker. It doesn't matter how symmetrical a person is, how V-shaped their torso, or how low their waist-to-hip ratio. Close relatives are a no-go zone.

Of course, unless you've lived in a cave all your life, this isn't going to be news. But it would seem extremely odd to our asexual alien scientist. Even if the alien broadened its reading horizons and started studying evolutionary psychology, the puzzle wouldn't immediately dissipate. That's because, at least at first glance, it looks as if incestuous mating would be favored by kin selection (the process that gave rise to the exploding carpenter ants we met in Chapter 2).[25] If an individual, A, has a child with, say, a full sibling, not only does she pass on some of her own genes, she also helps a close relative to pass on some of his. The inclusive fitness benefit of doing so would be equivalent to producing an offspring of one's own *plus* a niece or nephew: one-and-a-half for the price of one. Now, if both siblings instead had a child with someone else, A's inclusive fitness outcome would be the same: She'd end up with one offspring plus one niece or nephew.[26] However, if having a child with her sibling meant that the sibling would have an *extra* child – one he wouldn't otherwise have had – then A would get a bigger fitness boost by having a child with her sibling than she would by having a child with a non-relative. Any gene that encouraged such a tendency would have a strong chance of being selected. What's the catch?

The catch is *inbreeding depression*. This refers to the fact – long suspected and now well documented – that mating with close relatives drastically increases the chances that any offspring will have genetic defects.[27] The main problem is *deleterious recessives*. These are recessive genes which, if expressed, are highly detrimental to fitness: They kill their owners or severely compromise their health and wellbeing. Most people have some deleterious recessives, but most of the time, they're not expressed. This is because recessive genes, by definition, are only expressed when you get a double dose of them: one from each parent. In the normal course of events, that's not very likely, because any given deleterious recessive is exceedingly rare. Thus, although most people have *some* deleterious recessives, sexual partners rarely have the same ones… *if*, that is, they're unrelated. If they're related, on the other hand, the probabilities shift dramatically. Relatives share a large fraction of their genes, as a result of their recent shared ancestry. And because almost everyone has some rare deleterious recessives, if two relatives have a child together, they're almost certain to give that child a double dose of at least some of those fitness-harming genes, even though the genes are rare in the general population. The end result is that the offspring of relatives typically have lower fitness than the offspring of non-relatives.

So, we've got our smoking gun: a selection pressure that would favor any genes that reduced the chances that their owners would mate incestuously. And as we'd expect, when we take a panoramic view of the animal kingdom, we find animals steering clear of incest in a wide variety of ways. These fall into two main categories: *dispersal* and *disgust*.

Dispersal is probably the more common anti-incest adaptation. In many species, when the young hit puberty and their sexual desires come online, they experience a strong wanderlust which motivates them to leave the family group. This automatically reduces their chances of encountering, and therefore potentially mating with, relatives. Sometimes, both sexes disperse; often, only one does. In most monkeys, for example, the males disperse; in most apes, it's usually the females.

Dispersal is a common anti-incest adaptation, but it's not the only one. In many species (including many species that disperse), individuals continue to interact with at least some other-sex relatives throughout their adult lives. Where that's the case, a second incest-avoidance adaptation is common: Animals develop a sexual aversion to probable relatives. This may mean individuals that one was reared with or that one cared for when they were young. Or it may mean individuals that are unexpectedly similar to oneself in terms of smell or any other trait where relatives are more alike than non-relatives. Thus, chimpanzees and bonobos avoid mating with their mothers or sons, and lemurs avoid mating with individuals whose scent is too similar to their own.[28] In short, many animals have evolved to use rearing association and similarity as cues to kinship. Just as rotten meat dampens the desire to eat, so these kinship cues dampen sexual desire.

Note that, although incest aversion is widespread, it's by no means universal. Indeed, in some species, incestuous mating is commonplace; termites and naked mole rats are the best-studied examples. But although incest does happen, among large, long-lived animals it's rare. Incest avoidance is the rule.

Does this rule apply to us? In other words, do humans possess evolved psychological mechanisms designed to reduce the chances of incestuous mating? Given that most people feel an acute disgust at the mere idea of mating with kin, it certainly seems possible. However, spelling out exactly what these mechanisms might be is not such an easy task. Just as we can't directly perceive how good a person's genes are, we can't directly perceive whether or not they're kin. Strictly speaking, then, we couldn't have an evolved tendency to avoid mating with kin. What we could have, though, is an evolved tendency to avoid mating with individuals possessing markers that, throughout the course of our evolution, were statistically associated with kinship. Several such markers have been proposed, and they're very like those documented in other species.

The first is *early-life cohabitation*. For most of our history as a species, children typically grew up with parents, siblings, and an assortment of other relatives. In such circumstances, a good anti-incest rule of thumb would be: "Avoid mating with anyone you spent a lot of time with as a child." The first person to argue that humans evolved to obey such a rule was a Finnish sociologist by the name of Edvard Westermarck. In his 1891 book *The History of Human Marriage*, Westermarck proposed that people who grow up together tend not to find each other sexually attractive, and that this tendency is an adaptation crafted by natural selection. In scientific lingo, growing up together – and especially spending a lot of time together in the first five or six years of life – results in *negative sexual imprinting*. In everyday lingo, growing up together puts people permanently in the friend zone. This phenomenon is now called *the Westermarck effect*.

The best evidence for the Westermarck effect comes from studies looking at people who were raised together but who weren't actually relatives. This includes research on Israeli kibbutzim. As mentioned in Chapter 3, in many kibbutzim, the children were housed and raised together, like a vast colony of siblings. Based on traditional social psychological principles, we'd expect that these individuals would routinely fall in love with one another and marry. After all, they spent a lot of time together, they had similar backgrounds, and they weren't related so there'd be no reason for anyone to object to their union. In reality, however, such marriages were virtually non-existent. One study found that, of nearly 3,000 marriages involving kibbutznik, only fourteen were between people raised together on the same kibbutz. Furthermore, of these fourteen, only five lived together prior to the age of six, and of these five, none lived together for more than two of their first six years. People who grow up together on kibbutzim are often *emotionally* close to each other; they just don't find each other sexually attractive. They love each other but not in *that* way – consistent with the Westermarck hypothesis.[29]

The kibbutz case study is a convincing natural experiment. Like all natural experiments, however, it's an uncontrolled one, and thus we shouldn't rest our case on this single example. Fortunately, it's not the only one. A second relates to traditional arranged marriages in Taiwan and parts of mainland China. Before the practice was outlawed in the mid-1900s, arranged marriages were common, and took one of several forms. Among these were the major marriage and the minor (or sim-pua) marriage. In the major marriage, the betrothed met each other only when they were full-grown adults. In the sim-pua marriage, in contrast, the girl was adopted into the boy's family, and the pair grew up together like siblings. On paper, this arrangement made sense, especially for poor families: The girl's family got someone to raise their child, and the boy's family got a bride for their son and a spare pair of hands around the house. But sim-pua marriage turned out to be a bad idea in precisely the way that Westermarck would have predicted. Sim-pua marriages had more marital problems, more infidelities, fewer children, and a higher rate of divorce than other forms of marriage. One study found, for instance, that seventeen out of nineteen sim-pua couples refused point blank to consummate their marriages, even in the face of significant pressure from their parents.[30] In other words, like so much in human life, the pattern emerged *despite* social pressure, not because of it.

A less well-known finding is that sim-pua brides who were reunited with their biological families later in life often found that they were sexually attracted to their brothers or other family members. Because they hadn't grown up together, they didn't experience the normal sexual aversion to kin that most people develop. Their heads understood that they were related, but their hearts did not. The same phenomenon has been documented as well in the West. Sometimes, when siblings who were separated at birth meet again as adults, they discover – usually to their horror – that there's a sexual spark between them. And occasionally, despite being related, they fall in love and start families.[31] Cases like these show just how much work the Westermarck effect does in the ordinary course of events.

But Westermarck isn't doing *all* the work. Because kin are so genetically similar to one another, they tend to look (and act and smell) similar too. That being the case, similarity itself constitutes an additional cue to kinship. As we'll see in the next chapter, humans appear to have an evolved tendency to favor individuals who are similar to themselves, which was presumably shaped, at least in part, by kin selection. It turns out, however, that people don't favor similar others in every possible way. As the evolutionary psychologist Lisa DeBruine has shown, self-resembling individuals are viewed as more trustworthy but less "*lust-worthy*."[32] More generally, kinship cues seem to upregulate closeness and altruism but downregulate sexual attraction.[33] This is an important discovery. Without the guidance of an evolutionary perspective, there's no particular reason we'd expect it. On the contrary, based on standard social psychological principles, we'd expect that similarity would turn up the volume on every positive social feeling. Only an evolutionary perspective predicts the fine-grained pattern of reactions: the fact that similarity turns up the volume on trust and liking, but turns *down* the volume on sexual interest.

In spite of its predictive success, it's fair to say that not everyone is sold on the evolutionary explanation for incest aversion. The obvious alternative, of course, is that we *learn* our sexual aversion to kin. And it's easy enough to dream up plausible-sounding just-so stories about how anti-incest norms could have emerged through cultural rather than genetic selection. One could argue, for example, that people who adhered to the anti-incest norm had more surviving offspring, and that those offspring then tended to learn that norm from their parents. In this way, the anti-incest norm spread because it enhanced people's fitness, despite not being encoded in the genes. Alternatively, one could argue that groups that taught their members to avoid incest were less plagued by genetic illness, and thus that more of their members survived and thrived. As a result, these groups grew and spread more rapidly than those that didn't have that teaching. For either reason, or perhaps for both, the anti-incest meme spread like wildfire through the human population. Or so it might be argued.

But is it really plausible to think that our distaste for incest is purely a product of parental hectoring and cultural pressures? Steven Pinker, for one, doubts it. "Do brothers and sisters avoid copulating because their parents discourage it?" he asks.

> Almost certainly not. Parents try to socialize their children to be more affectionate with each other ("Go ahead – kiss your sister!"), not less. And if they did discourage sex, it would be the only case in all of human experience in which a sexual prohibition worked.[34]

Compare incest to premarital and extramarital sex. Parents and moralists have discouraged the latter activities for as long as parents and moralists have existed. Sometimes it works; sometimes it doesn't. But when it does work, it's because people manage to resist temptation, not because they've developed a sexual aversion to sex outside wedlock. In contrast, most people aren't remotely tempted to engage in incestuous mating; they're acutely disgusted by the prospect. And this is the case despite the fact that parents and moralists rarely bother preaching the evils of sex with relatives. Parents worry about their teenage offspring having sex with *non-relatives*, not with each other – and that's because it's the non-relatives that they're most likely to have sex with.[35]

One response to this argument would be to concede that, sure, it sounds plausible if we restrict our gaze to the West. But that's just being ethnocentric. In many non-Western cultures, incest has nothing like the stigma it does in the Western world. Consider, for instance, the traditional

Egyptian, Hawaiian, and Incan civilizations. Historical records show that brother–sister marriage was neither proscribed nor demonized in these cultures, and indeed that it was reasonably common, especially among the royal classes. This isn't to say it was a good thing, of course; inbreeding depression doesn't just magically go away because people don't know about it or worry about it. But these ancient marital practices show us that, in at least some cultures, incest was not the big bug bear it is for us now. And if that's true in any culture, then the whole notion of evolved incest adaptations is cast into serious doubt.

The problem is, though, that it's not true – or at any rate, not true in a way that has the anti-adaptationist implication. Brother–sister marriage was not a culture-wide phenomenon in any of these ancient civilizations; it was largely restricted to the upper echelons of society.[36] And there's a reason for that: The marriages were primarily political alliances rather than love matches. Their function was to keep wealth and power concentrated in the hands of the family group.[37] There's no evidence that the siblings involved were attracted to each other, and in some cases, there's evidence against it. Cleopatra, for instance, married two of her brothers, but she didn't have children with them, and the men she *did* have children with – Julius Caesar and Mark Antony – were unrelated to her. To be sure, married siblings did sometimes have children together; to cite a famous example, the Egyptian Pharaoh Tutankhamun was the product of sibling incest, and he himself had two children with his half-sister (both of whom were stillborn). But again, this kind of thing was never widespread, which suggests that brother–sister marriage was a product of social pressures overriding our anti-incest adaptations, rather than being evidence that the anti-incest adaptations don't exist.

A final reason to reject the Nurture Only theory of incest avoidance is that, as we've already seen, many nonhuman animals are just as allergic to incestuous mating as we are. This includes species as diverse as capuchin monkeys, elephants, fur seals, and cockroaches.[38] Not only that but some animals exhibit something very similar to the Westermarck effect; mice, for instance, avoid mating with individuals they were raised with, even when scientists sneakily swap the pups around so that – like simpua brides – they're actually raised with non-relatives.[39] Incest avoidance has even been documented in plants. Most plants (unlike most animals) are hermaphrodites: They have male *and* female reproductive equipment, and thus can theoretically pollinate themselves. In a sense, self-pollination is the most extreme form of inbreeding there is, equivalent to having kids with a clone. Some plants do, in fact, self-pollinate. But many do not;

they have inbuilt mechanisms designed to reject their own pollen, and often also the pollen of relatives.[40] Inbreeding avoidance in plants and nonhuman animals clearly isn't a product of learning or culture. It's a product of natural selection. When we find the same thing in our own species, the default assumption ought to be that it's a product of natural selection for us as well.

The evolutionary explanation for incest aversion is an example of the deep explanatory power of a good scientific theory. The selection pressures that, in many nonhuman species, result in members of one sex or the other abandoning the family group at puberty turn out to be the same selection pressures that, in our species, result in people finding the prospect of incestuous mating distasteful (to put it mildly). Without the gene's-eye view of evolution, who would have guessed that these phenomena were connected? Just as Newton's theory of gravity links the fall of an apple to the orbit of the moon, the gene's-eye view of evolution links facts about the world we would never have imagined had a common origin. The theory is explanatorily satisfying in just the way that the best theories in science always are.

Love and Its Discontents

So far, you've imagined that you were a non-reproductive worker insect, that you were a hedgehog, and that you were a neo-Darwinian demon, hell-bent on spreading your genes. Now imagine that you're the parent of a teenage boy or girl. Imagine, moreover, that you've just read about a powerful new drug that young people in your area are taking. The drug can be instantly addictive: A kid might sample it just once at a party and – bang! – they're hooked. They're so deeply hooked, in fact, that if they think the supply will be cut off, they might try to kill themselves – by drinking poison, perhaps, or perhaps by stabbing themselves with a dagger.

Understandably, a drug like this would be something you'd be very concerned about. Well, I don't want to alarm you, but there really is a drug exactly like this. It doesn't always have such catastrophic effects, but it certainly did for Romeo and Juliet. Shakespeare's star-crossed lovers met at a party, sampled the drug together, and then, when it seemed that the drug would be torn away from them, committed suicide rather than go on without it. The drug, of course, is love. And although Romeo and Juliet were fictional, plenty of non-fictional persons have had their lives turned upside down by this legal high. Love has many adverse side effects, from poor concentration to obsessive thinking. It's a common cause of divorce, as when a husband or wife falls in love with someone else. And it can provoke a wide range of pathological behaviors, from stalking to suicide to murder. The clinical psychologist Frank Tallis once suggested that being in love is the closest most people come to mental illness.[41] Even the most hopeless romantic would have to concede, at the very least, that they see what he was getting at.

Of course, the hopeless romantic would immediately want to point out that love also has many positive effects. It can turn ordinary people into heroes, inspire self-sacrifice worthy of a non-reproductive worker insect, and fuel creative achievements spanning from embarrassing teenage poetry to the majestic Taj Mahal. Love is a double-edged sword, and it's not at all obvious whether, on balance, the sword has brought more joy or more misery to the world. Neil Gaiman captured the mixed character of love beautifully in his *Sandman* comic series, when his character Rose Walker said:

> Have you ever been in love? Horrible, isn't it? It makes you so vulnerable. It opens your chest and it opens up your heart and it means someone can get inside you and mess you up. You build up all these defenses. You build up a whole armor, for years, so nothing can hurt you, then one stupid person, no different from any other stupid person, wanders into your stupid life ... You give them a piece of you. They didn't ask for it. They did something dumb one day, like kiss you or smile at you, and then your life isn't your own anymore. Love takes hostages. It gets inside you. It eats you out and leaves you crying in the darkness, so a simple phrase like "maybe we should be just friends" or "how very perceptive" turns into a glass splinter working its way into your heart. It hurts. Not just in the imagination. Not just in the mind. It's a soul-hurt, a body-hurt, a real gets-inside-you-and-rips-you-apart pain. Nothing should be able to do that. Especially not love. I hate love.[42]

Strong words! Maybe, then, before going any further, we should make sure we know what we're talking about. Love is just one word, after all, but it's not just one thing. There are various strains of love, from the love of family to the love of friends to the love of king and country. The strain we're putting under the microscope at the moment – the one that killed Romeo and Juliet – is *romantic love*, also sometimes known as *passionate love*. The following is a list of the typical features (or symptoms) of this love subtype. It's based largely on the work of the anthropologist Helen Fisher and the psychologist Dorothy Tennov.[43]

- *Intrusive Thinking.* When people are in love, they can't stop thinking about the person they're in love with. As Fisher observes, being in love is like having someone camping permanently in your head. In some ways, it's similar to a clinical obsession.

- *Aggrandization of the Beloved.* When we're in love, we idolize the other person and fail to notice their flaws – or if we do notice, we don't care. As Fisher points out, no one in your life is a hundred times better than anyone else, but we act toward our lovers as if they were. George Bernard Shaw put it well: "Love," he wrote, "consists in overestimating the difference between one woman and another." (And of course the same applies to men.)

- *Looking for Clues.* In the early stages of romantic love, people are hyperattentive to any possible evidence of the other person's feelings toward them. They replay and dissect each interaction and every furtive glance. "Does she feel the way I do?" Five minutes later: "Does she *still* feel the way I do?"

- *Intense Energy.* When people are in love, notes Fisher, they can walk all night and talk till the sun comes up. Sometimes they can't sleep; sometimes they forget to eat. Love is a little like a hit of cocaine. (More precisely, cocaine activates some of the same brain regions associated with being in love and other pleasurable psychological states.)

- *Mood Swings.* Love takes you on a roller coaster ride of emotions, from giddy pleasure to soul-crushing despair. When a relationship is going well, you feel like the luckiest person alive. When it hits a rocky patch, or disintegrates entirely, it feels like the worst thing that's ever happened to you – or to anyone else in the history of the world.

- *Sexual Desire.* Sexual desire is one of the central ingredients of romantic love, and the ingredient that most clearly distinguishes romantic love from the other love subtypes. As we'll see later, sexual desire usually goes hand-in-hand with a desire for sexual exclusivity, if not for oneself, then at least for one's mate.

- *Involuntariness.* Love is beyond our control. People can fall in love against their better judgment and despite their best intentions. That's why it's called "falling." Stendhal put it best: "Love," he said "is like a fever which comes and goes quite independently of the will."[44] Anyone who's been in love, and especially anyone who's been in love but not wanted to be, knows that we don't have complete control of our minds (that is, our minds don't have complete control of themselves).

Another common feature of romantic love – one which continually takes people by surprise, no matter how often they experience it – is that it tends to be fleeting. The sizzle often fizzles, as they say. And the sizzle-related fizzle is not unique to the West. One woman from the !Kung tribe of the Kalahari Desert, for example, observed that "When two people are first together, their hearts are on fire and their passion is very great. After a while, the fire cools and that's how it stays."[45] It's a familiar refrain.

But the end of the crazy, can't-keep-your-hands-off-each-other phase doesn't necessarily herald the end of love. Sometimes romantic love matures into a distinct form of love, which psychologists call *companionate love*. Actually, it's not quite right to say that romantic love and companionate love are distinct states; there are no clear boundaries between them. The main difference isn't the type of feelings involved but rather their relative prominence. Sexual desire tends to be an important ingredient in both forms of love, but in companionate love, it's less central. Meanwhile, other ingredients, such as intimacy and commitment, steal a larger share of the spotlight in companionate love than romantic.[46]

Companionate love is a less exhilarating form of love than romantic love, but in many ways, it's more real. With romantic love, or at least early-stage romantic love, we often don't really know the person we fall in love with. For whatever reason, this biological mechanism fastens itself to a particular person, like a dog with a bone, and we then cycle through a predictable sequence of feelings for them. In Ovid's *Metamorphoses*, Pygmalion sculpted a woman out of ivory and promptly fell in love with the sculpture. In a sense, we all do this every time we fall in love: We have a mental image of another person, project it onto someone who happens to get our hormones fluttering and happens to be available (ideally), and then fall in love with this image of our own creation. In effect, we fall in love with someone that we don't particularly know: a semi-stranger. And it's often rather hard to get to know this person, because every time we're with them, we start hallucinating – hallucinating that we're in the presence of someone whose every word and deed is transcendently perfect.

But the hallucinations don't last forever. In time, we start to get to know the real person hidden beneath our mental image of them – and occasionally, if we're lucky, we come to love that person too. This next-level love is often a more genuine, less delusional form of love. One of the most powerful descriptions of it that I've come across is in Louis de Bernières' novel, *Captain Corelli's Mandolin*:

> Love is a temporary madness. It erupts like an earthquake and then subsides. And when it subsides you have to make a decision. You have to work out

whether your roots have become so entwined together that it is inconceivable that you should ever part. Because this is what love is. Love is not breathlessness, it is not excitement, it is not the promulgation of promises of eternal passion. That is just being "in love" which any of us can convince ourselves we are. Love itself is what is left over when being in love has burned away, and this is both an art and a fortunate accident. Your mother and I had it, we had roots that grew towards each other underground, and when all the pretty blossoms had fallen from our branches we found that we were one tree and not two.[47]

Love, then, is a complex and multifaceted phenomenon. Where did it come from? Unlike some of the topics we've looked at, it's easy enough to see how romantic love relates to reproduction, and thus easy enough to see why love might initially have evolved. Getting the details right isn't so easy, though. Somerset Maugham once wrote that "Love [is] only the dirty trick nature played on us to achieve the continuation of the species."[48] Maugham would have failed my evolutionary psychology class! (See Chapter 2.) A dirty trick it may well be, but love did not evolve for the good of the species. It evolved for the good of the genes giving rise to it. What Maugham should have said – admittedly, it's less catchy – is this: "Love is only the dirty trick nature played on us to achieve the continuation of the genes contributing to the development of the capacity to fall in love."

How does love achieve this lofty goal? At one level, it's obvious. As we've seen, one of the central ingredients of romantic love is sexual desire: New lovers have a hard time keeping their paws off each other. The sexual connection is a rather unambiguous sign that one of the primary functions of love is achieving pregnancy. This often isn't *our* aim when we fall in love, but it is one of the main reasons that romantic love evolved.

If it were just that, though, there'd be no reason for love to last longer than a few months at the most, and certainly no reason for it to last beyond the point where the woman got pregnant. Human love affairs would be like chimpanzee "consortships," where a high-status male and a fertile female go off together for a few days, mate their brains out, and then return to the group and to business as usual, with no lingering bond between them and no expectation of paternal investment. Some human relationships do follow this trajectory, but most don't and the reason isn't hard to grasp: Human beings, unlike chimps, have profoundly dependent offspring. Because our young require so much care and attention, humans evolved the capacity to form pair bonds. Pair bonds not only boost the chances of pregnancy (or did in our natural, contraception-free environment), but also provide a setting for dual-parent care of the young.[49] Love is the glue that holds the parenting partnership together. In the ancestral environment in which it evolved, love typically kept the couple in one another's orbit for long enough for the man to provision and protect the woman during the later-stages of pregnancy, and then to provision and protect her and their new child during the difficult early months and years.[50] Sometimes love lasted longer than that; sometimes it didn't. As Helen Fisher sums up the situation, love enables us to tolerate another human being for long enough to produce a child or two – but not necessarily to tolerate them forever.[51]

This much seems uncontroversial. Beyond this point, however, the mysteries start to grow and multiply like mold in a petri dish. Why, when we fall in love, do we fall so hard? It's not immediately obvious how this would be in our best interests. Aside from anything else, it makes us vulnerable to exploitation by Lotharios and gold diggers. If love *is* in our best interests, why does it often wear off so quickly, even despite our best wishes? Conversely, why does love sometimes *fail* to wear off when a relationship is doomed, dying, or dead? An alien scientist would naturally predict that, after the collapse of a love affair, a well-designed Darwinian machine would immediately resume its search for a suitable mate. The alien would *not* predict that the "machine" would spend months or sometimes years pining for lost love and wallowing in futile self-pity – after all, life is short and there are plenty more fish in the sea. The wallowing hardly seems adaptive. Moreover, as Romeo and Juliet remind us, people sometimes kill themselves over a failed relationship, which is about as *mal*adaptive as it gets. What's going on?

Let's start with the question of why love doesn't always last.[52] One way to shed light on this sad state of affairs is to view the process of falling *out* of love as a late-stage form of mate choice. Seen through this lens, mate choice is not just a one-time event, which ends the moment one enters a relationship. Instead, every moment one remains in a relationship is, in effect, an additional instance of mate choice. Taking this perspective, many facts about the death of love begin to seem less puzzling. First, wherever the law permits it, women are more likely than men to initiate a divorce or breakup.[53] Viewed as a form of mate choice, this becomes yet another example of the fact that women are choosier than men about their mates. And it makes good sense; not only do women invest more than men into each individual offspring, but they have a shorter reproductive lifespan, and thus each tick of the reproductive clock is a bigger deal. Second, men and women differ, on average, in their reasons for ending a relationship. Across cultures, women are more likely than men to cut the marital knot if their partner becomes unemployed.[54] This fits well with the finding that women place more weight than men on a mate's resources and status. In contrast, men are more likely to end a relationship when the woman starts getting older.[55] This fits well with the finding that men place more weight than women on a mate's appearance and youthfulness.

So, the fact that love doesn't always last isn't too hard to fathom from an evolutionary point of view. But what about the fact that love sometimes lasts well past its use-by date, that people sometimes mourn the death of a relationship for longer than makes any sense, and that people sometimes go as far as to kill themselves over a love affair gone wrong? Perhaps surprisingly, the easiest part of this puzzle to solve is the last part: love-related suicide. *Most* people don't kill themselves over lost love, and although it does happen, it isn't something that humans specifically evolved to do, in the way that we did evolve to laugh and cry, fight or fly. It's just an occasional, tragic by-product of the negative feelings that people experience in the wake of a breakup or the breakdown of a romantic relationship. The real question for the evolutionary psychologist is why we have these feelings in the first place, and why they sometimes last so much longer than would seem to be adaptive. There are several possibilities, but one is that the feelings weren't so maladaptive in earlier times. People often tell each other that "There are plenty more fish in the sea," which seems to imply that we often act as if there aren't. But maybe we act that way because, when love first evolved, there really *weren't* a lot of other fish in the sea. For most of our evolutionary history, we lived in small groups with a limited pool of potential partners and a distinct absence of nightclubs and Internet dating sites. In such a world, selection may have favored a tendency to latch onto mates and potential mates with a vice-like grip, at least until other, more suitable options appeared. Our emotions are calibrated for this ancient lost world, not for the world we now inhabit. The excessive tenacity of love may, in other words, be a result of evolutionary mismatch.

Now admittedly, all of this assumes that romantic love is a product of evolution to begin with, and not everybody accepts that. Some argue that love is a product of Western civilization, and that the notion that it's a human universal is merely a manifestation of Western ethnocentrism. Love, they argue, was invented by Shakespeare, medieval knights, or individualistic societies, and is nourished and sustained today by fairy tales, Hollywood movies, and cringe-worthy Valentine's Day cards. Some go further and argue that love is a patriarchal, capitalist conspiracy. The conceptual artist Jenny Holzer, for instance, suggested that "Romantic love was invented to manipulate women." Similarly, Angie Burns proposed that, "From a feminist perspective, romantic love was, and is, seen to obscure or disguise gender inequality and women's oppression in intimate heterosexual relationships."[56]

But all these ideas strain credulity, to put it politely. When people fall in love for the first time, the experience is often very different from what they'd expected. Some are surprised that love makes them so vulnerable and possessive, others that it so radically rewrites their plans and priorities, and others still that the course of love so often runs so very far from smoothly. If love is just a matter of acting out a cultural script, stamped into our heads by fairy tales and rom-coms, why would love as we experience it deviate so profoundly from love as we think it will be?

And why, as far as we can tell, is romantic love found in all cultures? That's right; contrary to stubborn anthropological myth, people everywhere fall in love. One line of evidence for this claim comes from the anthropologists William Jankowiak and Edward Fischer, who scoured the anthropological research on 166 historically independent cultures, noting down any evidence of romantic love that they came across: romantic poetry, elopement, all the usual symptoms. Their conclusion? Romantic love was unambiguously present in around 89 percent of cultures – the vast majority. Note that love wasn't necessarily *absent* in the remaining 11 percent; it simply wasn't clear either way. If the ambiguous cultures had all been clustered together, then maybe we could argue that love was a contagious meme which had spread to most of the world but hadn't yet reached that region. Instead, though, the ambiguous cultures were dotted round the globe at random, and romantic love was clearly present in all or nearly all cultures from every world region: Africa, Europe, Asia, the Americas, Polynesia. This suggests that love actually is universal, and that if we just looked a little bit harder, we'd find it in every culture.[57]

In support of this assertion, another study using a very different method found evidence for romantic love in an even higher proportion of cultures. Jonathan Gottschall and Marcus Nordlund analyzed thousands of traditional folk stories from cultures around the world, again looking for telltale signs of romantic love. They restricted their survey to stories that predated contact with the West – stories, in other words, that couldn't have been "tainted" by Western individualism or Shakespearean sonnets. Once again, the findings strongly suggested that romantic love is a human universal: Love was unambiguously present in seventy-eight of the seventy-nine cultures in the sample. Notably, the cultures in question covered all the major human population groups, including sub-Saharan Africans, Australian Aborigines, South Asians, East Asians, and Native Americans.[58]

The question all these findings raise is a straightforward one: If romantic love is an invention of Western culture, why is it found in every geographical region, historical period, and ethnic group? The simplest and most plausible answer is that romantic love is *not* an invention of Western culture. Instead, the *idea* that romantic love is an invention of Western culture is itself an invention of Western culture, and a rather implausible one at that. Human beings were falling in and out of love for hundreds of thousands of years before we ever had Hollywood blockbusters or knights in shining armor. We're just that kind of animal – the kind that falls in love from time to time.

The Green-Eyed Monster

Different people have different ideas about the best way to spend one's honeymoon: where to go, how long to go for, what to do when one arrives. But one thing that nearly everyone agrees on is that it's best if the bride and groom spend their honeymoon together, and best if they *don't* spend their time behind bars. Sadly, though, that's exactly what a newly wed Massachusetts couple ended up doing back in 2010. The couple had been driving around after their wedding when they spotted an old flame of the groom, leaving work with her son. The bride was at the wheel and, after rolling down the window and shouting some sexually loaded insults at her new husband's ex, decided she'd try to run her down. Luckily, the ex and her son were able to get out of the way before the car collided with them, and the bride hit a fence instead. The newlyweds were arrested at the scene and spent their first night as husband and wife locked up in separate jail cells, while their story became viral news and a source of mirth all around the world.

Like the sexorcism case in Chapter 3, the specifics of this story are surprising – that's why it went viral – but the emotions underlying it are as familiar as one's face in the mirror. Most people "get" jealousy. Even before we experience it ourselves, most of us understand how upsetting it would be if a girlfriend or boyfriend, husband or wife, got involved with somebody else. And yet to our asexual alien scientist, jealousy would be another relationship mystery, every bit as puzzling as love. We don't usually mind our friends having other friends. Why, then, are most of us so worried about our sexual partners having other sexual partners? We don't usually mind if our sexual partners have a nice meal without us. Why, then, are most of us so worried about our sexual partners having a nice sexual encounter without us? Things would function so much more smoothly if we didn't worry. So, why don't we just do that?

According to some, we can and we should. Popular books such as *Sex at Dawn* and *The Ethical Slut* inform us that jealousy is not a part of human nature, but rather a cultural invention.[59] It's not found in our closest living relatives, the chimps or bonobos, and it's not found in other big-brained social animals such as dolphins. There are even plenty of human cultures where jealousy is a stranger.[60] This includes the Inuit of the Arctic Circle, among whom tribal chiefs will sometimes offer their guests one of their wives for the night. Likewise, in the traditional societies of the Musuo of China, the Siriono of South America, and the Samoans of the South Pacific, the desire for sexual exclusivity is as foreign as it would be to our alien scientist. Sexual exclusivity is a Western fetish and jealousy a Western neurosis. Indeed, even within the Western world, a lot of people refuse to tow the jealousy line. To take just one example, the actress Shirley MacLeane once told an interviewer "I've never really had sexual jealousy." The sooner the rest of us join MacLeane and cast off the shackles of our sexual possessiveness, the better it'll be for everyone.

That's what some people say, but how plausible is it? Could jealousy really just be a social invention, like money or a seven-day week? When people first experience the green-eyed monster, they're often quite taken aback by its intensity. If they're merely following a script picked up from the local culture, why the surprise? And are there really cultures where people are immune to the torments of jealousy? The idea is dubious in the extreme.[61] True, anthropologists have made bold claims about the absence of jealousy in various non-Western societies. But these claims invariably crumble on close inspection. William Stephens argued in the 1960s, for example, that among the Marquesas Islanders of Polynesia, adultery was considered perfectly acceptable. His evidence was that the Marquesas didn't have any social sanctions against it. What he'd overlooked, though, is that despite the absence of formal sanctions, men in this society would often beat or even kill their wives if they suspected them of infidelity.[62] Along similar lines, Margaret Mead once claimed, on the basis of a handful of interviews, that traditional Samoan societies were entirely free of the scourge of jealousy. But later, more in-depth research revealed that jealousy was a common cause of violence in Samoa.[63]

What about Inuit wife sharing? This does seem like a counterexample to the idea that sexual possessiveness is a human universal... but only if we assume that sharing one's wife was no big deal to the Inuit. That's not doing the custom justice, though; the whole point was that it was a hugely generous gesture. And it was generous because the Inuit, like human beings everywhere, are possessive of their spouses and lovers. How do we know? Because among the Inuit, male sexual jealousy is a common cause of spousal homicide – just as it is among all peoples.[64]

No doubt, there are individual exceptions. But they're few and far between. That's why when Shirley MacLeane announces that she's never really experienced jealousy, it makes headlines around the world. Most people have. Like it or not, jealousy is a constant companion of love. It's an unwelcome guest at the table – a guest we can never quite banish, try though some of us might. That's not what we'd expect if jealousy were just a social invention.

But if it's not just a social invention, where did jealousy come from? Here, evolutionary theory provides a clear answer and the only truly convincing answer there is. As we saw in Chapter 2, the function of any emotion is to motivate behavior. What kind of behavior does jealousy motivate? The obvious answer – obvious, at any rate, to anyone familiar with animal behavior – is that jealousy motivates *mate guarding*. Mate guarding is a common category of behavior in the animal world. And like any recurring behavior, it's designed to solve a certain set of adaptive problems.

The most basic adaptive problem that mate guarding aims to solve is one faced only by males: the problem of ensuring that a female you've mated with has *your* kids rather than someone else's. That's the primary purpose of the chimpanzee consortship I mentioned earlier. This brief mating arrangement is initiated by the male, not the female. By whisking her away from the group, the male isolates the female from other males while she's fertile, thereby increasing the chances that he'll be the one who impregnates her. For obvious reasons, any genes that help give rise to such a tendency have a good chance of being selected. For the same obvious reasons, male behavior along similar lines is found right throughout the animal kingdom. From water striders to warblers to wallabies, males in a diverse range of species stick close to fertile females, and fight to keep rivals at bay.

In species where the male contribution to the production of offspring goes beyond merely fertilizing the female's eggs, new adaptive challenges arise. The first pertains again only to males. The more a male invests in offspring, the more important it is for him to ensure that those offspring are really his. For male chimpanzees, the worst-case scenario is wasting time mating with a female who another male has already impregnated. But for males in species with high paternal investment, the worst-case scenario is spending months or years caring for another male's off-spring – and therefore that male's genes. This is a much worse worst-case scenario, and it greatly amplifies the adaptive benefits of mate guarding.

Paternal investment also creates new adaptive challenges for females. In species where the male contribution to baby-making is limited to pro-viding sperm, there's little reason to fight over males – a single male, given half a chance, can easily impregnate all the females in the group. Thus, among chimps, elephant seals, and indeed most mammals, males guard females but females don't guard males. In contrast, in species where males help care for the young, male investment becomes a scarce resource, and scarce resources are worth fighting over. You don't want your mate looking for a better option and you don't want another female poaching a good male away from you. Thus, in pair-bonding species, we often find *mutual mate guarding*. Gibbons are a good example. These agile little apes form pair bonds that typically last a lifetime. But gibbons know that the flesh can be weak, and thus engage in what's called *same-sex repulsion*: Males chase away other males, and females chase away other females. Gibbons *are* largely monogamous, but not because they have exclusively monogamous inclinations. Gibbon monogamy, it seems, has to be enforced.[65]

The landscape is similar for most pair-bonding species. Take birds. As we know from Chapter 3, pair-bonding is more common among birds than among any other family of animals, and for a long time, people thought that birds were entirely faithful. In the 1986 film *Heartburn*, Meryl Streep's character complains to her father that her new hus-band is having an affair. The father responds, rather heartlessly, "You want monogamy? Marry a swan." But about the same time, scientists figured out how to do DNA paternity testing, and quickly discovered that birds are not such models of piety after all. Although around 90 per-cent of birds are *socially* monogamous (that is, they live in male–female pairs), only around 10 percent of these species are *sexually* or *genetically* monogamous. In the other 90 percent – swans among them – a certain fraction of the youngsters is sired by males other than the social father.[66] Admittedly, sharp-eyed moralists had long suspected that birds weren't quite the innocents they made themselves out to be. In the third century, for instance, Saint Cornelius observed that "The habits of birds are, in many points, contrary to the commandments of the Church." But for many, the DNA evidence was a wake-up call. In an instant, we lost our innocence about our feathered friends. People wept openly in the streets.

And things only went downhill from there. Sex scandals soon plagued other animals formerly assumed to be bastions of sexual purity. Gibbons, prairie voles, shingleback skinks – all turn out to be prone to the occasional (or not-so-occasional) "extramarital" dalliance.[67] Social monogamy, it seems, is a lot more common than strict genetic monogamy. Don't get me wrong; some species really are genetically monogamous. Azara's owl monkeys are a case in point. An eighteen-year study of these pair-bonding South American primates failed to turn up a single instance of a youngster sired by a male other than the female's partner. Thus, if you want monogamy, you shouldn't marry a swan, or a gibbon, or a shingleback skink; apparently, you should marry an owl monkey. With most other species, though, there are no guarantees – which is why mate guarding is found in such a great diversity of animals.

What about the animal we're most interested in: the human animal? The first thing to say here is that the same evolutionary logic that explains mate guarding in other species applies as well to our own. For men, the key issue is paternity. A well-designed parental male will tend to end up investing in his own offspring, rather than the offspring of his good-looking next-door neighbor. And one way to help ensure that this happens is to be easily moved to jealousy – jealousy that leads you to keep a wary eye on your partner and the good-looking neighbor, to do what you can to keep them apart, and to abandon your partner if she strays, or else make life so unpleasant for her that she never does it again. (I'm not condoning any of this, incidentally; I'm just trying to explain it.) You don't have to understand the evolutionary logic of your jealousy in order for it to do its job; you only have to feel jealous. Any genes that incline you in that direction will automatically find themselves copied into more new bodies than rival alleles that incline you to think, "Hey, I'm an enlightened guy; I don't mind if my partner sleeps with other men." That the selection pressures tilt this way explains the fact that, as Margo Wilson and Martin Daly once put it, "men lay claim to particular women as songbirds lay claim to territories, as lions lay claim to a kill, or as people of both sexes lay claim to valuables."[68]

But it's not just men who lay claim to their mates. Women are sexually possessive too; that's why they sometimes try to run over their partner's former lovers. The two-way jealousy found in *Homo sapiens* makes good sense given that humans are a pair-bonding, biparental species. Indeed, it's further evidence that we *are* a pair-bonding, biparental species; if we weren't, women would be as devoid of jealousy as the average female chimp. Notice, however, that although both sexes are subject to the green-eyed monster, the evolutionary function of the monster is somewhat different for women than men. The main function for men, as we've seen, is to avoid getting cuckolded and raising another man's child. That can't happen to women. What can happen, though, is that the man a woman is with might fail to stick around and help care for their shared offspring. And one circumstance where that becomes likely is when the man falls in love with someone else. If, in our evolutionary past, a man ran off with another woman, leaving his partner holding the baby, the kid would often have had a reduced chance of surviving and thriving.[69] That would be bad for the man's fitness as well as the woman's. But if the man could have a kid with the other woman as well, it might be worth the risk: He might end up with two kids instead of one. For his original partner, in contrast, there'd be no silver lining but only clouds. As such, a well-designed pair-bonding female would seek to avoid that outcome if at all possible. And one way she could do *that* would be to be just as easily moved to jealousy as any man.

So, the selection pressures shaping men and women's jealousy were probably somewhat different. Does that mean, though, that jealousy itself differs between the sexes? Many evolutionary psychologists argue that it does. The argument goes something like this. If a man's wife has sex with another man, she could potentially fall pregnant to him, and the husband could end up raising the other man's child. That represents a huge fitness cost to the husband. If, on the other hand, a woman's husband has sex with another woman, that's clearly not great, but – as long as it's just sex – it's not quite as big a deal, evolutionarily speaking: She's not going to end up raising the other woman's kid. Thus, men should be particularly upset about sexual infidelity – more so than women. *Emotional* infidelity is a different story. If a woman's husband becomes emotionally involved with another woman, he could end up abandoning her and ceasing to invest in their children. That in turn could lower the chances of the children surviving, at least in premodern conditions. As noted, this represents a fitness cost to both parents; however, for the reasons outlined above, the cost is greater for the woman. Furthermore, even if the children did survive, the woman's childcare workload would presumably increase, making it less likely that she'd be able to have another child any time soon. For both reasons, women should be particularly upset about emotional infidelity – more so than men.[70]

Consistent with this line of thought, several decades of research show that, on average, the sexes differ in the predicted directions: Men tend to be more upset than women about sexual infidelity, whereas women tend to be more upset than men about emotional infidelity. These differences have been documented using a variety of methods in a variety of cultures, including at least one hunter-gatherer tribe.[71] However, without denying that the sex differences are real, my own view is that their importance has been somewhat overstated.[72] In self-report studies, the size of the differences depends on how you ask the question. If you use a forced-choice format – that is, if you ask people to choose which would upset them more, sexual or emotional infidelity – the sex differences are large: Men are generally quite evenly split but a significant majority of women choose emotional infidelity as the more upsetting option.[73] But if you use a free-response format – that is, if you ask people to rate how upset they'd be about sexual infidelity, and separately how upset they'd be about emotional infidelity – the differences are notably smaller.[74] Indeed, what stands out most clearly is not the sex differences but the fact that *both* sexes tend to get extremely upset about both aspects of infidelity: sexual and emotional.[75] There *are* average differences, but they're swamped by the commonalities. So, although the selection pressures shaping jealousy were different for women and men, the emotion of jealousy itself – the proximate mechanism shaped by these selection pressures – is largely the same for both.[76]

This whole discussion raises another, rather uncomfortable issue: How common is infidelity in our species? The short answer is that it's less common than it is in the soap operas, but it does happen at non-trivial rates. One way to get a conservative fix on the numbers is to look at the non-paternity rate in our species. How many people's father is not the guy whose name is on their birth certificate? The best-known estimate puts the figure at 10 percent, and estimates range as high as 30. Most people find these numbers surprising – and they're right to be surprised. More than likely, the estimates are extreme *over*estimates. Most come from studies looking at people for whom the non-paternity rate is likely to be much higher than it is in the general population. Many estimates, for instance, are based on data from professional paternity testing services. The problem with this is that the men who use these services usually already harbor doubts about the paternity of their ostensible offspring. Taking this into account, we might want to argue that the human non-paternity rate is unexpectedly *low*. Even among the men who most strongly suspect that their offspring are not their own, only a third are right at the most.

If we limit ourselves to studies looking at more representative samples of people, we find that the non-paternity rate is closer to 1 percent than to 10 percent or 30. Furthermore, at least in the West, this has been the case for at least the last few centuries, suggesting that the lower rate isn't just a product of modern birth control.[77] Unfortunately, though, the 1-percent estimate is less interesting to drop at parties or the pub, which is presumably why the higher estimates are better known: It's bad news but it's great gossip. Humans are no owl monkeys, certainly, but we're relatively faithful nonetheless. On reflection, this shouldn't come as too great a shock to the system. If the non-paternity rate were as high as some commentators claim, then male parental care would be a losing strategy. And if that were the case, men would not have evolved to be parental.

But if infidelity is so rare, why are people so prone to jealousy? Aren't we all more jealous and suspicious than we need to be? The answer is probably yes. The irony, though, is that if people *weren't* more jealous than they needed to be, they'd probably do less mate guarding and their partners would be a little more likely to stray – in which case, they'd need to be excessively jealous after all. People's mild paranoia regarding their mate's fidelity plausibly functions as a reverse self-fulfilling prophesy, helping to bring about its own falsity. Thus, the fact that infidelity is relatively rare in our species doesn't imply that jealousy isn't needed. On the contrary, part of the reason it's relatively rare is that people are so naturally prone to the green-eyed monster.

And Baby Makes Three

My favorite children's book is a slim volume entitled *Love You Forever* by the Canadian author Robert Munsch. I recommend it even if you don't have children. It's a quirky story but touching too. It's so touching, in fact, that I only ever got to read it to my children once; they cried their eyes out and refused to ever hear it again. The book is about the relationship between a mother and her son. It goes like this. Once upon a time, there was a woman who had a little baby boy. Sometimes, when the baby slept, she would tiptoe into his room, cradle him in her arms, and sing him a little song:

> I'll love you forever; I'll like you for always.
> As long as I'm living, my baby you'll be.

As the baby grew into a toddler, and the toddler into a boy, he would sometimes get into trouble, as juvenile primates sometimes do. But even

when his mother got mad at him, as soon as he went to sleep, she would
pick him up and sing:

> I'll love you forever; I'll like you for always.
> As long as I'm living, my baby you'll be.

Eventually, the son grew up and left home, and his mother grew old
and frail. But every now and then, the son would drive across town as his
mother slept, pick her up from her bed, and sing:

> I'll love you forever; I'll like you for always.
> As long as I'm living, my Mommy you'll be.

By this time, the son had a child of his own, a cute little baby girl. And
as soon as he got home from visiting his mother, he would tiptoe into her
room, cradle her in his arms, and sing:

> I'll love you forever; I'll like you for always.
> As long as I'm living, my baby you'll be.

And that's it! I don't know what it is about this simple story, but it
seems to have a powerful effect on a lot of people. Perhaps not coincidentally, it captures in its net some of the central themes of human life: the
endurance of family relationships, the inevitability of aging, the transi-
tory nature of human existence – and most importantly, the deep bond
between parents and their children.

This brings us to our next topic. Woody Allen once said that "Only
two things in life are important: One is sex, and the other isn't all that
important." A lot of people seem to think that that's the central message
of evolutionary psychology: "It's all about sex!" But is this a fair
assessment? The answer is probably yes – *if* we're talking about most fish,
most insects, and most other small, short-lived animals. When it comes to
human beings, on the other hand, the answer is an unequivocal no. Sex
is just the start for us. Humans – like all other mammals and all birds –
lavish enormous amounts of attention on their young, and this is every
bit as important to our reproductive success as finding a mate, falling in
love, and fending off rivals. Furthermore, with the exception of the occasional eternal bachelor, this is true not only of females in our species but
of males too. It's clear enough why humans are wired up this way. Our
offspring are entirely dependent at birth, and take much longer than any
other animal to reach maturity. As a result, women *and* men evolved not
simply to make babies but also to look after them and love them. The
human mind is not just a mating mind; it's a parenting mind as well.[78]

What psychological mechanisms has natural selection built into our brains to push us into the parenting niche? There are various candidates, but to introduce a particularly important one, consider a simple question: Why do we find babies cute? At first, this sounds like a silly question, up there with William James's "Why do we smile, when pleased, and not scowl?" and Jared Diamond's "Why is sex fun?"[79] The natural response is to say: "*Of course* we find babies cute; babies *are* cute. Are you a robot?" But just as an alien scientist wouldn't find adult humans any more attractive than fertile chimpanzees, nor would it find human infants any cuter than human adults or than any other animal. We'd probably all look quite distasteful to a psychologically healthy alien. Similarly, to a chimpanzee, our babies might look crumpled, bloated, and ugly, and our children bulbous-headed and weirdly hairless. They might look as unappealing to a chimp as hairless cats and naked mole rats do to us.

What does this tell us? Consider the concept of "visible light." When we describe light as visible, we're not actually talking about a property of light. We're talking about a property of our visual apparatus, namely that it's sensitive to some frequencies of electromagnetic radiation but not others. In the same way, when we describe babies as cute, we're not actually talking about a property of babies; we're talking about a property of ourselves. We find babies cute because we're designed to find them that way. If our babies looked like grotesque little monsters relative to our current standards, then that's what we'd find cute, and the babies we actually have would be the grotesque little monsters.

So, one mechanism that natural selection built into us to make us parental is a tendency to find babies cute. More precisely, selection built into us a tendency to find cute a certain cluster of features. This includes big eyes in the center of the face, a big bulbous head, chubby cheeks, and a button nose. Collectively, this cluster is known as the *Kindchenschema* or *baby schema*. The more a baby resembles the *Kindchenschema*, the more adorable we find it, and the more we want to protect and lavish care on this tiny, helpless creature. Again, this is true not only of females in our species but of both sexes. I was once at an evolutionary psychology conference in Japan, and during one of the talks, the speaker showed a slide of a baby chimpanzee, explaining that adult male chimps provide little in the way of care for the young. It occurred to me as he spoke that the male human beings in the audience, myself included, probably had a stronger parental reaction to that cute little chimpy face than would the male members of its own species. On average, women may be *more* responsive than men to cute baby faces, especially among people who don't yet have kids. However, most men are far from immune to the *Kindchenschema*.[80]

Our love of infantile faces has a number of interesting side effects. One, which I touched on in Chapter 2, is that we don't just find our own babies cute; we find babies of other species cute as well. Roughly speaking, other animals' cuteness is proportional to their resemblance to human offspring. It's rare to hear people say "Aww, what a cute baby worm" or "That's the most adorable baby scorpion I've ever seen."[81] But most people are suckers for baby mammals, which are far more closely related to us and thus far more similar. Not only that, but we're suckers for adult mammals that have baby-like faces: flat-faced cats, for instance, as opposed to pointy-faced opossums. Our love of these furry nonhumans is probably not an adaptation in itself; more than likely, it's spillover from the *Kindchenschema*. It's also a big part of the explanation for one of our strangest habits: our habit of keeping pets. People have many kinds of relationships with their pets; pets can be friends or protectors, servants or tools. Often, though, pets play the role of surrogate children. Where that's the case, pet keeping is a by-product of our unique parental instincts.[82]

The tendency to find babies cute is an important cog in the machine of our parental psychology. In itself, though, finding babies cute is not enough. If you were designing a parental animal from scratch, you couldn't just leave it at that; you'd also need to make sure that perceptions of cuteness and feelings of love translated into tangible care. How might you do that? Most people's first thought is that you should equip the animal with a tendency to enjoy caring for its young, just as you'd equip it with a tendency to enjoy sex and food. Surprisingly, though, this doesn't seem to be the path that natural selection has taken. If it were, people with children would presumably be happier than those without, and the more children they had, the more overjoyed they'd be. It's not obvious, however, that this is the case. Indeed, a lot of research points in the opposite direction. As the economist Tim Harford summed it up, if you want to maximize your personal happiness, the optimal number of children to have is zero... and if you absolutely must dabble in procreation, the least depressing number is two.[83] Don't take these numbers too seriously; the exact estimates vary from study to study, and there's always plenty of variation from person to person. The stable finding, though, is that having kids doesn't reliably make people happy, and that for a fair number of people, it leads to a slump in happiness. Parents swear blind that they love having kids and that they love spending time with them. But careful research suggests that many parents enjoy their time with

As a general rule, caring for kids is a good way to do that only if the kids in question are one's own biological offspring. This leads to a prediction: In parental species, natural selection will generally favor discriminating investors – individuals that don't just care for any old child but that favor their own progeny over the rest of the world.

This turns out to be true of most nonhuman animals. But is it true of us? This question was at the heart of some of the earliest and most enduring research in human evolutionary psychology – research conducted by two of the field's founders, Martin Daly and Margo Wilson. (Full disclosure – and slight boast: I did postdoctoral research with Martin and Margo at McMaster University in Canada.) The spark that lit the fuse on Daly and Wilson's research was a series of seminars exploring E. O. Wilson's then-recent book, *Sociobiology: The New Synthesis*. One day, in one of the seminars, someone pointed out that, based on the gene's-eye view of evolution, we'd expect that parents would typically have a stronger attachment to their genetic offspring than to stepchildren. Do they, though?

It certainly seemed possible. A common character in fairy tales from around the globe is the wicked stepparent: the unloving, abusive, or even murderous parental substitute. The stories of *Hansel and Gretel*, *Snow White*, and *Cinderella* are the best-known examples in the Western canon, but similar stories can be found wherever in the world we look. These stories aren't realistic, of course, but perhaps part of the reason they're so widespread is that they capture and caricature something real – something about the relationship between stepparents and their stepchildren. That, at any rate, is what Daly and Wilson began to wonder. The received wisdom at the time, however, was that there was no truth whatsoever in the fairy tales, and that they simply reflected malign and unfounded stereotypes about stepparents. It's a sensitive subject. Most non-fictional stepparents are not abusive; like most people, they're decent human beings just trying to do their best. At the same time, though, evolutionary principles give us a strong reason to think that, in general, the bond between parents and their genetic offspring will be stronger than that between stepparents and their stepchildren. If the experts claimed otherwise, thought Daly and Wilson, then maybe the experts were wrong.

So, they started looking into it. To their surprise, and despite the experts' confident pronouncements, they quickly discovered that there was essentially no research on the topic. If they wanted an answer, they'd have to do the research themselves. But how? Questionnaires wouldn't be up to the task. Stepparents might be reluctant to admit that they valued their stepchildren less than they would their own kids, and in some cases, they might not even realize it. What Daly and Wilson needed was a reliable, deception-proof barometer of the strength of the bond between parents and their children or stepchildren. And that's when they had a brainwave: They could look at public records of child abuse, child neglect, and child homicide. Such records are usually highly reliable, especially in the case of the homicides, which are nearly always recorded. And these records could serve as the relationship barometer that Daly and Wilson were looking for. The rationale is as follows. Emotional closeness acts as a brake on aggression. The closer two people are, the less likely it is that either will harm or kill the other. As such, records of abuse, neglect, and homicide are, in effect, reverse measurements of closeness within different categories of relationship. If stepparents are overrepresented among the minority of people who abuse, neglect, or kill the children in their care, this would suggest that, on average, stepparents are less close to their stepchildren than parents are to their own kids – not just among the abusers but in the general population.

And that's exactly what Daly and Wilson found, again and again, in study after study.[85] The phenomenon is now known as the *Cinderella effect*, and in some cases it's shockingly large. In several nations, for instance, Daly and Wilson discovered that children were around a hundred times more likely to be killed by a stepparent than by a biological parent. Not 100 *percent* more likely (i.e., twice as likely), but 100 *times* more likely. That's a huge difference. Of course, the absolute numbers are still miniscule. Child homicide is mercifully rare, and most caregivers would never harm their wards. Nonetheless, on the rare occasions it does happen, stepparents are far more likely to be the perpetrators.

To be absolutely clear, Daly and Wilson weren't suggesting that killing unrelated children is an adaptation in our species, in the way that it very obviously is in some others: gorillas, langur monkeys, and lions, to name but a few.[86] The suggestion instead was that child homicide in humans is a maladaptive, occasional by-product of psychological mechanisms that, in the normal course of events, lead people to value their own children above all others. Supporting this interpretation, the Cinderella effect is seen not only in the homicide stats, but also in less extreme forms of behavior. For example, on average, stepparents spend less money on food for their stepchildren, spend less time helping them with their homework, and less often take them to the doctor or dentist.[87] Subtle instances of discrimination such as these are vastly more common than stepchild homicide, which would be strange if the relevant adaptation were specifically about killing stepchildren. On top of that, it seems unlikely that stepchild homicide would boost one's reproductive prospects in almost any imaginable human society. Even if the perpetrator managed to avoid punishment or banishment, who'd want to partner up with a child killer? The most plausible interpretation is that the Cinderella effect is not about killing stepchildren per se, but rather reflects an evolved tendency to favor one's biological offspring.

Over the years, Daly and Wilson have addressed various attempts to explain away the Cinderella effect, or to explain it in non-evolutionary terms. Is it just a Western phenomenon? No. The preferential treatment of genetic offspring has been found in every society where researchers have looked for it, including small-scale, non-Western societies such as the Aché of South America and the Hadza of Tanzania.[88] Is it just that stepparents are more likely to be young, poor, or unmarried, or that stepfamilies tend to be larger and therefore more stress-ridden? No. Stepfamilies don't actually differ much from genetic families in most of these ways, and in any case, the Cinderella effect remains even when controlling for such factors. Is it just that stepparents tend to be more violent or abusive in general? No. When people have stepchildren *and* genetic children in the same home, they're much more likely to harm the stepchildren. Is it just that stepparents usually arrive on the scene later, and thus have less time to bond with their stepchildren? No. The Cinderella effect appears even when stepparents and stepchildren have known each other for many years, and often even from birth.[89] Is it just that people are more likely to report mistreatment by stepparents? No. The most extreme form of mistreatment – homicide – is almost always reported or detected, but the Cinderella effect for homicide isn't smaller than that for other forms of mistreatment; it's much larger.[90]

In short, Daly and Wilson have made a compelling case that humans everywhere favor their own kids over other people's. But how strong is this favoritism? If we take the homicide stats at face value, we might assume that the answer is "very." As mentioned, people are up to a hundred times more likely to kill a stepchild than a genetic child. But is this a good indicator of the typical level of own-child favoritism in our species? Probably not. If our alien scientist made a close study of human child-rearing habits, the first thing it would notice would *not* be that stepparents treat stepchildren radically worse than their own. The first thing it would notice would be that humans, unlike many other animals, commonly care for children who aren't related to them. And a little historical rummaging would soon persuade the alien that this is nothing new for our species.[91] Certainly, stepfamilies tend to be somewhat less harmonious than intact biological families. But biological families aren't perfectly harmonious either, and the thing that would stand out most to our alien observer is not our antagonism toward stepchildren but rather the fact that we engage in such high levels of stepparental care. How can we reconcile the alien's observations with the undeniable enormity of the Cinderella effect for homicide and abuse?

The answer, I suspect, is that homicide and abuse are rare and extreme behaviors, which represent a total breakdown of the bond between those involved. As a general rule, when you look at extreme behaviors, differences between groups are much larger than for behavior in the normal range. Thus, although people are much more likely to kill or abuse stepchildren than genetic children, when it comes to less extreme, more common forms of own-child favoritism, the Cinderella effect is considerably smaller. Genetic parents and their children do tend to be somewhat better bonded than stepparents and their stepchildren. However, for most people most of the time, the gap is nowhere near as large as the homicide and abuse stats would seem to suggest.

That being the case, maybe we should be asking a different question: not "Why do we favor our own offspring?" but instead "Why don't we favor them more?" One suggestion from Daly and Wilson is that, in the evolutionary past of our species, caring for stepchildren could sometimes boost our ancestors' reproductive success.[92] It wouldn't have done this directly, in the way that caring for one's own offspring would have, but it might have done it indirectly. Specifically, it might have functioned as a way to obtain a sexual partner. The sexual partner in question wasn't the stepchild, of course; it was the stepchild's mother or father. For most of the lifespan of our species, there was no reliable contraception, and romantic relationships were only occasionally of the till-death-do-us-part variety. As such, most of the potential mates in our ancestors' social circles had children from prior relationships.[93] For women in particular, these children were a central part of their lives. Thus, women – and to some extent men – may have preferred suitors who helped them to care for their pre-existing young. If the helpers ended up having more children of their own as a result, their willingness to help would have been selected. This appears to have happened in a number of nonhuman species, including many birds, and there's no obvious reason that it couldn't have happened in our own species. So, although people are clearly *less* willing to invest in stepchildren than genetic children, they're a lot more willing than the average gorilla, langur monkey, or lion. The Cinderella effect is real, but humans are a surprisingly stepparental species nonetheless.

Maximizing Returns on Parental Investment

It seems reasonable to conclude that the psychological mechanisms underpinning human parental behavior are sensitive to genetic relatedness: We tend to value our own children more than we value others. But humanity's evolved parental psychology isn't sensitive *only* to relatedness. Other factors come into play. Perhaps the most important is *reproductive potential* (also known as *reproductive value* or *RV*). Reproductive potential refers to the number of children that an individual is likely to have in the remainder of his or her life.[94] Individuals with high reproductive potential are likely to have lots of children; individuals with low reproductive potential are likely to have few or none. As we've seen, reproductive potential plays an important role in mate choice. For men especially, the traits that people find attractive in a mate are often statistical indicators of reproductive potential and fertility. But reproductive potential also helps to determine what we find "attractive" in babies and children – and indeed in all relatives. Along with genetic relatedness, this variable helps to regulate how much we value these individuals and how willing we are to make sacrifices for them.

Various factors shape a person's reproductive potential, but the most important is age. Reproductive potential peaks at sexual maturity and then declines steadily through the adult years. A person who's just hit puberty has higher reproductive potential than one who's just celebrated their fortieth birthday, for the simple reason that the younger individual has more time left to have children. In a traditional forager society, the average sixteen-year-old might have four children in the future; the average forty-year-old would be lucky to have one. Thus, as age goes up, reproductive potential comes down.

That, at any rate, is what happens after people reach sexual maturity and become capable of having children. What happens before that? Some might guess that reproductive potential is uniformly high throughout childhood and only starts to drop after puberty, when the reproductive clock starts ticking. But they'd be wrong. What actually happens is that reproductive potential starts out low at birth, climbs rapidly in the first year of life, and then continues to climb at a more sedate pace until reproductive maturity is achieved. In other words, in childhood, reproductive potential rises with age. The reason for this is that the younger children are, the less likely it is that they'll survive to reproductive maturity. Those who *don't* survive obviously won't have any offspring, and this pulls down the average number of future offspring for their age group. Thus, whereas the average sixteen-year-old in a traditional society might be expected to have four offspring, the average newborn might be expected to have just two. They'd have four if they reached sexual maturity, but they've only got a 50 percent chance of doing that. Ergo, their expected number of future offspring, statistically speaking, is two.

Why are younger children less likely than older ones to survive to sexual maturity? It's partly because younger children are more vulnerable, less robust, and less competent. This is especially so in the first year out of the womb, at least in premodern conditions, which is why the increase in reproductive potential is particularly rapid in the first year of life. But vulnerability isn't the only reason that younger children are less likely to make it to adulthood. Another is simply that they have a longer stretch of time to traverse to get there, and thus more opportunity to walk off a cliff, get struck by lightning, or get eaten by a crocodile. The upshot is that, even if younger children were no more vulnerable than their older peers, they'd still have lower reproductive potential.

What does any of this have to do with parental care? The answer to that question, like so much else in biology, is most easily expressed in economic terms. The optimal investment strategy for a parental organism is to invest preferentially in those offspring that are best able to turn the investment into a fitness advantage. In many cases, the offspring best able to do that are those with high reproductive potential. To see why, let's conduct another thought experiment.[95] Earlier in the chapter, I asked you to imagine that your one goal in life was to have as many children as possible. This is a useful way to start thinking about which traits natural selection is likely to favor. However, as we saw in Chapter 2, natural selection isn't just about children. It's ultimately about genes, and if you want your genes to survive, you need your children to have children of their own. In a word, you need grandchildren – and the more the merrier. So, let's assume that your overriding goal is not simply to have as many children as possible but to have as many grandchildren. With that as the backdrop, imagine that you're sharing a house with two of your children. One is sixteen; the other is forty. (I guess you must be quite old.) One night, you wake up to discover that the house is on fire. Your children are sleeping in separate rooms at opposite ends of the house, and you realize, to your utter dismay, that you're only going to have time to save one.

Bearing in mind that your one goal in life is to maximize the number of grandchildren you end up with, which of your children should you save?

It's a horrible question, but the answer is clear: You should save the sixteen-year-old. Doing so would almost certainly mean more grandchildren, because the sixteen-year-old probably hasn't started having kids yet whereas the forty-year-old has quite possibly finished. The sixteen-year-old, in other words, has higher reproductive potential, and that's why that's who you should save.

Now imagine a second scenario. Again, you find yourself in a burning building with two of your sleeping children, and again you can save only one. This time, however, one of the children is a ten-year-old, and the other is a newborn baby. Which of *these* two children should you save? In the last example, you saved the younger child because the younger child had more miles left on the reproductive clock. But in this situation, you should save the older one: the ten-year-old. Why? Because the ten-year-old has a greater chance of reaching sexual maturity and thus a greater chance of giving you grandkids. Nothing's guaranteed, of course; maybe the *newborn* would ultimately have given you more. But you don't know that, and in the absence of any reason to suspect it, saving the ten-year-old would be your best bet. Again, this is because the ten-year-old has higher reproductive potential.

This isn't to say that parents should *always* favor the child with the highest reproductive potential. The child's needs have to be factored in as well, and an infant's needs are much greater than a ten-year-old's. Without parental input, the ten-year-old has at least some chance of surviving; the infant, in contrast, has none. As such, helping the infant would generally give a bigger boost to the parent's lifetime fitness, and in the normal course of events, an optimal parental investor would prioritize the infant. However, if for some terrible reason, the parent had to choose between the two, or could invest heavily in only one, then in strictly evolutionary terms, the ten-year-old would be the better choice. Reproductive potential would be the tie breaker.[96]

That's the theory. How does the reality match up? One way to address this question is to turn again to the homicide records, treating them once more as a reverse indicator of attachment and interpersonal valuation. If parents really do place more value on children with greater reproductive potential, we'd expect that parents would more often kill infants and toddlers than they would older children. Sure enough, when Daly and Wilson looked at the data, they found precisely that. An obvious counter-explanation would be that older children are larger, stronger, and more mobile, and thus that they're harder to kill. But the counter-explanation doesn't work. Although *parents* are less likely to kill older children, non-relatives are not; they're *more* likely to. This makes sense in sociobiological terms: A child's reproductive potential has no direct bearing on the fitness of non-relatives, and thus there's no reason that non-relatives would value an older child more. The pattern doesn't make sense, however, in terms of the older-kids-are-harder-to-kill hypothesis. It contradicts it.[97]

Age is not the only indicator of reproductive potential. Another is health. Let's return again to the burning building. Imagine this time that you have to decide whether to save a healthy child or a child of the same age but with debilitating health problems. Which choice would give the bigger boost to your lifetime tally of grandchildren? The answer is as obvious as it is sad: More than likely, saving the healthy child would give you the bigger boost, because the healthy child has a better chance of surviving to maturity, attracting a mate, and having healthy children of its own. Certainly, when it comes to low-cost help, people may help the sickly child more, just because the sickly child needs more help. However, for costlier, more evolutionarily significant forms of help, parents may tend to prioritize the healthy child over the unhealthy one. When it comes to the crunch, they may value the healthy child more.[98]

Again, that's the theory. But again, sadly, the facts and the theory seem to line up well. On average, parents respond more favorably to healthy babies than to unhealthy ones.[99] They're less likely to abuse or neglect healthy children than children with disabilities or congenital abnormalities.[100] And they tend to grieve the death of a healthy child more intensely than the death of an unhealthy child, despite the fact that most people consider such favoritism to be almost unforgiveable.[101] These various strands of evidence all point in the same direction. Parents tacitly place more value on the lives of healthy children (high reproductive potential) than they do on the lives of less healthy ones (low reproductive potential).

Needless to say, this is heart-breaking. But natural selection is heartless. That's why it's always a mistake to assume (as a surprising number of people do) that if a trait is favored by natural selection, it must be good, right, or proper. Often, the reverse is true: Doing the right thing means going against natural selection, and trying to right nature's wrongs. Evolutionary theory explains much that is bad about the world; it doesn't imply that the bad things are actually good, and it doesn't provide a template for how we ought to live our lives.[102] (For more on this issue, see Appendix A.)

Not So Fast!

> Monogamy is the Western custom of one wife and hardly any mistresses.
> —*H. H. Monro*

In the course of the chapter so far, I've sketched out a picture of the human reproductive cycle involving one man and one woman, pair-bonding, and biparental care. Lest there's any doubt, I'm not suggesting that this is the way things always work out; humans are nothing if not flexible. What I would suggest, though, is that the psychological dispositions underlying pair-bonding and biparental care – mate preferences, romantic love, parental affection – have their roots in natural selection, and that they define human beings' *primary* baby-making system. To round out the discussion, let's consider some of the criticisms that have been leveled at this view.[103]

1. "*You seem to be suggesting that humans are naturally monogamous, but that's not true. Romantic relationships rarely last a lifetime, and infidelity is ubiquitous. In trying to maintain lifelong, exclusive relationships, we're all just trying to press a square peg into a round hole.*"

This criticism stems from a misunderstanding. Some opponents of evolutionary psychology seem to think that, according to evolutionary psychologists, the natural human mating system is sexually exclusive, till-death-do-us-part monogamy, embedded in a 1950s-style nuclear family. I'm not quite sure where this idea comes from, but it's not what evolutionary psychologists claim. Most, I believe, would agree with each of the following statements:

  i. Pair bonds sometimes last for life, but most of the time they don't. We know this because most people fall in love more than once in their lifetimes.

  ii. Infidelity isn't exactly ubiquitous, but it's also not hugely uncommon. As a species, we're socially monogamous, rather than sexually or genetically monogamous. (Note, though, that most individual relationships *are* sexually and genetically monogamous.)[104]

iii. As we'll soon see, humans sometimes engage in mating arrangements other than pair-bonding, including polygyny and casual sex, and these arrangements come just as naturally to us as pair-bonding does.

All these facts and findings are inconsistent with the claim that humans are designed solely for lifelong, exclusive monogamy, and that any deviation from that path is some kind of evolutionary mistake. Like I say, though, that's not a claim that evolutionary psychologists have ever made. It's certainly not what I'm claiming. My view is not that exclusive, lifelong monogamy is our one true mating system; it's that pair-bonding is an *important part* of our reproductive repertoire. The fact that pair bonds don't always last for life, and that they're not always entirely exclusive, doesn't undermine that view. On the contrary, these facts fit neatly into an evolutionary model. In *most* pair-bonding species, pair bonds are temporary and infidelity is an occasional part of life. It's no surprise, then, that we find the same thing in our own species. The real surprise for the evolutionist would be if pair bonds always lasted a lifetime and if no one ever cheated.

2. *"Nice try, but even with those provisos, you're still mischaracterizing our species. Humans are a polygynous animal, not a pair-bonding one. Left to our own devices, some men take several mates while others are relegated to bachelorhood. That's our natural mating pattern: harem polygyny."*

The idea that humans are naturally polygynous is a common one. The biologist David Barash, for instance, wrote that "A Martian zoologist, reporting on the species *Homo sapiens*, would have no doubt: Human beings are mildly polygynous by nature. Like other polygynous mammals, we exhibit all the hallmarks."[105] The hallmarks, as I see it, are threefold. First, on average, men are larger than women. This is true of most polygynous animals; as we saw in Chapter 3, for example, male gorillas and elephant seals dwarf the females of their respective species. Monogamous animals, in contrast, dance to a different tune: The males and females are much more similar in size.[106]

Second, polygyny is common across the cultures of the world. According to George Murdock's authoritative *Ethnographic Atlas*, a whopping 83 percent of human societies are polygynous.[107] Even Western culture, where polygynous marriage is outlawed, has polygyny in its backstory. In the Old Testament, for instance, King Solomon had 700 wives and 300 concubines. Other Biblical figures married polygynously as well, though not on quite the same scale as Solomon.

Third and finally, the fact that polygynous marriage *needs* to be outlawed in the West is telling. If people didn't naturally lean toward polygyny, outlawing it would presumably be unnecessary. Furthermore, despite the fact that it *is* outlawed, people still engage in de facto polygyny: Rich, high-status men sometimes have one official wife but also one or more mistresses. This suggests, yet again, that rather than being a *product* of culture, polygyny reflects something in human nature that survives even when culture tries to stamp it out. (We'll take up the question of Western monogamy norms again in Chapter 6.)

What should we make of all this? Let me concede right off the bat that harem polygyny is a persistent trend in our species, and that this mating arrangement seems to come naturally to human beings. Despite that, though, it's misleading to classify us as a polygynous species. First, it's worth emphasizing that the sex difference in size doesn't settle the issue. The human size difference is nowhere near as large as that found in highly polygynous species such as gorillas and elephant seals, but instead falls awkwardly within a range consistent with occasional harem polygyny, serial monogamy, generalized promiscuity, or some combination of all of these. It's possible to argue that the sex difference in size underestimates how polygynous we are; one might argue, for instance, that the difference in upper-body muscle mass is a better indicator of past polygyny, and that this is much larger than the overall size difference.[108] But it's also possible to argue that the size difference *over*estimates our polygynous tendencies; one could argue, for instance, that the sex difference in size (*and* in upper-body muscle mass) is due in large part to the fact that human males are adapted for hunting.[109] My verdict on the size difference? Inconclusive.

Second, although it's true that most societies are classified as polygynous, this doesn't mean that polygyny is common. Technically, it means two things: first, that polygyny is permitted in these societies, and second, that there's at least one polygynous relationship. In most so-called polygynous societies, the majority of relationships – usually the vast majority – are *not* polygynous; they're pair bonds.[110] This makes us very different from animals like gorillas. Most male gorillas either have a harem or they don't have a mate. In contrast, most male humans who have more than zero mates have only one. For that reason, pair-bonding is likely to be deeply inscribed in our evolved nature.[111] Certainly, humans have a polygynous streak. However, the idea that we're a polygynous species *rather than* a pair-bonding one vastly overstates the centrality of polygynous mating in our species. Polygyny is not our one true mating system, any more than monogamy is.

3. *"Doubly wrong! Humans aren't monogamous or polygynous. In our natural state, we're a promiscuous animal. Life for our prehistoric ancestors was a sexual free-for-all. It's only since the advent of agriculture that people have started insisting on long-term, exclusive relationships. But such relationships don't come naturally to us; they're purely products of repressive cultural norms and coercion."*

Many animals, from chimps to dolphins to rabbits, have promiscuous mating systems. The one that really grabs people's attention, however, is the bonobo.[112] Bonobos are famous for engaging in lots of sex with lots of partners (despite having center partings – it's one of the miracles of nature). Indeed, these frisky apes' antics make the Playboy Mansion look wholesome. We're not just talking boy-on-girl action. Bonobos engage in boy-on-boy action, girl-on-girl action... even family-member-on-family-member action. The only combination that's never been witnessed is mother-on-son action. It's not all fun and games, though. Sex plays an important role in bonobo social life, especially among the females: Bonobos use sex to forge bonds within the group and defuse social tensions. Admittedly, bonobos' reputation as sex maniacs is probably somewhat overblown.[113] These amiable critters seem to be a lot more sexually maniacal in captivity than in their natural habitat, and most of their sexual mania consists of brief "copulatory gestures," rather than full-blown intercourse. Still, there's no question that bonobos have multiple sexual partners and that they don't form exclusive sexual bonds, monogamous or polygynous. Bonobos, in other words, are a promiscuous species. And so, according to some commentators, are we. In the absence of corrupting civilization, humans are basically bonobos without the fur coats: free-loving hippies devoid of sexual hang-ups and possessiveness.[114]

For many people – more men than women, I imagine – this is an appealing vision. Is it realistic, though? On the one hand, casual mating does seem to be part of the natural sexual repertoire of *Homo sapiens*. People can be sexually attracted to individuals they don't love or want a relationship with, and casual sex is found in every culture for which we have data.[115] But to say that casual sex is part of our repertoire is very different from saying that it's our sole evolved mating system, and that everything else is just a repressive cultural imposition. And not only is it different, the latter is almost certainly false. To start with, the idea is implausible on its face. Can we really believe that our pair-bonding proclivities come entirely from culture – that we're bonobos that got confused and mistook ourselves for pair-bonding birds? Is it just a huge coincidence that culture has replicated the full package of traits found in most pair-bonding species, including mutual mate choice, mutual jealousy, and biparental care? It seems unlikely. It seems especially unlikely given that there's an obvious reason why humans would evolve this package of traits whereas bonobos (and chimps) would not: Our offspring are vastly more dependent than theirs and require more investment over a longer period of time. Given that difference, it would be decidedly odd if humans had the same mating system as bonobos or chimpanzees.

Theoretical problems aside, various lines of evidence militate against the pure-promiscuity hypothesis. The most important concerns a peculiar phenomenon known as *sperm competition.*[116] In promiscuous species, females often mate with more than one male within the space of a single reproductive cycle. For animals with internal fertilization, this means that females often end up with several males' sperm in their reproductive tract at the same time. In that context, any male that produces more sperm than the rest has a better chance of impregnating the female. This works on a simple principle, namely that the more lottery tickets you buy, the better your chances of winning. Males that produce more sperm have more offspring, and therefore males evolve over the generations to mass produce sperm. As any business magnate will tell you, if you want to mass produce something, you need a large "factory." Thus, males in promiscuous species tend to have large testicles relative to the size of their bodies. Male chimps, for instance, have testicles nearly as large as their brains. Males in monogamous species, on the other hand, don't need to mass produce sperm, because females in these species only rarely mate with more than one male within a single reproductive cycle. Thus, monogamous males have much more modestly proportioned testicles. The same holds for males in polygynous species – gorillas, for example. Although harem-holding males mate with more than one female, each female mates with only one male: him. As a result, the mighty silverback gorilla, with his terrifying muscles and harem of subordinate females, has only tiny testicles.

What about us? Where do we fall in the testicular hierarchy? If you've read a little in this area, you might have heard that humans fall somewhere in between polygynous gorillas and promiscuous chimpanzees. Many take this to mean that our species has traveled some way down the road toward chimpanzee promiscuity. This isn't entirely false, but it is somewhat misleading. Though men have larger testicles than polygynous gorillas, human testicles are very close in size to those of pair-bonding gibbons, and nowhere near the magnitude of chimp or bonobo testicles.[117] As the primatologist Frans de Waal remarks, "both chimps and bonobos are far more promiscuous than we are. Our testicles reflect this: they are mere peanuts compared to our ape relatives' coconuts."[118] And although humans are hardly goody-two-shoes, especially in early adulthood, our sexual exploits really aren't in the same league as our coconut-wielding cousins. Jane Goodall once wrote about a female chimpanzee who mated with more than fifty males in a single day. This is uncommon among human beings, even in LA. Human testis size suggests that sperm competition is relatively modest in our species. And this argues rather strongly against the idea that we're an exclusively promiscuous species.[119]

(An aside: Traditionally, scholars learned about prehistoric societies solely via the methods of archaeology and paleoanthropology. It's a fascinating implication of the new Darwinian vision of life that we can now surmise things about the lifestyles of our ancestors – including things that people usually try to keep secret – by looking at anatomical features such as testicle size.)

4. *"You argue that humans are a biparental species: a species in which both sexes naturally help care for the young. But men differ enormously in how much they invest in their kids, and in some cultures, they barely invest at all. This suggests that male parental care is a social invention rather than an evolved propensity – a bit like forcing a tiger to be a vegetarian."*

It's true that men are more variable in their parental participation than women. But the vegetarian-tiger scenario is a non-starter for several reasons. The most important is that men appear to have hormonal mechanisms designed to facilitate pair-bonding and male parental care – mechanisms similar to those found in other biparental species. For instance, in many birds, adult males have two main modes of operation: mating mode and pair-bonding mode. In mating mode, they spend their time seeking out mates and fighting other males. In pair-bonding mode, they establish pair bonds, settle down, and help raise a clutch of kids. These different modes of operation are switched on and off primarily by testosterone: In mating mode, testosterone levels run high; in pair-bonding mode, they plummet.[120] I'm simplifying, of course, but there is a clear trend in this direction.

We find this trend in birds, but we also find it in humans. The average bachelor has more testosterone than the average man in a serious relationship and more than the average man helping to care for kids.[121] This may be partly due to the fact that high-testosterone guys are more likely to steer clear of long-term relationships and less likely to want anything to do with kids. But that's not all that's going on. Studies tracking the same men over time show that most men's testosterone levels are dialed down when they form pair bonds, and dialed down even more when they become fathers.[122] Are men *socialized* to secrete less testosterone under these circumstances? Presumably not. More than likely, the tendency is part of our basic evolved nature. Like male birds, men's pair-bonding and parental tendencies are modulated by hormones in ways pre-programmed by natural selection.[123]

Certainly, men vary a lot in terms of how involved they are in childcare, particularly when we compare across cultures. But there are several things to say about that. First, men in all cultures are considerably more involved in childcare than most male mammals. This is especially true when we count not just direct care of children but also indirect care – bringing home the bacon, so to speak.[124] Second, some of the variation in men's parental involvement may itself be explicable in evolutionary terms. I mentioned in Chapter 3 that variables such as paternity uncertainty and mating opportunity cost help to explain average levels of male parental care in our species. But the same variables may also help to explain differences among individual men in the time and effort they devote to parenting. As a general rule, men are less inclined to care for kids who are unlikely to be their own – kids who look less like them than like the mailman, for instance.[125] In addition, men are less inclined to care for kids when the mating opportunity costs of doing so would be high – when, for example, they're especially attractive to women.[126] One explanation for this pattern is that human males have an evolved tendency to lean toward mating mode when paternity certainty is low and mating opportunity costs high, but to lean toward pair-bonding mode when the cards fall the other way.[127]

If this is right, then it begins to explain not only differences in paternal care among men within a given culture, but also average differences in paternal care between cultures. In cultures where promiscuous mating is common, and where paternity certainty is therefore low, men tend to engage in less childcare than men in less promiscuous cultures.[128] Likewise, in societies where women outnumber men, and where men therefore have more sexual opportunities, men again invest less in their young.[129] Critics sometimes argue that individual and cultural differences in male parental care undermine an evolutionary perspective on this topic. It may turn out, however, that an evolutionary perspective helps to explain those very differences.

What's Our Natural Mating System?

So, are human beings naturally monogamous, polygynous, or promiscuous? In light of the above discussion, I hope you agree that this is the wrong question. In a sense, we're all of these things; in another sense, we're none. But there's a better way to frame the whole issue. Human mating systems and childcare arrangements are not directly dictated by natural selection. Instead, selection has bequeathed us a diverse palette of evolved desires and drives, and these are compatible with a variety of socially molded mating systems. For example, both sexes naturally fall in love and are possessive of their mates, and both sexes naturally love their kids. As a result, a common reproductive arrangement in our species involves largely exclusive pair bonds coupled with biparental care. But we also have other desires and drives, and thus other arrangements are possible. People have a desire for sexual novelty, and men are less inclined to care for kids that probably aren't their own. Thus, another common arrangement in our species involves casual sexual relationships coupled with relatively low levels of paternal investment. Similarly, many men desire multiple mates, and many women desire a man with wealth and status. Thus, yet another common arrangement involves rich or high-status men partnering up with more than one woman – polygyny, in other words.[130] Pair-bonding is our *primary* mating system, at least in as much as that it's the most common. But polygyny and casual mating are not aberrations; they're central elements of the human reproductive repertoire.

To be clear, this isn't to say that pair-bonding, polygyny, and casual mating are all naturally trouble-free. On the contrary, none of them are. This is partly because what's best for one person in a relationship is not always best for the other, which means that relationships almost always involve a tug of war between those involved.[131] But even if people's interests aligned perfectly, no mating arrangement would ever be perfectly satisfactory, for the simple reason that human beings have multiple, incompatible desires. Pair bonds satisfy the desire for an intimate long-term relationship, but may leave some people with an unfulfilled desire for sexual variety – probably more men than women. Casual mating may satisfy the desire for sexual variety, but may leave some people with an unfulfilled longing for an intimate long-term relationship – perhaps more women than men. Open relationships or polygamy may enable some people to satisfy both desires, but can bring new troubles of their own, jealousy being the most obvious example. In short, it's not a matter of figuring out which mating pattern is human beings' natural and ideal one – the one that will finally let us live happily ever after. Instead, it's a matter of choosing which of the naturally occurring relationship types we prefer, each of which will satisfy some aspects of human nature but not others. This is the irritating reality of the human condition: Whatever we do, we're left with unfulfilled desires. Human beings are chronically conflicted animals. And that's because that's what selection made us.

5

The Altruistic Animal

The Trouble with Altruism[1]

In 2011, the Australian state of Queensland suffered extreme flooding. As with any disaster, this one left many tales of heroism in its wake. Among the most poignant is the story of thirteen-year-old Jordan Rice. Jordan had been out shopping with his mum, Donna, and his younger brother, Blake. They were in the car heading home when, out of the blue, they found themselves caught in the middle of a flash flood. Unable to drive any further, and unable to get to dry land, the three scrambled onto the roof of the car and then sat there, stranded in the middle of a violent torrent of water. Fortunately, some bystanders saw what had happened. One man – Warren McErlean – tied one end of a rope to a post, and the other around his waist, and then pushed his way through the rapidly rising waters to the car. He reached for Jordan, but Jordan pulled away, begging him to save his little brother first. McErlean complied: He picked Blake up and carried him quickly to safety. Before he had time to rescue the others, however, a sudden surge of water flipped the car. Jordan and his mum were swept away and killed.

By putting his brother ahead of himself, Jordan lost his life. From an ethical point of view, this represents the height of moral action. From a Darwinian point of view, on the other hand, it's initially quite perplexing. Evolutionary theory seems to imply that the only organisms that will prevail in the harsh Darwinian struggle for existence are those that look out for number one, and thus that the whole world will be populated with self-interested, self-serving organisms. But Jordan was no such organism; he risked his own life to save someone else, and died in the process. And although his was an extreme and extraordinary case, Jordan was no freak of nature – no moral equivalent of a two-headed gorilla. Self-sacrifice of one sort or another is nearly as common as breathing.

It's not just common in our own species, either. Self-sacrifice is found across the length and the breadth of the animal kingdom. Let's run through some examples. A particularly memorable example is the broken-wing display found in many birds. If a bird has a nest-full of chicks, and a predator starts getting too close for comfort, the bird will lure the predator away in a particularly risky manner. It will fly out into the open so the predator can see it, and then flutter around erratically, so that it looks like it's got an injured wing. An easy meal! The predator starts heading toward the malingering bird, and thus away from the nest and the chicks. At the last minute, the malingerer is miraculously healed and makes a quick getaway… if it's lucky. Sometimes, though, like Jordan, the bird doesn't make it. It loses its life in the quest to save another.

Admittedly, the broken-wing display is a clear-cut case of parental care, and parental care isn't especially tough to explain in evolutionary terms. But other cases of animal self-sacrifice are not as easily waved away. One such case, also found in birds, is called *helping at the nest*. What happens is that a reproductively capable adult – an adult that could just as well have its own offspring – instead joins an existing family and helps the breeding pair to care for *their* offspring. Helpers do often go on to have kids of their own. Presumably, though, they have fewer kids overall than individuals who don't help even for a while. How, then, does helping at the nest persist?

Another example of non-parental altruism is the widespread phenomenon of *alarm calling*. In various group-living species, including Belding's ground squirrels, prairie dogs, and vervet monkeys, group members that spot a predator – a hawk or a prowling coyote, for instance – immediately issue a loud warning call.[2] This is good for anyone who hasn't yet noticed the predator, because it gives them a chance to scurry away before it's too late. But it's not so good for the caller. The best-case scenario is that it burns up the caller's time and energy without conferring a personal advantage. The worst-case scenario is that it attracts the predator's attention, thereby increasing the caller's chances of falling prey to the predator. Given the costs, it seems reasonable to suppose that, on average, callers will have fewer offspring than non-callers, and thus that the tendency to issue alarm calls will slowly evaporate from the population. But it doesn't; it persists. How? And how did it get selected in the first place?

Helping at the nest and raising the alarm are nice things to do. But some animals take this kind of niceness to a whole new level. Bees are a good example. When a predator approaches the hive, a horde of workers streams out and starts stinging the intruder. In most bee species, the sting gets stuck in its target like a fishing hook. Unfortunately for the bee, the hook isn't designed to detach after use. This means that not only the sting but the whole bee gets stuck to the intruder. When it tries to escape, or when the intruder tries to brush it off, the sting stays put and the bee is ripped apart, usually soon dying as a result. In effect, stinging the intruder is suicide – something we wouldn't expect natural selection to favor.

Is the bee protecting its offspring in the nest, like the bird performing its broken-wing display? No. Most bee species – and most wasps, ants, and termites – are *eusocial*: They have a large caste of individuals which don't produce offspring of their own, but instead help the reproductively active minority to produce their offspring. The worker bees are such a caste. And *that's* the real mystery. Given that the workers don't reproduce anyway, their suicidal defense of the nest doesn't actually make a difference to their lifetime reproductive success. Whether they die now or die later, they won't be passing on any genes to any offspring. Imagine a gene that causes its owners to help *others* to survive and reproduce but which prevents them from doing so themselves. Who will inherit that gene? It's not a trick question; the answer is that no one will. So how could such a gene keep itself afloat in the gene pool?

All these examples raise the same question: If natural selection favors ruthless self-interest, why is there so much niceness in the world? Why do people and other animals risk their own necks to help individuals other than themselves? Evolutionary biologists call this *the problem of altruism*. In everyday life, the word *altruism* refers to acts that are intended to increase another's welfare, without any ulterior motive on the part of the altruist. If I help you because I want you to be better off, that's altruism; if I help because I'm hoping it'll somehow benefit me, that's not. Biologists, however, have a different definition in mind. An altruistic act for them is one that benefits the recipient of the act at some cost to the actor – not in terms of pleasure or happiness or any other subjective metric, but in terms of reproductive success.[3] Intentions are not part of the equation; an altruistic animal *might* have an intention to help but might just as well not. The problem of altruism is the problem of explaining how altruism in the biologists' sense could evolve.

For advocates of the blank slate view of human nature, this might not seem like much of a problem, at least if we confine our attention to our own species. Human beings, according to blank slaters, are not naturally selfish, or naturally altruistic, or naturally anywhere in between. We're cultural animals, equally capable of learning to be any of these things or anything else. As I've said before, this view does have a superficial plausibility; humans are the consummate learners and lean more heavily on culture than any other animal. Yet for all its intuitive appeal, the blank slate view faces serious objections. First, if we really were blank slates, it would be just as easy to teach people to be indiscriminately altruistic as to teach them to favor themselves and their kith and kin. It's not. We know it's not because people have been trying to persuade each other to be more altruistic for millennia (sometimes to further their own selfish interests, ironically enough). It hasn't worked! We're still only partial altruists.

Second, it's extremely implausible to think that natural selection could produce an animal equally capable of learning to be entirely selfless as to be somewhat self-interested. It's almost as implausible as thinking that natural selection could produce an animal equally capable of learning to love or hate being savaged by wolves. Organisms with no natural bent toward self-interest would be far too vulnerable to exploitation: It would be all too easy for less scrupulous organisms to mold them into willing slaves. If our capacity for culture made us that vulnerable, then selection would quickly have eliminated the capacity for culture. For both these reasons, we can safely reject the blank slate hypothesis. How, then, can we explain our undeniable altruistic streak?

Thicker than Water

Assume for a moment that our alien scientist was a bit lazy. Instead of observing our species directly, the alien decided to steal a social psychology textbook and learn about us from that. Would its subsequent report to the Great Galactic Council give an accurate account of the social life of *Homo sapiens*? Sadly not. It's an unfortunate fact but in a number of ways, most social psychology textbooks present a distorted view of our species. Most notably, they focus almost entirely on interactions among strangers, and thus largely ignore the fact that kin occupy a central place in most people's lives.[4] This is no minor oversight. Kinship is important not only in the West or even only in our species; it's important right across the living world. In species that live in groups composed of both kin and non-kin, individuals usually favor the former over the latter. This is just as true of our own species as of any other, and yet many psychologists and psychology textbooks overlook that fact.[5] As a result, the lazy alien would be misled about our social nature, would botch its report to the Great Galactic Council, and would probably be vaporized.

What accounts for psychologists' kin-blindness – a blindness so profound it would lead to the vaporization of lazy aliens? Part of the answer is that many psychologists, as I've already mentioned, have an empty space in their brains where their knowledge of evolution should be. They know little about other animals and little about the nature of the evolutionary process. This impairs their understanding of their own species. To be fair, the importance of kinship *is* slowly dawning on the field. But the stimulus to this awakening came not from within psychology but from another discipline entirely: evolutionary biology. Ground zero for this slow-motion revolution was the great British genius, William D. Hamilton. Hamilton's big breakthrough was his *kin selection theory*.

To get a firm grasp of this theory, we need to do some creative thinking. In previous chapters, I asked you to imagine that your one goal in life was to have as many children or grandchildren as possible, and then to contemplate what you might do to achieve that goal. As noted, this is a useful way to start thinking about which traits and tendencies natural selection is likely to favor. But there's another way to approach this task which gives us an even more accurate picture. Rather than putting ourselves in the shoes of a fellow human being, we put ourselves in the shoes of a gene – a little snippet of a coil of DNA, which can shape the organism in which it resides and whose only goal in life is to increase its representation in the gene pool. If *you* were such a snippet, how might you set about achieving your expansionist aim? Of course, just as people don't actually strategize about how to pass on their genes, genes don't actually strategize about how to get *themselves* passed on. The exercise is useful, though, because any genes that act *as if* they strategize in this way – any genes, in other words, that have effects that increase their chances of getting passed on – are likely to be selected. As such, the thought experiment provides a useful method of identifying which traits have a good chance of evolving.[6]

So, you're a gene and you want to increase your representation in the gene pool. What are you going to do? The most obvious tactic would be to boost the survival and reproductive success of the organism in which you reside – your "vehicle," as Richard Dawkins puts it.[7] Thus, you could band together with other genes to install in your vehicle a fear of snakes, or a lust for fertile mates, or a proneness to sexual jealousy. But there are other options you might consider instead. In certain circumstances, you could increase your representation in the gene pool by boosting the survival and reproductive success not of your own vehicle but of *other* vehicles. In short, you could achieve your ultimate goal in life by making your vehicle altruistic.

To see how this could work, imagine first that you have a foolproof method of identifying which other organisms in the local environment definitely possess copies of you. Armed with such a superpower, one way you could get yourself passed into the next generation would be to persuade your vehicle to help those organisms to have offspring. From the perspective of an ambitious young gene like yourself, it would make little difference whether your vehicle had offspring or whether one of the other vehicles did. Both possess copies of you, so either outcome would increase your representation in the gene pool to exactly the same degree. There *is* a complication: If you were the *only* gene that your vehicle and the other vehicle shared, your vehicle's altruistic behavior wouldn't advantage any of the other genes in your vehicle's genome; for them, it'd be wasted effort. That being the case, some of those other genes would probably try to prevent your vehicle from behaving altruistically (or to put it in literal terms, any gene that suppressed your activity would probably be selected).[8] But let's leave that aside for the moment. The important point at this stage is that, as long as you're certain that another individual has copies of you, it could well be worth your while to help that individual to survive and reproduce.

But what if you're not completely certain? What if you *suspect* that the other individual has copies of you, but there's at least some chance that you're wrong? Your first thought might be that you should veto the plan to help; after all, if you help someone who *doesn't* have copies of you, you'll be helping a rival allele. But don't be too hasty. In certain circumstances, altruism could help you get ahead in the gene pool, even when there's some doubt about whether the potential recipient of that altruism possesses copies of you. All that's necessary is that the recipient is *more likely than chance* to possess those copies – more likely, that is, than a randomly chosen individual.

To understand why, first imagine that you decide to make your vehicle *indiscriminately* altruistic: You adjust the knobs and dials of your vehicle's brain in such a way that your vehicle helps any other individual it encounters. What are the chances that the recipients of your vehicle's altruistic aid will possess copies of you? Well it all depends how common you are in the population at large. If, say, 50 percent of the population have copies of you, and your vehicle helps other individuals entirely at random, then the odds that the recipients of your help will have copies of you is, naturally enough, 50 percent. Thus, by making your vehicle indiscriminately altruistic, your vehicle will end up helping you 50 percent of the time. That sounds pretty good – until you realize that your vehicle will *also* end up helping competing alleles the other 50 percent of the time. Assume there's just one competing allele. That means that your vehicle will help *you* half the time, but help the rival allele the other half. The net effect of your vehicle's efforts will be to have no effect at all on the relative frequencies of you or the rival allele. And this turns out to be the case no matter what the population frequencies happen to be: 60:40, 40:30:30, whatever. The implication is that indiscriminate altruism won't actually help you. In fact, it will harm you, because it'll use up time and energy that could otherwise be spent doing something that definitely would help: boosting your own vehicle's fitness.

But what if you had some way to get your vehicle to channel its altruism toward individuals who were *more likely than chance* to possess copies of you? Then, everything changes. Suddenly, your vehicle would be helping you at a rate higher than your own population frequency – more than 50 percent of the time, to stick with the previous example. And that means, as a matter of simple logic, that your vehicle would be helping rival alleles at a rate *lower* than *their* population frequencies. The population frequencies of the available alleles would no longer be unaffected by your actions. *Your* population frequency would increase; that of competing alleles would fall. And with that, we reach our first major conclusion:

- A gene that promotes altruism could potentially be selected *if* its vehicle preferentially helps individuals that are more likely than chance to possess copies of that very gene.

A promising start. But how could you get your vehicle to reliably target such individuals? There are several ways, but the most important is that you could get your vehicle to target its genetic relatives. Kin! Your vehicle's kin are more likely than anyone else in the world to possess copies of you, as a result of their recent shared ancestry. As such, by programing your vehicle to be nice to its relatives, you could potentially increase your representation in the gene pool. Importantly, your vehicle's kin are also more likely than anyone else in the world to possess copies of *every other gene* in your vehicle's genome. For that reason, those other genes would probably have no objection to your nepotistic masterplan, and some might even jump on the bandwagon and help you to achieve it.

This brings us to one of the most important building blocks of Hamilton's theory: *the coefficient of relatedness*, also known as *r*. This is a metric which expresses the degree of genealogical relatedness between two organisms. There are several definitions of *r*, but for present purposes, the most useful is that *r* represents the probability that two organisms share any given gene as a consequence of recent shared ancestry. That's quite a mouthful, so let's break it down into bite-size pieces. The simplest case concerns relatedness between parents and offspring. Parents pass on half their genes to each of their children. That means that there's a 50 percent chance that any given gene in the parent's genome will be copied into any given child it produces. And this means in turn that relatedness between parent and child is 50 percent – or to put it in technical terms, $r = .5$. To work out *r* for grandparents and grandchildren, we simply perform this operation twice. There's a 50 percent chance that the grandparent passed a given gene to the parent (its child), and then a further 50 percent chance that the parent passed the same gene to its own child, the grandchild. Thus, relatedness between grandparents and their grandchildren is 50 percent of 50 percent, or 25 percent ($r = .25$).

Things get more complicated for non-descendant kin. For half-siblings (who share one parent), there's a 50 percent chance that the shared parent passed a given gene to one sibling, and a further 50 percent chance that the shared parent passed the same gene to the other. Thus, the chances that *both* inherited a given gene from the shared parent is 25 percent, and relatedness between half-siblings is .25. For full siblings (who share two parents), there's a 25 percent chance that both inherited a given gene from one shared parent, and another 25 percent chance that both inherited it from the other shared parent. Add these together and you find that there's a 50 percent chance that both inherited a given gene from *either one* of the two shared parents. Thus, relatedness for full siblings is .5.

Without going into all the details, it's possible to crank out relatedness values like this for any family members you care to name. Here are the relatedness values for some of the most important categories in the social world of *Homo sapiens*:

- First-degree relatives (offspring, parents, full siblings): $r = .5$.
- Second-degree relatives (grandparents, grandchildren, half-siblings, nieces and nephews, aunts and uncles): $r = .25$.
- Third-degree relatives (e.g., first cousins): $r = .125$.
- Non-relatives (e.g., mates, in-laws, stepfamily, friends, acquaintances, and strangers), $r \approx 0$.

The concept of relatedness looks simple enough, but it's easily misunderstood.[9] A common misunderstanding runs like this:

> Hang on a minute. You're saying that parents and kids share 50 percent of their genes and that non-relatives share none. But I've heard that *all* human beings share more than 98 percent of their genes, and that humans share more than 94 percent of their genes with chimpanzees. Hell – we even share 50 percent of our genes with bananas! Isn't this a contradiction?[10]

The answer is no, it's not a contradiction, because the coefficient of relatedness is *not* a measure of shared genes. It's a measure of the probability that two individuals share any particular gene *by descent*: that is, either because one inherited it from the other (as in offspring from a parent), or because both inherited it from one of their most recent common ancestors (as in siblings from a shared parent or cousins from a shared grandparent). Simplifying somewhat, if they *don't* share the gene by descent, the probability that they share it anyway – that is, the probability that a gene in individual A's genome is also present in individual B's – is equal to the population frequency of that gene. This means, in effect, that relatedness is a measure of the probability that two individuals share a given gene *over and above* the population frequency of that gene. In other words, relatedness represents the probability that two individuals share a given gene *over and above chance*.

This is an important point, because it ties the concept of relatedness to our earlier conclusion that altruism can be selected when the target of that altruism is more likely than chance to possess any genes helping to create the altruistic tendency. In a sense, relatedness is an index of the probability that altruism will pay off for the genes giving rise to it. And this means that, if you were a gene trying to increase your representation in the gene pool, one way you could do it would be to wire up your vehicle to be altruistic toward its genetic relatives – and especially its *close* genetic relatives.

But a sensible gene wouldn't settle on this strategy yet. Relatedness, it turns out, is not the only relevant variable. We also need to factor in the benefits of altruism to the recipient and the costs of altruism to the altruist. To see why, consider the case of identical twins. Identical twins are an exception to the rule that first-degree relatives are related at the .5 level; these natural-born clones share all their genes, and thus r for identical twins is 1. This means that, if you were a gene seeking safe passage into the next generation, it wouldn't matter to you one iota whether your vehicle had a child or your vehicle's identical twin did. Either way, you'd have a 50 percent chance of being copied into the child's genome. So, who should your vehicle help – its twin or itself? It depends how much good the help would do. Imagine that your vehicle could either have three extra offspring of its own or help its identical twin to have an extra two. In that case, it'd clearly be better for you if your vehicle had the three extra offspring and didn't help the twin. Self-interest would prevail. But what if those numbers were reversed? What if your vehicle could have two extra offspring of its own or help its identical twin to have three? In *that* case, it would be better if your vehicle helped the twin. Altruism would get the upper hand.

This much is clear enough. Now, though, consider what would happen if the potential recipient of your vehicle's largesse was not an identical twin, but a run-of-the-mill first-, second-, or third-degree relative. When would you want your vehicle to help *that* individual? At first, it might seem the answer is never. Sure, your vehicle's relatives are more likely than a random stranger to have copies of you. But your vehicle *definitely* has copies, so why shouldn't your vehicle help itself every time? The reason is that if the benefits of altruism to the recipient are sufficiently large, they can outweigh the costs to your vehicle, even taking into account the fact that the recipient might not possess copies of you by descent. This was one of Hamilton's key insights. It's captured in a simple formula called *Hamilton's rule*, which the evolutionary psychologist Oliver Curry described as the $E = mc^2$ of evolutionary psychology.[11] Hamilton's rule states that altruism can be selected when...

$$br > c$$

What does this mean? Well, we already know what r means; it's the degree of relatedness between altruist and recipient. b, on the other hand, is the *benefit* of altruism to the recipient, and c is the *cost* of altruism to the altruist. These benefits and costs are measured in terms of Darwinian fitness; thus, b is the number of extra offspring the recipient of the altruism has as a result of receiving help, whereas c is the number of off-spring that the altruist *doesn't* have as a result of helping. Putting it all together, what Hamilton's rule tells us is that a gene promoting altruism can be selected when the reproductive cost of altruism to the altruist is less than the reproductive benefit to the recipient, but with the benefit to the recipient scaled back in proportion to the degree of relatedness between recipient and altruist – in other words, the probability that they share the gene by descent.

OK, so you're a gene and you now understand Hamilton's rule. How would you use your newfound knowledge to get ahead in the genetic rat race? Recall that, in the identical twin scenario, you decided that your vehicle should give up two offspring of its own in order to help its twin to have three. But what if the potential recipient of your vehicle's altruism was a regular .5 sibling? Then the landscape changes completely. There's only a 50 percent chance that the sibling possesses copies of you by descent. Thus, from your perspective, the average expected gain from your vehicle helping its sibling wouldn't be three extra offspring; it would be 1.5. That's less than the cost of altruism to your vehicle (the two extra offspring of its own it would have otherwise). As such, you'd be better off if your vehicle was *not* altruistic. In this situation, you should advise your vehicle to look after number one, and leave its sibling to fend for itself.

But what if your vehicle could give up two extra offspring of its own in order to help its sibling to have *five* extra offspring? Well, that's a different story. The expected benefit to you would still be halved; it would be 2.5 offspring instead of five. But that's still more than the two extra offspring your vehicle would have if it didn't help. Thus, in *this* situation, you should urge your vehicle to help its sibling.

Now obviously genes can't actually micromanage our moment-to-moment decisions in this way. What they can do, though – and what kin selection theory predicts that they will – is help to set up brains that make these kinds of decisions for themselves. The optimal Hamiltonian brain would be sensitive to information about the three terms of Hamilton's rule – relatedness, benefits, and costs – and would use this information to calibrate its altruistic output. This wouldn't involve conscious, deliberative reasoning of the sort you're using to read these words; it would involve nonconscious, automatic processing of the sort underpinning visual perception. To put it another way, Hamiltonian altruists wouldn't sit down with pen and paper and work their way through Hamilton's rule; they'd simply find that they had a stronger urge to help other individuals when relatedness was high, benefits substantial, and costs low, and a weaker urge to help when these values were reversed. Needless to say, no single gene could give rise to a psychological disposition so complex. However, any gene that nudged its vehicle in this direction could potentially be selected, and working together, large teams of genes could very well create such a disposition.[12] Hamilton's math shows that, in a wide variety of circumstances, genes like these could make a living in the competitive world of the species' gene pool. As we'll see in the next section, that's exactly what seems to have happened.

Family Man

Hamilton's theory was a revolution in evolutionary biology. In a single stroke, it explained most of the altruistic behavior seen in the animal world, including most of that described at the start of the chapter. Consider helping at the nest in birds. Why do reproductively capable adults sometimes help to raise chicks that aren't theirs? Kin selection suggests an answer: Although the helpers aren't the chicks' parents, maybe they're related in some other way. Turns out that's true; the helpers are usually the chicks' uncles.[13] Next, consider alarm calling in Belding's ground squirrels. Why does the caller put its life on the line to warn others of danger? Again, kin selection suggests an answer: Maybe it benefits the caller's kin. Right again; the squirrels are much more likely to raise the alarm when relatives are in the vicinity, and much, *much* more likely when they're close relatives.[14] Finally, consider the suicidal nest defense of honeybees, and the reproductive altruism of eusocial animals in general. Once again, kin selection makes mincemeat of this initially mystifying phenomenon. Eusocial colonies, as I explained in Chapter 2, are vast families of thousands or even millions of siblings, some small fraction of which go on to found new colonies. As such, protecting the nest is an indirect way of protecting the genes that made you do it, even if you never have kids of your own.

With our kin selection goggles on, a general pattern suddenly becomes glaringly obvious. Across the living world, organisms are more altruistic toward kin than non-kin, and more altruistic toward closer kin than more distantly related kin. And not only are they more altruistic, they're less likely to harm close kin. Among Belding's ground squirrels, for instance, close relatives fight less than distant relatives. Similarly, among wolf spiders, mothers are less likely to cannibalize their own spiderlings than to cannibalize spiderlings that aren't related to them.[15]

Here's something you might be wondering about, though: How do spiders and squirrels and other languageless animals know who their relatives are? It's a lot easier to see *why* selection would favor kin-directed altruism than to figure out *how* it might actually implement it. But we do have some idea. Roughly speaking, animals evolve to follow simple rules-of-thumb that, in the species' ancestral past, led them to favor kin over non-kin most of the time. One common rule is "Be nice to anyone you lived with as a juvenile." This works well in any species in which the young are reared with kin. Belding's ground squirrels, for example, tend to be nicer to individuals they were nursed with as pups. This is the case even when evil scientists sneakily swap the pups around so that they're actually reared with non-relatives rather than siblings – persuasive evidence that the relevant variable at the proximate level is early-life cohabitation, rather than relatedness per se.[16]

It's not *just* early-life cohabitation, however. Among Belding's ground squirrels, females usually mate with several males within a single reproductive cycle. As a result, litters usually contain pups from several different fathers, and any given pup will have some full siblings and some half-siblings among its littermates. Despite this, females are more cooperative and less aggressive with full sisters than with half-sisters.[17] Early-life cohabitation can't explain this, because the females cohabited with both types of sisters. The squirrels seem to be following a second rule as well: "Be nice to individuals who are similar to you." This strategy is known as *phenotype matching*. (The phenotype is the collection of observable features of an organism, shaped by the interaction of its genes with the local environment.) Phenotype matching works because the more similar an individual is to oneself, the more probable it is that that individual is a relative.

(Attentive readers will notice that early-life cohabitation and phenotypic similarity are also used as kinship cues in another context: incest avoidance. In a range of species, statistical indicators of relatedness appear to upregulate altruism but downregulate sexual attraction. As mentioned in Chapter 4, this fine-grained pattern of responses is difficult to explain without the gene's-eye view of evolution.)

There's little doubt now that kin selection has had a profound impact on the nonhuman animal world.[18] What about us? Has kin selection helped to sculpt the clay of human nature? Casual observation suggests, at the very least, that it's possible. For one thing, kin selection offers an explanation for the case we kicked off the chapter with: the heroism of Jordan Rice. Simply put, Jordan risked his life for a sibling, and siblings are as closely related to each other as parents are to their children. But kin selection also offers an explanation for the millions of little ways in which people extend greater kindness to kin than to non-kin in everyday life. Parents spend thousands upon thousands of dollars on their children every year, but if they also give a few hundred dollars to charity, this is seen as uncommonly generous.[19] Grandparents rarely insist on being paid to babysit their grandchildren; non-relatives usually do. People often say things like "They're such close friends she sees her as a sister," but rarely "They're such close sisters she sees her as a friend." And in nuclear families, if the parents split up, family members generally maintain every other bond within the group – children with mother, children with father, siblings with siblings – even when the parents can no longer stand to be in the same room as each other. The fact that kinship bonds so often persist in this awkward circumstance is a testament to their durability. Non-kin bonds are rarely so robust.

These observations are suggestive, but anecdotes alone can't clinch the win for kin selection. Fortunately, in the decades since Hamilton first proposed his theory, scientists have assembled various lines of more rigorous evidence that human behavior does indeed fall into conformity with the theory's predictions. One line of evidence comes from anonymous surveys asking people about their relationships with the people in their social circles, and how much help they give to or receive from these individuals. People's responses in these surveys are almost exactly what other animals' responses would be if other animals could fill out questionnaires. As a general rule, people report being closer to kin than to non-kin, more helpful to kin than to non-kin, and more willing to help kin than non-kin in the future. Not only that, but among kin, people are closer to and more willing to help close kin than more remotely related kin.[20]

Importantly, these trends are most pronounced for costly, evolutionarily significant help: donating a kidney, for instance, as opposed to just comforting someone who's down in the dumps.[21] Some of my own research has dealt with this issue. In one series of studies, I asked Canadian students how much low-, medium-, and high-cost help they'd given to various categories of kin and non-kin in the last two months. I found that as the cost of help went up, the share of help that people gave to kin increased, whereas the share they gave to non-kin nosedived.[22] In other words, the more it mattered, the more nepotistic people became. These findings have since been replicated by other researchers in other parts of the world, including Carey Fitzgerald in the United States and Ming Xue in rural Tibet. The Tibetan replication is particularly noteworthy, because it suggests that the cost-of-help effect isn't unique to the Western world, but reflects something deeper in human nature.[23]

As usual, self-report studies are a good first step, but there's always a nagging worry that people's responses might be less than accurate. Before we can rest our case, we need to look at other sources of data. Public records are a particularly useful source: They're often highly reliable, they capture real-world behavior, and they weren't collected specifically to test evolutionary predictions, which means that experimenter bias is less of a concern. One line of research in this vein focuses on inheritance records, looking for the fingerprints of kin selection in the ways that people divide up their estates in their wills. As any non-alien would predict, people tend to leave more of their wealth and belongings to kin than to non-kin, and more to close kin than to distant kin. One study found, for example, that nearly half of all bequests were to offspring and siblings ($r = .5$), that around 10 percent were to nephews, nieces, and grandchildren ($r = .25$), and that less than 1 percent were to cousins ($r = .125$).[24] People rarely leave money or belongings to friends, even though they're often just as close to their friends as they are to their relatives.[25] For the big decisions in life, genetic closeness seems to trump emotional closeness.

There is one exception to the nepotistic rule, though, and that's spouses. People almost always leave a large chunk of their estates to their wives or husbands, despite the fact that their wives and husbands are (usually) non-relatives.[26] And this, of course, is just one example of a broader trend, namely that people tend to have extremely close bonds with their long-term mates, and are normally more than willing to extend them significant help. The deep bond between long-term mates isn't directly explicable in terms of relatedness, but it is still explicable in evolutionary terms. Although mates are not generally related to each other, they *are* both related to any offspring they produce together. This means that mates have overlapping fitness interests in the shape of their shared offspring: those they have already and those they could have in the future. Anything one does to boost a mate's fitness – anything, that is, that increases a mate's chances of bearing and rearing kids – is very likely to boost one's own fitness as well, for the simple reason that their kids are your kids and your kids are theirs. It's no surprise, then, that long-term mates often treat each other, in effect, as honorary kin.[27]

Another line of research based on public records looks at police statistics on violent crime, abuse, and homicide. As we saw in Chapter 4, Martin Daly and Margo Wilson are the pioneers of the research in this area, and they use the police statistics as a "reverse assay" of closeness and altruism: The more violence there is within a given type of relationship, the less close the relationship typically is and the less altruism there's likely to be within it. Daly and Wilson's central finding is that, like mother spiders and Belding's ground squirrels, human beings are more likely to harm or kill non-relatives than relatives. This is true even when the relatives and non-relatives in question all live under the same roof, and therefore have similar levels of contact and interdependency. In one study, for instance, Daly and Wilson found that people were more likely to kill cohabiting spouses and other non-relatives than they were to kill cohabiting children, parents, or other related individuals.[28] As well as supporting the predictions of kin selection theory, this finding sheds further light on the nature of the relationship between long-term mates. The fact that people are more likely to kill spouses than kin suggests that, although people *can* be as close to mates as they are to genetic relatives, love can more easily turn to hatred among the former than the latter. In other words, a mate's status as an honorary relative can quickly be revoked. This makes good sense in the light of kin selection theory. Mates, after all, aren't kin, and thus their evolutionary relevance can change almost as quickly as the quality of their relationship.

Of course, humans can't peer into one another's genomes and directly perceive who their kin are, any more than nonhumans can. Strictly speaking, then, people couldn't have evolved to favor kin. What could have happened, however, is that we evolved to follow certain implicit rules that, in our ancestral past, led us to favor kin over non-kin nine times out of ten. What might these rules be? First, like many other animals, humans seem to be especially willing to lend a helping hand to individuals they were raised with, or those they helped to raise. Second, also like many other animals, humans seem to be especially willing to help individuals who resemble them more closely than the average person in their social environment. To use the vernacular, early-life cohabitation and phenotypic similarity seem to function as kinship cues in our species, just as they do in many others. A responsiveness to these stimuli is plausibly a part of our species-typical mental architecture.[29]

It's worth acknowledging that many of the research findings related to human kinship are utterly unsurprising. No one is the slightest bit taken aback, for example, to learn that will makers leave more of their estates to relatives than to non-relatives. Most of us are good "intuitive evolutionary psychologists" (an ironic fact given how controversial evolutionary psychology has been over the years). We typically see acts of altruism toward non-relatives as noble in the extreme, but comparable acts among kin as par for the course. To take just one example, the Carnegie Hero Fund Commission, which honors people who risk their lives to save others, doesn't even consider cases where people save relatives, except when this results in death or serious injury.[30] And most of us wouldn't blink at this. It's as if we tacitly accept the overlap of genetic interests among kin, and see rescuing kin as little more heroic than rescuing oneself. *Dying* to rescue kin might be going beyond the call of duty, but anything short of that is almost indistinguishable from self-interest. Given this mindset, it's only natural that we find evidence for our blood-relative bias unsurprising. But the fact that it's unsurprising is not a mark against kin selection theory. On the contrary, it's the things that seem most obvious to us that most cry out for an explanation – after all, they wouldn't seem so blindingly obvious if they weren't so deeply ingrained in our psyches and in our societies. The fact that we don't feel that these things require an explanation is, paradoxically, exactly why they do.

Why, though, should we favor the kin selection explanation over a rather obvious competitor: the idea that we simply *learn* to favor kin over non-kin? I'll give you several reasons. First, the importance of kinship is a human universal. Wherever in the world we look, people's affinity for kin stands out like a sore thumb – or, more to the point, like an *opposable* thumb: a species-wide attribute with an obvious adaptive rationale.[31] Not only that, but our affinity for kin persists even when we try to eliminate it. This was the case in the Israeli kibbutzim we discussed in Chapter 3: In many kibbutzim, the powers-that-be attempted to eradicate people's special affection for their own children, but that special affection just kept on popping back up like a Bobo doll. And this isn't the only example. Another comes from a polygynous Mormon community studied by William Jankowiak and Monique Diderich. Men in this community often had several wives, and thus siblings and half-siblings often ended up living together in the same home, like a mixed litter of Belding's ground squirrels. Despite growing up together, however, and despite a community ethos which downplayed differences in relatedness among siblings, people were still closer to their full siblings than their half-siblings.[32] The blood-relative bias runs deep – deeper than non-evolutionary theories can readily explain.

Admittedly, the fact that a trait is universal doesn't necessarily mean that it has an evolutionary origin; if it did, we'd have to conclude that drinking Coca-Cola and yakking on mobile phones are products of natural selection. But cross-cultural universality sits a lot more easily with an evolutionary explanation than it does with a Nurture Only one. We've got a good understanding of how Coke and mobile phones came to be universal despite not being innate. Until advocates of the Nurture Only view can offer us an equally good explanation for the universality of our kin bias, the default prediction from the Nurture Only perspective is that, across cultures, people's behavior toward kin and non-kin will vary randomly with respect to kin selection theory: In some cultures, people will favor kin, in others they'll favor non-kin, and in others still they'll show no bias in either direction. That's not what we see, though; people everywhere favor kin. This represents a major predictive failure for the Nurture Only view.

A further nail in the Nurture Only coffin is the fact that kin altruism is not unique to our species. As we've seen, it's found in other mammals, in birds, and in species as remotely related to us as honeybees and carpenter ants. And that's just scratching the surface. Kin altruism is even found in plants. For example, when American sea rocket shares soil with sibling plants, it grows its roots less aggressively and less competitively than it does when it shares soil with non-relatives. In other words, just as Belding's ground squirrels and human beings are less aggressive toward relatives, so too is American sea rocket.[33] And even *that's* just scratching the surface. Kin altruism is also found in bacteria.[34] This means that a lot of kin altruism – indeed, the vast majority of the kin altruism taking place on this planet – is invisible to us. But it's there, and it shows us that kin selection is profoundly important, not only in complex multicellular organisms such as ourselves and plants, but right across the living world. The fact that kin altruism is so ubiquitous – vastly more so than Coke or mobile phones – tells us something important. It tells us that William Hamilton, in positing his kin selection theory, identified an extremely deep principle in nature: the biological equivalent of the laws of thermodynamics.

The ubiquity of kin altruism also tells us something about the origins of our own nepotistic inclinations. No one would dream of explaining kin-directed altruism in Belding's ground squirrels, American sea rocket, or bacterial molds in terms of socialization or cultural norms. We explain it, without hesitation, in evolutionary terms. Is it plausible, then, when we see the same pattern in our own species, to explain it in completely different terms, as a product solely of learning or culture? The short answer is no, it's not plausible. It's *possible*, I suppose. But in the absence of strong evidence that it's actually the case, the default presumption should be that humans are continuous with the rest of nature and thus that our nepotistic streak has an evolutionary origin, just as it does for every other organism. The burden of proof falls squarely on the shoulders of the Nurture Only theorist.

And what a burden it is! Anyone who wants to deny that human kinship behavior has anything to do with kin selection has an arduous task ahead of them. They've got to explain why human beings are mysteriously exempt from a selection pressure that runs like an unbroken thread throughout the animal kingdom and all the way down to plants and bacteria. They've got to explain why, at some point in our species' past, kin altruism ceased to be adaptive for us, and thus why selection wiped out the nepotistic streak that presumably existed in our pre-human ancestors. And they've got to explain why, despite this, people in every culture still manage to act in ways consistent with kin selection theory. Why would culture recapitulate biology in this manner, given that kin altruism supposedly ceased to be adaptive in our lineage?

Kin selection theory is one of the great theories in the history of science. It places human beings within an explanatory framework that embraces the rest of the biological world – a framework that links suicidal nest defense in bees and patterns of root growth in plants with the loving bonds and quiet self-sacrifice found in human families. I'm willing to go out on a limb and say that no psychologist or social scientist who lacks a detailed understanding of kin selection theory, and an awareness of the importance of relatedness throughout the living world, can claim to be an expert on behavior. Furthermore, any student of psychology or the social sciences who isn't taught Hamilton's ideas in some depth has been short-changed and should immediately ask for a refund.

But kin selection theory shouldn't exhaust psychologists' exposure to evolutionary ideas. Although Hamilton's theory explains most of the altruism we see in other animals, it explains only a sliver of what we see in our own species. To expand on an earlier point, the thing that would stand out most to an alien scientist about our species is not the fact that we help kin more than we help non-kin; most social animals do that. It's the fact that we help non-relatives as much as we do. In this respect, as in so many others, humans are freaks of nature. We're freakishly kind to non-relatives.

We've all heard stories about extraordinary human kindness. One of my favorites concerns the rock star Elvis Presley. As a boy, Elvis decided to give his bike to a neighborhood kid who was too poor to get a bike of his own – an incredibly generous act by any standard. The kid's parents felt that they couldn't accept such an expensive gift and insisted on returning the bike... whereupon Elvis immediately went and gave the bike back to the kid. Another favorite: In 2012, an NYPD cop spotted a homeless man sitting shoeless on the pavement on a freezing New York night. The cop didn't just keep on walking and he didn't just tell the man to move on; instead, he went into a nearby store and bought him a pair of boots. Some tourists snapped a photo of the cop helping the homeless man put the boots on, which then went viral on social media. Finally, on a lighter note, a US couple pledged to legally change their surname to Van Squigglebottoms if they could raise a million dollars for Oxfam – all of which, contrary to Darwinian imperatives, would be channeled to non-relatives. It's hard to think of parallels to any of this behavior in other species. And yet in our own, it's common. For all the terrible things we do to each other as well, humans are capable of incredible kindness – kindness that can bring a tear to the eye.

Heart-warming anecdotes aside, there's now solid scientific evidence that humans get a hedonic kick – a warm glow – from helping other people, even when the people in question are unrelated to them. The psychologist Elizabeth Dunn is one of the big names in this area. In one clever study, Dunn gave participants a small sum of money. Half were told to go out and spend it on themselves, the other half to spend it on someone else: a stranger. Most people, if you ask them, guess that the first group would end up feeling happier than the second. But most people turn out to be wrong; participants who spent the money on others were the happiest.[35] Money *can* buy you happiness, it seems, but it helps if you spend it on other people. A cynic might say that this just gives us a selfish reason to act unselfishly. However we cut it, though, from the perspective of tooth-and-claw Darwinism, it's a puzzle that we're set up this way. How can we explain the fact that human beings like to help people who are no more likely than chance to share their genes?

The Evolution of Mutual Back Scratching

One important answer comes from the evolutionary biologist Robert Trivers – the same Robert Trivers who cracked the mystery of sex differences in Chapter 3. Trivers is another figure like Hamilton: Any psychology student who doesn't study his work in detail should demand a refund. Steven Pinker went as far as to describe him as one of the great thinkers in the Western intellectual tradition, thus placing him alongside such luminaries as Plato, Aristotle, Descartes, Wollstonecraft, Mill, and Darwin. And one of Trivers' landmark contributions is a theory about how altruism could evolve among non-relatives. It's called *reciprocal altruism theory*.[36]

To understand how the theory works, we first need to draw a distinction between altruism and cooperation. As we saw earlier, an altruistic act is one that benefits the recipient of the act but inflicts a cost on the actor. A *cooperative* act, in contrast, is one that benefits both parties. A widespread view of evolution casts it as a mountain of competition set against a molehill of cooperation. But nothing could be further from the truth. As with kin altruism, cooperation is ubiquitous in the living world, visible at any level of magnification you care to choose. Lions, killer whales, and army ants cooperate to hunt prey; cells cooperate to form multicellular organisms; organelles cooperate to form cells; and genes cooperate to form genomes – which then give rise to organelles, cells, and multicellular organisms, which in turn cooperate to pass on those genes. Don't get me wrong; there's plenty of competition in the natural world as well. Indeed, the cooperation evolved *through* competition: Cooperative entities exist today because their ancestors outcompeted their non-cooperative counterparts. Still, the conception of life as a Hobbesian war of all against all is misleading. Life is a messy mix of war and cooperation, the nasty and the nice.[37]

Unlike altruism, it's easy to see how cooperation could be adaptive, even among non-relatives. People (or lions or army ants) that excel at teamwork often do better for themselves than they would if they went it alone. If five hunters manage to bring down a giraffe, all of them end up with more meat in their bellies than they would if each brought down, say, a turtle. As such, cooperators generally do better and have more offspring than non-cooperators. Cooperation evolves. But what about altruism? What about the fact, for instance, that human hunters often share their bounties with people who weren't involved in the hunt? How could such generosity of spirit arise?

That's where reciprocal altruism comes in. A lot of apparent altruism, argues Trivers, is really cooperation smeared across time. If we look at a single isolated incident, we see individual A helping individual B, and wonder how such noble self-sacrifice could evolve. However, as soon as we step back and take in a broader slice of the timeline, we realize that A and B, whether they know it or not, are actually swapping favors: A helps B this time, but later the roles are reversed and B helps A. As with hunters working together to bring down a giraffe, A and B both do better than either would if they put the same efforts into purely self-interested pursuits.

The reason for this, though, is slightly different than in the case of the giraffe hunters. Imagine that I have two portions of meat but that you have none. Imagine also that eating the first portion will give me two units of fitness, whereas eating the second will give me only one – I'll already be full, after all, and thus I'll get diminishing returns from stuffing my face any further. You, on the other hand, haven't eaten, so if I give you my second portion, you'll get two units of fitness from it. That's slightly worse for me – I'll get two units in total instead of three – but it's better for you and it's better overall: Between us, we'll get four units of fitness from the meat, rather than the three we'd get if I gorged myself and let you starve.

Next imagine that, at a later date, *you* have two portions of meat and *I* have none. If you now eat one portion and give me the other, we'll again get two units of fitness each, rather than three for you and none for me. That's slightly worse for you, but it's better for me and again it's better overall: We'll squeeze four units of fitness out of the meat instead of just three. But there's something else to notice at this point. Although sharing is worse for the sharer at the time, by swapping sharing duties, both of us will do better in the end than we would have if neither of us had shared. Specifically, both of us will get a total of four units of fitness from trading favors, whereas we'd both get only three if we'd horded our windfalls. In other words, by taking turns at being generous, we'll both be better off. And of course this doesn't apply only to meat. Reciprocity works its magic whenever just two conditions are met: first, that the benefits of helping to the recipient are greater than the costs to the altruist, and second, that recipient and altruist trade places from time to time.[38]

It sounds like a foolproof plan: Reciprocators do better than non-reciprocators, and thus reciprocators should fill the world with their progeny. Unlike cooperation, however, reciprocity is as rare among the animals as snowballs in summer. If it's such a great idea, why aren't polar bears, grasshoppers, and ladybugs routinely trading favors? The reason is that there's a major roadblock to the evolution of reciprocity, which biologists call *the problem of cheating*. Recall that, in the meat example, when you and I take turns sharing, we're both better off than we would be if neither of us did; each of us gets four units of fitness instead of three. But there is a way that one of us could be better off still: We could rip the other one off. I, for example, could take the meat you offer me when I have none, but then just eat both my portions when you come home empty-handed. Sure, the second portion would do me less good than it would do you. But it would still do me *some* good. I'd get three units of fitness from eating both my portions, as well as the additional two units from eating the meat you so kindly gave me. That would give me five units of fitness in total, instead of the four I'd get if I reciprocated your generosity like a chump. Meanwhile, you'd get only two units of fitness – less than the three you'd get if you just ate all your meat when you had it and didn't risk sharing. The upshot is that individuals who exploit others' generosity – *free riders* – do better than would-be reciprocators. If there were ever a population of reciprocators, they'd be vulnerable to invasion by free riders. Like suicidal lemmings, the reciprocators would get less and less common until one day, eventually, the free riders would inherit the Earth.

If that were the last word on reciprocity, no one would ever reciprocate. People do reciprocate, however, which tells us that – if reciprocity has an evolutionary origin – natural selection must have equipped us with some kind of defense against getting cheated. According to Trivers, that's exactly what happened.[39] Human beings, he argues, have a set of emotions and preferences that lead us to establish mutually profitable reciprocal relationships, all the while avoiding getting too severely short-changed by the free riders of the world. First and foremost, we dislike it when people take more than they give. We get angry with them. This probably isn't just something we've learned to do; it seems unlikely that people could just as easily learn to cherish being cheated as to despise it. Instead, argues Trivers, the reaction is part and parcel of human nature. Just as disgust motivates us to avoid spoiled food, anger at a person who cheated us motivates us to avoid getting cheated again. We might refuse to help the cheater next time; we might cut our social ties with them; or we might punish them in some way to dissuade them from cheating us in the future. Such actions limit the damage done by free riders and cheats.

Other familiar emotions also help make us natural reciprocators. When someone does something good for us, we feel gratitude. This motivates us to return the favor later, thereby continuing the mutually beneficial cycle of reciprocal exchange. If, for some reason, we fail to return a favor, we feel guilt or shame. These emotions motivate us to mend our ways before we activate other people's anti-free-rider adaptations – that is, before they get mad at us – or, if they're already mad at us, the emotions motivate us to make amends. Consistent with this suggestion, we seem to be particularly prone to guilt when we've been caught doing wrong, or when there's a good chance that we will be. (As H. L. Mencken said, "Conscience is the inner voice that tells us someone might be looking.")[40] Finally, when we come across a person in need, we feel sympathy – and the greater the need, the more sympathy we feel. This motivates us to channel our help toward those who'll benefit from it the most, thereby earning the maximum gratitude for the smallest outlay of effort. Robert Wright put it well: Sympathy, he said, is investment advice.[41]

At first glance, this makes it sound like people are constantly scheming and seeking payback for any help we give to others. But that's not the idea. We don't help people in need because we're deliberately trying to find the cheapest way to earn the most gratitude. Well, sometimes we might, but most of the time, we just do it because we care. Why, though, do we care? From the standpoint of reciprocal altruism theory, it's because caring helped our ancestors to establish profitable relationships of reciprocal exchange, without any advanced planning and without any awareness that that's what they were even doing. The same applies to the other emotions involved in reciprocity. Just as love and jealousy lead us to forge romantic relationships, emotions such as anger, gratitude, guilt, and sympathy lead us to forge reciprocal alliances (in plain English, "friendships") and to defend ourselves against cheats and free riders (in plain English, "assholes"). If this is right, then some of the most basic human emotions, and some of our most cherished moral impulses, have their origin in the economic logic of mutually beneficial trade.

The Trading Animal

Trivers' theory sounds reasonable enough, but how can we be sure it's not just a plausible-sounding just-so story? At this stage, I don't think we *can* be as sure as we are in the case of kin selection or parental investment theory. Still, in various ways, the theory of reciprocal altruism has been tested and challenged, prodded and poked, and so far it's passed with flying colors. At the very least, it's a theory we need to take seriously.

To begin with, the underlying logic of the theory has been tested with computer simulations. The basic idea is that researchers set up colonies of reciprocators and non-reciprocators inside virtual worlds, push the go button, and then see how well each strategy fares. The political scientist Robert Axelrod got the ball rolling on this line of research.[42] Axelrod invited academics from all over the world to submit computer programs which would play against other computer programs in a round-robin tournament. The programs, he explained, would repeatedly encounter one another, and with each encounter, would have to choose to either cooperate or defect. Cooperation (or helping) would be the equivalent of sharing your meat; defection would be the equivalent of hording it. Once both programs had made their choices, they'd be awarded points as follows.

- Receiving help but not helping in return (i.e., free riding): five points.
- Mutual helping: three points.
- Mutual defection: one point.
- Helping but not receiving help in return (the so-called *sucker's payoff*): zero points.

The winner, said Axelrod, would be the program that accumulated the most points by the end of the tournament. Given this set up, what should a virtual animal do? For a single interaction, the answer is clear: Defect.

Don't share. This is the best option no matter what the other player does. If the other player defects as well, you'll get one point instead of zero. If the other player cooperates, you'll get five points instead of three. Thus, in a "one-shot game," the best move for both players is always to defect, and cooperation could never get off the ground. The tragedy is that both players would do better if they both cooperated: They'd get three points each instead of the one point they end up with when they both defect. Ironically, then, in a one-shot game, the rational pursuit of self-interest makes both parties worse off. This is a hypothetical example of a real-world phenomenon – a phenomenon the ecologist Garrett Hardin called *the tragedy of the commons*.[43]

But what if the virtual animals get to encounter each other again and again, like real animals often do? In that case, the best course of action is no longer quite so obvious. And that's why we need the computer simulations. Back to the main story. After issuing his invitation for programs to compete in his tournament, Axelrod soon had a diverse collection of contestants, each reflecting its creator's best guess about which strategy would rack up the most points. Some were very simple, others highly complex. Among the simple ones was a program called Always Cooperate. This program was a regular Mr. Nice Guy; no matter what the other player did the last time they met, Always Cooperate... always cooperated. Some call this strategy Doormat. Another simple program was called Always Defect. This was Doormat's evil twin; no matter what the other player did last time, this program always defected. Alongside the simple programs were various more complex ones – programs that tried to predict what other players would do based on their past behavior, and then use that information to calculate the most profitable course of action for each encounter.

On top of all that was a program called *Tit-for-Tat*. Tit-for-Tat was one of the simpler strategies; all it did was cooperate on its first encounter with any other player, and thereafter simply match that player's last move. If the other player cooperated on the last move, Tit-for-Tat cooperated now. If the other player defected, Tit-for-Tat defected right back. As Trivers observed, Tit-for-Tat acted on a variant of the Golden Rule: "First, do unto others as you wish them to do unto you, but then do unto them as they have just done unto you."[44] In this way, the program did roughly what Trivers argued that humans evolved to do: It established recipro-cally altruistic partnerships where possible, but defended itself against cheats by refusing to cooperate with them further.

Axelrod ran several versions of the computer tournament. In each, Tit-for-Tat emerged as the undisputed champion. It beat Always Cooperate; it beat Always Defect; it even beat the complex predictive strategies. This was a major shot in the arm for Trivers' theory. It showed in a rigorous way that reciprocal altruism really could work. Admittedly, later simulations revealed that Tit-for-Tat isn't *always* the best option; it all depends who one's up against.[45] But the clear message emerging from the mountain of research that now exists on this topic is that, in a wide variety of circumstances, reciprocity can trump unadulterated selfishness.

The computer simulations also shed light on how. Tit-for-Tat had three features that made it a force to be reckoned with.[46] The first was that it was nice. Tit-for-Tat never aimed for the top prize: It never tried to take from another but not return the favor. Despite that, it prevailed over nasty programs such as Always Defect, which did exactly that. One reason for this is that nice strategies do well when they're common; they happily trade favors with other nice strategies. Nasty strategies, in contrast, do well only when they're rare and there are thus lots of nice strategies around to exploit. As soon as the bad guys get too common, they spend most of their time running into one another, mutually defecting, and doing poorly as a result.

The second feature making Tit-for-Tat a success story was that, although it was nice, it wasn't *too* nice. Unlike Always Cooperate, Tit-for-Tat had a way of limiting the damage done by cheats: It simply stopped cooperating with them. Nasty programs like Always Defect got one big payoff from Tit-for-Tat, but then they were out in the cold. And though Tit-for-Tat took a one-time hit when interacting with these programs, it quickly racked up more points than them, just by patiently accumulating the lesser rewards of mutual cooperation. In the short-term, Always Defect did better; in the long-term, Tit-for-Tat did.

The third and final secret of Tit-for-Tat's success was that it was forgiving. If a former cheat turned over a new leaf and started cooperating, Tit-for-Tat didn't hold a grudge. It immediately began cooperating again too, thereby securing the rewards of mutual assistance. Grudge-holders, on the other hand, cut themselves off from those rewards and did worse for that reason. In sum, Tit-for-Tat prevailed because it was nice, because it defended itself against cheats, and because it was always willing to forgive a reformed cheater.

The Axelrod simulations show us that reciprocal altruism is a viable proposition. But has anything like a Tit-for-Tat strategy actually evolved in our species? As with earlier cases, casual observation is consistent with the idea. When we give someone a birthday present, we don't necessarily do so with the expectation that they'll give us one in return. But for how long would we continue giving presents to someone who never returned the favor? Probably not long, especially if the ingrate was a non-relative. We resent freeloaders, brand those who don't pull their own weight "parasites," and look down our noses at those who ride on others' coat-tails. The science writer Matt Ridley summed it up well:

> [R]eciprocity hangs, like a sword of Damocles, over every human head. He's only asking me to his party so I'll give his book a good review. They've been to dinner twice and never asked us back once. After all I did for him, how could he do that to me? If you do this for me, I promise I'll make it up later. What did I do to deserve that? You owe it to me. Obligation; debt; favour; bargain; contract; exchange; deal ... Our language and our lives are permeated with ideas of reciprocity.[47]

The importance of reciprocity isn't limited to any particular culture. Wherever you find human beings, you find them embedded in networks of reciprocal assistance.[48] Some suggest that this reflects the influence of a universal norm of reciprocity. But although social norms presumably *augment* our reciprocal tendencies, it seems unlikely that they explain them in their entirety. In everyday life, reciprocity is propped up not by abstract normative principles, but by emotions such as gratitude, guilt, sympathy, and anger. All neurologically normal human beings possess these emotions, and it seems highly unlikely that the emotions are inventions of culture – that culture somehow dragged them "into existence by the hair, out of the swamps of nothingness," to borrow a line from Nietzsche.[49] More plausibly, the emotions undergirding reciprocity are birth rights of the human animal – and so, therefore, is the reciprocally altruistic behavior these emotions help to inspire.

Moreover, the cross-cultural commonalities go deeper than the fact that people everywhere exchange favors. People everywhere also seem to be more concerned that non-relatives return favors than that relatives do.[50] The evolutionary anthropologist Raymond Hames discovered, for instance, that among the Ye'kwana of Venezuela – a tribal group living in the Amazon rainforest – people exchange help with relatives and non-relatives alike, but that the exchanges with non-relatives are more balanced and equitable.[51] This suggests that people are more willing to tolerate a lack of strict reciprocity when dealing with relatives. A later study found the same pattern among the Mayans of Belize.[52] Without an evolutionary perspective, this pattern is hard to explain. With it, it's a piece of cake. Altruism channeled to kin can boost inclusive fitness even when there's no rebounding benefit to the altruist, as a result of the fact that kin are more likely than chance to share genes. In contrast, altruism channeled to non-kin can only boost fitness if there *is* a rebounding benefit. As such, reciprocation is more important, evolutionarily speaking, when dealing with non-kin than kin.[53]

Although reciprocal exchange isn't nearly as widespread among the animals as kin altruism, it's not just a human speciality. The most persuasive research on nonhuman reciprocity has focused on vampire bats. These spooky flying mammals make their living by venturing out at night and siphoning blood from other animals, including occasionally humans. Sometimes they succeed; sometimes they don't. But when they don't, they're at risk of starving. And that's when things take an interesting turn. The vampire "haves" – those that succeed in finding blood on a given night – often regurgitate some of their bounty into the mouths of the vampire "have-nots," even when the have-nots are unrelated to them. As the biologists Gerald Wilkinson and Gerald Carter have convincingly demonstrated, the vampire haves are most likely to share blood with individuals who've shared with them in the past. Meanwhile, free-rider bats – bats that don't reciprocate – get frozen out of the game. Bizarre though it might seem, vampire bats seem to be playing their own real-world version of the Tit-for-Tat strategy.[54]

Other animals seem to be doing the same. There's evidence, for instance, of reciprocal exchange in various nonhuman primates, including macaques, baboons, and chimps.[55] There's also evidence of reciprocal exchange in animals as diminutive as rats and mice, and as distantly related to us as birds and perhaps even fish.[56] And not only do other animals reciprocate, the adaptive contours of animal reciprocity are similar to those in humans. First, wherever we find reciprocity, we find defenses against cheating. Ravens, for example, refuse to cooperate with other ravens who failed to cooperate with them in the past.[57] Second, in at least some species, reciprocity is more important when dealing with non-relatives than it is when dealing with kin. Thus, when vervet monkeys interact with non-relatives, they're more likely to help those who groomed them in the recent past; when they interact with *relatives*, on the other hand, the likelihood of helping is unrelated to the recency of grooming.[58] This is strikingly similar to what we see in human beings.

I mentioned earlier that reciprocal altruism is not as well-established as kin selection. Nonetheless, given the evidence available, it's not unreasonable to think that, for humans and nonhumans alike, reciprocity explains much of the altruism we see among non-relatives – much of the altruism that's *not* explained by kin selection. But does it explain it all? Some think it does. In the early days of the field, a common attitude was: "We've got kin selection; we've got reciprocal exchange – that's the problem of altruism wrapped up!" Unfortunately, though, this was probably a case of premature extrapolation. Kinship and reciprocity are important, but they're not the be-all-and-end-all when it comes to explaining altruism. We need to keep on digging.

Altruism as a Peacock's Tail

Another, very different way to tackle the problem of altruism is with sexual selection theory. As we saw in Chapter 2, sexual selection was Darwin's explanation for certain extravagant structures in nature, the best-known being the peacock's tail. According to Darwin, this pretty-but-awkward appendage is essentially a "babe magnet." It announces to the world that "I'm an extremely fit specimen of manhood – so fit, in fact, that I can afford to grow this useless tail. Mate with me and you'll have offspring as fit and as fine as I am, and your sons will be popular with the ladies." Some argue that altruism, and morality in general, send essentially the same message. These traits are human equivalents of the peacock's tail: ways in which men and women show off to attract mates. Just as a good tail is a costly signal of health and good genes, so too is good behavior and a generous spirit.

Geoffrey Miller has been the main cheerleader for this idea.[59] In his book *The Mating Mind*, Miller expounds on the link between moral virtue and sexual success:

> Murder, unkindness, rape, rudeness, failure to help the injured, fraud, racism, war crime, driving on the wrong side of the road, failing to leave a tip in a restaurant, and cheating at sports. What do they have in common? A moral philosopher might say they are all examples of immoral behavior. But they are also things we would not normally brag about on a first date.[60]

And for good reason. Consider Ebenezer Scrooge, the ruthlessly self-interested protagonist of Charles Dickens' *A Christmas Carol*. To some, Scrooge is the ultimate Darwinian machine: a ruthless social Darwinist with a take-no-prisoners, winner-takes-all attitude to life. But Scrooge, remember, was single. Charles Darwin, in contrast, was not; he had a long and happy marriage, and was, by all accounts, a kind and deeply moral man. Here's how his soon-to-be wife Emma described him in a letter to her aunt in 1838, shortly before she and Charles were wed:

> He is the most open, transparent man I ever saw, and every word expresses his real thoughts. He is particularly affectionate and very nice to his father and sisters, and perfectly sweet tempered, and possesses some minor qualities that add particularly to one's happiness, such as not being fastidious, and being humane to animals.[61]

Admittedly, the future Mrs. Darwin was probably in the first flush of romantic infatuation at the time she penned these words. But the testimony of friends and associates throughout his long life confirms that Darwin really was the anti-Scrooge. And importantly, he was also more reproductively prolific than Scrooge: He and Emma had ten children, seven of whom survived to adulthood. Scrooge had none.

Now, this argument does have one minor weakness: the fact that Scrooge didn't actually exist. But real-world evidence converges with the fiction and provides at least tentative support for the sexy-altruism hypothesis. In his massive cross-cultural study of human mate preferences, David Buss discovered that one of the traits that men and women all over the world most desire in a long-term mate is kindness.[62] More recently, Pat Barclay and colleagues found that altruists tend to have more sexual partners than the chronically self-interested, and that altruism is particularly good at elevating *men's* sexual success.[63] It is possible that, in our modern era, there's no longer any link between how nice people are and how many children they end up having. But before the advent of reliable contraception, altruists might not merely have had more sexual partners; they might also have had more offspring. And if that's the case, it raises an intriguing possibility, namely that since the preference for altruism first evolved, we humans have selectively bred ourselves into a nicer, more extravagantly generous species than we would otherwise have been.

The biggest victory for the costly signaling theory to date is in explaining a conspicuous form of altruism found among hunter-gatherers. In most hunter-gatherer groups, men engage in big-game hunting. They go out in small hunting parties in the hope of returning with a bonanza: a giraffe, a zebra, an impala perhaps. Like vampire bats, the hunters sometimes succeed and sometimes fail – actually, most of the time they fail. On the rare occasions they succeed, however, the spoils of the hunt are shared widely throughout the group. The hunters put forward all the effort and take all the risks, but then everyone enjoys the fruits of their labor. In other words, big-game hunting is a form of altruism. Admittedly, the hunters have little say over how the meat is distributed, which has led some anthropologists to argue that meat sharing is really tolerated theft rather than spontaneous gift giving.[64] The problem with this argument, though, is that the hunters don't have to go hunting in the first place. They know they'll be forced to share the meat, and yet they do it anyway. How can we explain this?

Let's work our way through the standard evolutionary psychological theories, and see if any help. Is it kin selection? No. Hunters only rarely manage to channel meat from big game selectively to their children or other relatives. They *do* channel other foraged food to kin and mates – small game and honey, for instance. But big game is usually a public good: Everyone gets a slice of the pie. Is it reciprocal altruism? No. Men who do little in the way of hunting, or who contribute little to the hunt, get just as much meat as the best hunters. This means that they're never able to repay their "debt" (at least not in the same currency). Is it for the good of the group? No. Hunters would secure more calories per bead of sweat expended if they targeted smaller, more easily caught prey. A steady supply of squirrels would represent a more economical use of group resources than the occasional giraffe or impala. And big-game hunting is a risky business. Why go for the large, dangerous bonanza when you could much more easily go for a big pile of squirrels?[65]

The standard theories fall short of the mark. How, then, can we explain big-game hunting? According to the anthropologist Kristen Hawkes, the answer is sexual selection. Big-game hunting is not primarily about calories; it's primarily about showing off. Vanquishing animals that could easily trample you underfoot and kill you is a pretty convincing display of strength, skill, and bravery. It's a cultural peacock's tail – a hard-to-fake signal of good condition and good genes. Men do it not to provision their wives or children but as a way of attracting mates and allies. That, at any rate, is its primary function, especially among younger men.[66] Consistent with Hawkes' conjecture, better hunters command more respect, and end up with younger wives, more sexual partners, and ultimately more offspring. This has been documented in a range of small-scale societies, including the Hadza of East Africa, the Aché of South America, and the Meriam people of Torres Strait in Australia.[67]

As always, the claim is not that when men hunt – or that when people donate to charity or help out at a nursing home – they're only ever trying to get laid. Male proboscis monkeys don't think "You know what? I think I'll grow a giant nose; females will love that." The nose just grows. Likewise, men and women don't think "You know what? I think I'll grow an altruistic disposition and a moral sense. Members of my preferred sex will find that irresistible." Like the monkey's nose, the altruistic disposition and the moral sense just grow, and the fact that these traits evolved to attract mates doesn't imply that that's people's secret goal whenever they act altruistically. As Miller observed, "A trait shaped by sexual selection does not have to include a little copy of its function inside, in the form of a conscious or subconscious sexual motivation."[68] Thus, to say that altruism is a product of sexual selection is not to say that it's necessarily motivated by the desire for sex. Most of the time, it's not; people are just nice. But the reason we're nice is that, throughout the course of our evolution, kind, loyal, and generous people attracted more romantic interest than their cruel, traitorous, or selfish peers, and therefore did better in the reproductive sweepstakes.

As with kin altruism and reciprocity, precedents for sexually selected altruism can be found in other animals. The best example is the Arabian babbler, a little brown bird that nests in large communal groups in arid areas of the Middle East. According to the Israeli biologist Amotz Zahavi, male babblers not only help unrelated individuals, they compete for the privilege of doing so. For instance, males compete to act as sentinels for the group, putting their lives on the line to keep a lookout for hawks and other predators. Similarly, males compete to provide each other food, sometimes even forcing it on unwilling recipients. What drives these odd antics? Zahavi's contention is that the hyper-helpfulness is a way of saying "I am awesome. Indeed, I'm so awesome that not only can I feed and protect myself, I can feed and protect you too, you inferior creature." In essence, the helpers are signaling their fitness to the rest of the group. The payoff for their efforts is attracting mates and deterring same-sex competitors.[69] Not everyone agrees with Zahavi, but the theory is certainly plausible, not just for Arabian babblers but for us.

Needless to say, not all human altruism is a product of sexual selection. Just for a start, parents don't care for their offspring merely to impress bystanders or outshine same-sex rivals. But the smart money says that sexual selection joins kin selection and reciprocity as one of the selective forces shaping our altruistic bent. The next question is: Should we add group selection to that list?

The Groupish Animal

The theories we've discussed thus far, especially kin selection and reciprocal altruism, have been hugely influential in evolutionary psychology. But not everyone is satisfied. Some argue that they don't go far enough. They explain most of the kindness and cooperation we see in other animals, but they barely scratch the surface when it comes to human kindness and human cooperation. People are routinely good to one another in ways that strain bread-and-butter sociobiological theories such as kin selection, reciprocal altruism, and sexual selection or costly signaling.

I'll give you three examples. The first is what we might call *extreme altruism*. The archetypal instance of extreme altruism is the soldier who dives onto a live grenade, giving up his life to save his brothers in arms. Most of the time, these so-called brothers aren't really brothers at all; they're non-relatives. That means it can't be kin altruism. And after you've dived onto a grenade, there's little chance that your benefactors will be able to repay the favor or that you'll suddenly be more magnetically attractive to potential sexual partners. That rules out reciprocal altruism and sexual selection, respectively. Compounding the mystery, extreme altruism among non-kin is virtually unheard of in other animals. It's almost exclusively a human activity. What makes us so different?

Extreme altruism tends to grab the spotlight, but other, more common forms of altruism are equally difficult to explain in terms of kinship, reciprocity, or costly signaling. In big cities, people routinely give directions or spare change to individuals they're not related to (so it can't be kin selection) and who they're unlikely ever to see again (so it can't be reciprocal altruism or costly signaling). Similarly, in group studies in the lab, people often give up potential prize money in order to punish individuals who fail to contribute to a public good, even when this couldn't benefit them personally but could only benefit the group.[70] Some call this mode of helping *strong reciprocity*.[71] It's an unfortunate name, given that, by definition, the help in question is never reciprocated. But it's also a theoretical conundrum. Why do we engage in such behavior?

Finally, human beings are abnormally cooperative and "groupish."[72] We live together relatively peacefully in groups of hundreds, thousands, or even millions of people. And not only do we usually manage to avoid killing each other, we cooperate and collaborate in a myriad of ways. We build pyramids and skyscrapers; we recycle and pay our taxes; we buy and sell and teach and learn. Moreover, we don't just teach our children that "blo r" or that "one good deed deserves another." We teach them to give without expectation of reward and to place the interests of the group above their own selfish interests. These teachings don't just fall on deaf ears; often the lesson sticks. Throughout history, millions have fought and died for their religion, their country, or some other sacred cause. Once again, the bread-and-butter sociobiological theories have a hard time squaring this circle. Why are human beings so much more cooperative and group-oriented than our current crop of theories predict that we ought to be?

For some, the answer seems obvious. Extreme altruism, strong reciprocity, and hypercooperativeness weren't selected because they were good for the individuals engaging in them; they were selected because they were good for the groups those individuals belonged to. In technical terms, these things are products of group selection, rather than individual selection.[73] As I mentioned in Chapter 2, group selection is a controversial topic in evolutionary biology – a topic like politics or religion that polite people avoid raising at dinner parties. The main sticking point for the idea is the question of whether selection among groups could ever override selection among individuals. Most biologists accept that it could happen in principle, given the right conditions, but doubt that these conditions have ever been met in reality. Still, some argue that they have, and that human beings have been strongly shaped by group selection. (Note that I'll limit the discussion here to *genetic* group selection. Cultural group selection is another story, which we'll pick up in Chapter 6.)

What are the necessary conditions for group selection to work? Many have been discussed over the years, but two in particular stand out. The first is that fitness differences between individuals within groups are small: Most group members have, say, three offspring and few have more or fewer. The second is that average differences in fitness *between* groups are large: Individuals in one group have an average of three offspring, for instance, whereas individuals in a neighboring group have an average of six. When both these conditions are met – when you have small fitness differences within groups but large fitness differences between them – selection is concentrated at the level of the group. Traits that benefit the group can be selected, even if they don't always benefit the individuals possessing them.[74]

According to group selectionists, that's precisely what happened in the human lineage: Between-group selection got the upper hand over within-group selection, and this led to the evolution of our hypercooperative nature. In case this sounds like an overly rosy picture of our species, note that, as far as the group selectionists are concerned, it's not all good news. Although group selection favors brotherly and sisterly love *within* groups, it also often favors antagonism toward *other* groups. This is especially likely if the context of group selection in our species was warfare between neighboring tribes, as a number of scholars contend.[75] Contrary to popular opinion, then, modern group selection is not a feel-good theory of universal love and understanding. As the moral psychologist Jonathan Haidt notes, group selection didn't make us saints; it made us team players, and team players are not always saintly.[76] Indeed, as the blood-soaked pages of history make all too clear, they're often anything but. Nonetheless, if we want to explain our hypercooperativeness and our willingness to put the interests of the group ahead of our own, we need to invoke group selection.

Or so say the group selectionists. But are they right? For various reasons, I suspect that they're not. To start with, there are other ways to explain our hyperprosocial tendencies, and to my mind, these are much more plausible. Let's start with extreme altruism: the fact that people sometimes throw themselves on grenades or make comparably heroic sacrifices to help unrelated comrades. It is true, as earlier noted, that this kind of behavior is not found in other animals. But the fact is that it's barely found in our own species, either. It's an extremely rare phenomenon; that's why it grabs our attention so forcefully on the rare occasions it happens. And we need to ask: Is the tendency to sacrifice one's life to help unrelated group members something that was *specifically selected* in our lineage? Almost certainly it isn't. The emotions that underpin extreme altruism – love for and loyalty to one's friends and allies, for instance – are very likely products of natural selection. Extreme altruism itself, however, is presumably just an occasional maladaptive by-product of these emotions. Most of the time, the emotions lead to less extreme, more evolutionarily adaptive behavior.

A similar argument can be made regarding strong reciprocity: the fact that people sometimes help unrelated strangers even when reciprocation and reputation enhancement are out of the question. As with extreme altruism, this type of behavior probably isn't an evolved tendency in itself; we presumably don't have a specific evolved motivation to help strangers we'll never meet again when no one else is watching. Instead, we have a relatively generalized willingness to help others, which usually produces adaptive behavior but sometimes doesn't. The deeper lesson here is that, when we're testing the water for an adaptive explanation, we shouldn't ask "How might this particular behavior in this particular circumstance enhance this person's fitness?" What we should ask is "How might the emotions and motivations underpinning this behavior enhance people's fitness in general?"

In fact, even the latter question needs tweaking. What we really need to ask is "How might the emotions and motivations underpinning this behavior have enhanced people's fitness in the kinds of environments in which these emotions and motivations evolved?" And when we ask *that* question, strong reciprocity starts to look even less anomalous. For most of our evolutionary history, we only rarely encountered people we'd never encounter again. Most of those who came into our social orbit were potential reciprocators, and most interactions had the potential to tarnish or enhance our reputations. In such a world, a relatively indiscriminate willingness to help other people might usually have been adaptive. These days, we live in a very different social world: a world in which we pass more strangers as we walk to work in the morning than most of our ancestors encountered in a lifetime. But our altruistic dispositions are not calibrated for this evolutionarily novel world. They're calibrated for the small-scale societies in which most of our evolution took place. The upshot is that people in modern societies may sometimes help one another even when reciprocity and reputational effects are out of the question – not because we evolved specifically to do this, but just because we didn't evolve not to. In short, strong reciprocity may be a product of evolutionary mismatch.[77] Trying to figure out why selection favored it may be as misguided as trying to figure out why selection favored obesity or postpartum depression (see Chapter 2).

What about our hypercooperativeness and groupishness: the fact that we cooperate to a degree unprecedented in the living world, and sometimes put the group above self or family? Well, there's no denying that we're groupish; the evidence is all around us. It's important to note, however, that groupishness is not necessarily a product of group selection. We need to draw a clear, bright line between traits, on the one hand, and the selection processes that produce them, on the other. Groupishness is a trait; group selection is a selection process. And although it seems self-evident to many that groupishness must be a product of group selection, it's possible – likely, in fact – that groupishness was selected because it advantaged groupish individuals, not because it advantaged the group as a whole. Consider pack hunting in whales, wolves, and lions. This is about as good an example of animal groupishness as you're likely to find outside the realm of the eusocial insects. As we've already seen, though, this groupish behavior is advantageous to each individual engaging in it, because each ends up with more food in its belly than it would if it acted alone. The moral of the story is that groupishness can advantage individual group members, and thus cannot be treated as direct evidence of group selection.

Certainly, human beings cooperate on a larger scale than any other animal than ants – and unlike ants, our cooperative endeavors commonly involve non-relatives. But is group selection the best explanation for these facts? Probably not. First, cooperation is, by definition, good for all involved, which renders group selection somewhat superfluous. Humans have clearly taken the cooperative trend much further than any other animal, in large part because language, intelligence, and cumulative cultural evolution permit more sophisticated forms of social organization than we find elsewhere in nature. However, the adaptive logic of our cooperation plausibly remains the same. Second, large-scale cooperation is often *explicitly* motivated by considerations of self-interest. When people work together en masse to build pyramids or fight wars, it's generally not because they have a desire to sacrifice their time or their lives for the sake of the group. Most often, it's because someone is either forcing them to do it or paying them a wage. In other words, they're acting on immediate self-interest, not on some conception of what's good for the group as a whole.

Once we start looking at things through this prism, something quickly becomes apparent. Humans don't actually look like we're strongly group-selected.[78] If we were, we wouldn't need tax collectors; people would be just as willing to pay their taxes as to pay for a holiday. We wouldn't need conscription; people would be just as happy to serve as cannon fodder as to eat, drink, and be merry. And we wouldn't need to worry about free riders trying to exploit us; as long as they belonged to our group, and as long as their gains outweighed our losses, we would welcome the opportunity to be exploited. It's fair to say, I think, that we're not that kind of species. Our group-interested behavior often has to be incentivized. This makes little sense on the assumption that we're strongly group selected, but perfect sense on the assumption that we're not. To be clear, this isn't to deny that we can choose to act for the good of the group, the species, or even life as a whole; obviously, we can. When we do, though, we're using our brains to accomplish something they didn't specifically evolve to accomplish – and as a result, we may have to fight harder against ourselves than we would if we just wanted to benefit ourselves and our nearest and dearest.

It is true, as the group selectionists claim, that in every society, people extol the virtues of group-wide altruism, and deploy a broad array of tactics to get people to subordinate their own interests to those of the group. But this is dubious evidence for group-selected altruism. We don't need preachers or teachers to tell us to be self-interested, after all. And if we did, this wouldn't be evidence that we're innately selfish; it would be evidence that we're not. In the same way, the fact that societies use all manner of tricks and tactics to get us to put the group before ourselves doesn't prove that we have an evolved tendency to prioritize the group. On the contrary, the fact that the tricks and tactics are necessary is evidence for the opposite. (Note that this hints at a very different explanation for our extreme prosocial tendencies: culture. I'll say more about this soon.)

So, we're not quite as groupish as we sometimes seem to be, and to the extent that we are groupish, other theories do a better job of explaining it. But the group selectionists still have one more card up their sleeves. This is the idea that group selection and inclusive fitness are *mathematically equivalent*.[79] The two approaches are simply different ways of construing the same phenomenon in nature. Just as you can describe the same glass of water as either half full or half empty, and be right either way, so too you can describe the same instance of evolutionary change in terms of inclusive fitness or group selection. There's still some debate about whether the theories really are mathematically equivalent.[80] However, if we assume for the sake of argument that they are, the implication is clear: Wherever an inclusive fitness explanation is valid, so too is an explanation framed in terms of group selection.

The equivalence argument throws a monkey wrench into the earlier criticisms of group selection. But there are several things to notice. First of all, many advocates of group selection do seem to view this selection process as something more than just another way of construing inclusive fitness.[81] Anyone who argues, for example, that group selection explains instances of altruism that inclusive fitness cannot is tacitly assuming that group selection is something over and above selection for inclusive fitness. That being the case, they can't then play the mathematical equivalence card. Conversely, if group selectionists *do* play the mathematical equivalence card, they must drop the claim that group selection goes beyond inclusive fitness, and accept that inclusive fitness provides a perfectly adequate explanation for our altruistic instincts.

Of course, this cuts both ways; the inclusive fitness advocate would equally have to accept that group selection provides a perfectly adequate explanation for human altruism. But then the question becomes: Which is the more useful way to frame things? Is it *ever* more useful to frame things in terms of group selection? I suggest that it's probably not. For one thing, doing so often involves an awkward and counterintuitive redefinition of what a group is. For example, reframing reciprocal altruism in terms of group selection involves treating any pair of reciprocators as a group – a two-person group that does better than two-person *non*-reciprocating groups.[82] This is a far cry from the traditional definition of a group as a relatively stable, geographically delineated cluster of individuals. To my mind, the fact that we need to adopt such a contrived and artificial definition to make group selection work is a real stumbling block for this approach. Inclusive fitness offers a more intuitive and tractable way to conceptualize things, even if it's not the only way. Perhaps for that reason, inclusive fitness has been much more empirically fertile than group selection, leading to discoveries as diverse and unexpected as parent–offspring conflict, maternal–fetal conflict, genomic imprinting, selfish genetic elements, and deviations from a 50:50 sex ratio in eusocial insects.[83]

To sum up, the situation is this: Either group selection is something over and above inclusive fitness, in which case it's probably had little effect on us, *or* it's equivalent to inclusive fitness, in which case it's just an awkward way of reframing insights that are better captured in inclusive fitness terms. Either way, group selection doesn't seem to be a particularly fruitful approach to understanding our hypertrophied altruistic tendencies.

Is Altruism Just Disguised Self-Interest?

Group selection seems to be a dead end. To the extent that people are more altruistic than the standard sociobiological theories predict, this is probably a product of culture, not of group-level selection. Still, it's reasonable to think that even if the standard sociobiological theories don't solve the problem of altruism in its entirety, they do take a giant bite out of it. Before moving on to new pastures, let's pause to consider what implications this might have for our understanding of human goodness. Specifically, let's ask whether the evolutionary origins of our altruistic streak imply that all altruism is, in reality, nothing but disguised self-interest.

This is certainly the conclusion that many have drawn, including many evolutionary biologists. The unstable genius George Price, for instance, whose work on the problem of altruism underpins modern group selection theory, concluded that an evolutionary approach to altruism pulls the rug out from under the concept. The behavior we consider altruistic, reasoned Price, is really just a strategy which genes employ to propagate themselves, and therefore isn't really altruistic at all. This conclusion helped to push Price into a deep depression that ultimately led to his suicide. Now obviously this was an extreme reaction. But less extreme reactions in a similar vein are commonplace. Randy Nesse, who helped found the field of evolutionary medicine, couldn't sleep for a week after he first read Richard Dawkins' book *The Selfish Gene*. The gene's-eye view appeared to him to imply that altruism is an illusion, and that at bottom we're nothing but selfish and self-serving automatons.[84] George C. Williams, one of the architects of the gene's-eye view and co-founder with Nesse of evolutionary medicine, had a similar reaction to the science he helped to found. He was haunted till the end of his days by the apparently dismal moral implications of altruism's lowly origin.

But does an evolutionary account of altruism actually have such dismal implications? There's an irony in the fact that so many people worry that, if the selfish-gene theory is true, human beings might ultimately be selfish or amoral; it seems to suggest, after all, that at least some of us ain't so bad. If we were, why would we lose sleep over the idea? In my view, the claim that evolutionary psychology debunks altruism is deeply misguided. To appreciate why, however, we need to distinguish two possible interpretations of the claim.

The first is that natural selection has made us selfish at the psychological level. From this perspective, we're never genuinely motivated to improve other people's lot as an end in itself; we only ever do it as a means to some other selfish end. Perhaps we do it to make others like us, perhaps to put them in our debt, or perhaps to impress bystanders. Whatever the reason, though, there's always a hidden agenda lurking somewhere in the background, if not at a conscious level then certainly at an unconscious one. The second interpretation is that, even if we really are altruistic at the psychological level – even if we sincerely desire to help other people – this tendency still evolved as a way to pass on the genes giving rise to it, and thus it's still ultimately selfish. It's selfish, we might say, *at the genetic level*.

Let's start with the first interpretation: the idea that all apparent altruism has an ulterior motive, and therefore that human minds are every bit as selfish as the selfish genes that give rise to them. It's certainly possible to concoct an argument to that effect. Consider our moral codes and maxims. At first blush, these look like evidence *against* the claim that humans are ultimately selfish; after all, our moral codes generally recommend unselfish behavior. On closer inspection, however, many moral memes turn out to involve subtle appeals to self-interest. For example, people say "You've got to give in order to receive." Sounds nice. But the unspoken message is that we should give *because it will benefit us*. Similarly, people say "He who lives by the sword shall die by the sword" (Matthew 26:52). Nice sentiment. Again, though, the unstated assumption is that we should avoid living by the sword *for our own good*. Once you open this can of worms, it's unnervingly easy to find more examples – and then more and more. "What you give is *what you get*." "You're only cheating *yourself*." "Judge not, *lest ye be judged*" (Matthew 7:1). "I served and I saw that *service is joy*" (Mother Teresa). In each case, the implicit rationale for goodness to others is gain for oneself. Even in the moral domain, self-interest rules the roost.

One might respond that, sure, this is true in the Western world, with its individualism, capitalism, and worship of the almighty dollar. But it's not intrinsic to human nature. We know it's not (so the argument might go) because we know that in collectivist cultures and socialist states, people are more concerned about the interests of the group than about their own selfish interests. This difference between the West and the rest permeates every aspect of the respective cultures – including their proverbs. According to a Western proverb, for instance, the squeaky wheel gets the grease: Don't be afraid to stand up, stand out, and fight for your rights! According to a Japanese proverb, on the other hand, the nail that sticks up gets pounded down: Don't make waves or make a fuss; blend in and be loyal to the group! Doesn't this prove that not all cultures prize the pursuit of self-interest, and that our deep-seated selfishness is merely a cultural habit?

Well, it does suggest that there's a cultural difference. But the difference is in the type of behavior recommended; it's not that one proverb appeals to self-interest while the other does not. *Both* appeal to self-interest. The Japanese proverb doesn't warn that the nail that stands up will diminish the good of the group; it warns that *the nail itself* will be disadvantaged – it'll get hit on the head with a hammer. That's bad for the nail, not for the group as a whole. Other collectivist proverbs prove the same point. Confucius, for example, observed that "He who wishes to secure the good of others, *has already secured his own*." Similarly, Laozi advised that "*If you would take*, you must first give." In short, collectivist values, like individualist values, are often maintained by appeals to self-interest. The intended counterexample just adds fuel to the fire.

There's a case to be made, then, that altruism and morality are only skin deep, and that beneath the shiny, altruistic surface of our public personas lurks hidden depths of selfishness. But although no one should deny that we sometimes nurse secret selfish motivations, it's easy to take this argument too far. Some cynical types insist that, deep down, people only really care about themselves. I don't buy it. A family friend from the UK was driving home from work one day when a nine-year-old boy ran out in front of his car. He slammed on the brakes but it was too late: He hit the kid. As he jumped out of the car to help, the boy's sister started screaming "You killed my brother! You killed my brother!" And he had. The family friend wasn't related to the boy and he didn't know the boy's family. It wasn't his fault so he didn't go to prison and wasn't punished in any tangible way. But it was still the worst thing that ever happened to him. He couldn't bear to stay in the UK, so immigrated to New Zealand and started his life again there. Most people would have a similar reaction. Look at your own reaction just to reading about it! When it really comes down to it, it simply isn't plausible that we only ever care about ourselves, and that when we act otherwise, that's all we're ever doing: acting. The occasional psychopath notwithstanding, people aren't so bad.

There's another tack that the "everyone is ultimately selfish" crowd often takes at this point. Sure, they argue, we do sometimes care about others, and this does sometimes move us to help them. But even then, helping is selfish because helping makes us feel good. We know that it will, and that's why we do it. In the final analysis, we do it for ourselves, not for the other person. This argument rings true for a lot of people, but it really shouldn't. Aside from anything else, it implicitly assumes a very strange definition of altruism, namely that altruism involves doing something good for someone else *and* not enjoying it. This last stipulation seems muddled to me; surely an altruistic person is one who *does* enjoy doing good for other people. A selfish person, in contrast, is one who doesn't enjoy doing good for others as an end in itself, but who only ever does it as a means to some other end that he does enjoy. There's a meaningful distinction between these two types of people (or rather, these two motives for helping), and presumably this is one of the distinctions that the words *altruistic* and *selfish* were originally designed to mark. If we insist that people who enjoy being altruistic aren't actually altruistic after all, we'll no longer have the verbal tools to draw that distinction, and we'll need new words to do the same job. Why reconfigure our natural language like this? What's the justification?

I think we can safely dismiss the idea that evolution made us entirely selfish at the level of mind and motives. But we're not out of the woods just yet. We've still got to deal with a second version of the selfishness thesis: the idea that altruism is a product of *genetic* self-interest. Certainly, argue proponents of this position, we do genuinely care about other people, some of the time at least. Nonetheless, the *capacity* to care about other people evolved purely as a way to pass on the genes giving rise to it. This implies that altruism isn't really about benefiting others; it's really about benefiting those genes. People might *look* like they're behaving altruistically, even to themselves. In fact, though, they're pursuing their own genetic interests. They're doing what's best for them *at the genetic level.*

I've tried to make this position sound as reasonable as I can, but I have to say that I don't find it very reasonable at all. It is true that, if altruism has an evolutionary origin, then the genes giving rise to it must have benefited from that behavior. By definition, those genes are selfish, to use Richard Dawkins' metaphor. The question is, though, whether it makes sense to say that, if the genes are selfish, then the organism possessing those genes must therefore be selfish as well. In other words, does the fact that the genes benefit metaphorically imply that the organism benefits literally in any meaningful way? And the first thing to say about *this* is that anyone who thinks that it does has surreptitiously strayed away from the original biological definition of altruism. The original definition, you'll recall, is that altruism is any act that benefits the recipient at a cost to the actor, with the benefits and costs measured in terms of reproductive success. Given that definition – which underpins more than half-a-century's worth of research in evolutionary biology – there's no question that altruism can evolve. We've seen many examples in this chapter, including helping at the nest, alarm calling, and cooperative brood care in insects. If we're now going to say that these apparently altruistic acts are actually selfish, we've tacitly abandoned the original definition of altruism – again, that altruism is an act that benefits the recipient at a cost to the actor – and pivoted to a new one, namely that altruism is an act that benefits the recipient at a cost to the actor *and* which doesn't benefit the genes giving rise to it. This rather awkward definition, or something very like it, is implicit in the idea that altruism is really genetic self-interest.

Now, there's no one-true meaning of any word, and thus we can define altruism this way if we want to. But why would we want to? Why would we want to ditch the established definition and adopt this new one instead? As far as I can tell, the entire rationale for doing so rests on the intuition that, if the genes benefit from a trait, the organism benefits too. As soon as we start scrutinizing this intuition, however, it crumbles like dust in our hands. In effect, it identifies the individual with its genes – or more precisely, the *interests* of the individual with the "interests" of its genes. And therein lies the problem. The individual is *not* its genes; the genes are merely one factor among many shaping the individual. The fallaciousness of treating the individual and its genes as isomorphic becomes clear when we apply the same treatment to some of the non-genetic causes of behavior – culture, for instance. Cultures can foster altruism in their members, and it's easy to imagine that this altruism could benefit the cultures that foster it. Would we want to say, though, that when someone acts altruistically as a result of their cultural training, they're acting selfishly at the cultural level? No. In this case, the error is obvious. There *is* no cultural level; there's the individual, and there's the culture, and the one may benefit while the other is disadvantaged. But the same applies to genes. As we've seen, genes can foster altruism, and this altruism can benefit the genes giving rise to it. Should we say, then, that when someone acts on an evolved inclination toward altruism, they're acting selfishly at the genetic level? No. There *is* no genetic level – not in any literal sense. There's the individual, and there are the genes, and the one may benefit while the other is disadvantaged. The notion of selfishness-at-the-genetic-level is merely an abstraction designed to capture the fact that the genes unpinning altruism can proliferate as a result of that altruism. However, it's a misleading abstraction if it tempts to us to think that the individual has a genuine vested interest in the frequency of those genes in the gene pool. To assume otherwise is to mistake the metaphorical interests of the genes for the literal interests of the individuals possessing them.

However we slice it, an evolutionary explanation for our prosocial instincts doesn't suck the altruism out of altruism, and doesn't imply that, deep down, we're all writhing snake pits of selfishness. Human beings often act altruistically, both in the everyday sense of the word and in the technical sense used by evolutionary biologists. Obviously, we're not angels. But we're not Machiavellian monsters, either. As is so often the case, humans are inbetweeners, falling somewhere in the gray area between the two extremes.

It's time to take stock. I hope I've persuaded you, over the course of the preceding chapters, that evolutionary psychology is an indispensable tool for understanding the human mind and behavior. It illuminates a wide range of phenomena, from sex differences and romantic relationships to parental care and altruism. As I see it, there's no longer any reasonable doubt that the basic constituents of the human mind – including various emotions, motivations, and social propensities – were crafted and honed by natural selection, and that their ultimate function is to enhance inclusive fitness. There is, however, still a serious question about *how much* of what we think, do, and feel will yield to this approach. Some have high hopes. As Claire El Mouden and colleagues write:

> How would an alien biologist that was capable of observing our behaviours and reading our minds sum up what humans maximize? … the alien would conclude that humans, along with all other organisms, are best described as striving to maximize their inclusive fitness over their lifetimes.[85]

If it were any other animal than us, I'd be inclined to agree. But human beings aren't like other animals. As the philosopher Daniel Dennett points out, our species is utterly unique among the flora and fauna of the Earth in that we pursue goals and ends other than simply having more grandchildren, nieces, and nephews than our neighbors. And that's because, although we are gene vehicles, we're not *just* gene vehicles. As soon as we evolved the capacity for culture, we became vehicles for something else as well. We became vehicles for memes.

6

The Cultural Animal

If an alien anthropologist had crashed in Europe in 1943, our intrepid observer would surely have called us the "fierce" people. But if, say, the landing site was Woodstock, New York, or San Francisco, California, in 1968, ET would likely have labeled us the "erotic" people. Local and historical context matters.

—*Michael Shermer (2004), p. 84*

A Virus in the Letterbox

Guess what? Earlier today, I received some good news. I don't want to say too much about it in case I jinx it, but... well, just between you and me, I got a letter today that was sent to me for good luck. The letter has been around the world nine times already. I'll receive good luck within four days of getting it – providing I send it on. This is no joke. I need to send copies to people I think need good luck. I should *not* send money, because fate has no price. I must not keep the letter. It must leave my hands within 96 hours. I'm usually skeptical about this kind of thing, but an RAF officer followed the instructions and received $70,000. I could do with $70,000. Meanwhile, Joe Elliot received $40,000, but then lost it because he didn't send the letter on and thereby broke the chain. And in the Philippines, Gene Welch lost his wife just six days after receiving the letter. He'd failed to circulate it. Before his wife died, though, Welch received $7,755,000. That probably eased the blow a little.

The letter originated in Venezuela and was written by Saul Anthony Decroup, a South American missionary. Since the letter must make a tour around the world, I need to make twenty copies of it and send them to friends and associates. Then, after a few days, I'll get a pleasant surprise. This is true *even if I'm not superstitious.* Constanton Dias received the letter in 1953. He asked his secretary to make twenty copies of it and send them out. A few days later, he won two million dollars in the lottery. Carle Dadditt, an office employee, received the letter but forgot it had to leave his hands within 96 hours. He lost his job. Later, after finding the letter again, he mailed out the requisite copies, and a few days after that, he got a better job. Dalan Fairchild received the letter and, not believing, threw it away. Nine days later *he died.* I'd be crazy to ignore this. It really seems to work.

OK, obviously it doesn't work. As you probably know, I'm referencing that infamous, irritating cultural virus, the chain letter. The above paragraphs are based on a letter that circulated in the latter decades of the twentieth century, and which was the subject of some fascinating research by the physicist Charles Bennett. Bennett and several colleagues collected thirty-three different versions of the letter, and quickly established that some were "parents" of other, slightly different versions. With a bit of jiggling, they were able to arrange the letters in a family tree, much like your own family tree or the family tree of the primates. They could then trace how the letters had changed and mutated over time.

In one lineage, the RAF officer won $470,000, rather than the paltry $70,000 I mentioned above. (I could *really* do with $470,000.) In another lineage, Gene Welch originally received $1,755 before his wife's death, but then in an offspring version, he received $50,000 in a lottery which his wife won before she died. In both cases, the bigger prize might have inspired a few more people to suspend their disbelief and send out copies of the letter. In yet another lineage, the letter began with a Biblical passage: "Trust in the Lord with all your heart and he will light the way." This presumably aided the letter's transmission in some circles, but impeded it in others.

I haven't spelt it out yet, but the parallels with biological evolution should be clear. Bennett put it like this:

> These letters have passed from host to host, mutating and evolving. Like a gene, their average length is about 2,000 characters. Like a potent virus, the letter threatens to kill you and induces you to pass it on to your "friends and associates" – some variation of this letter has probably reached millions of people. Like an inheritable trait, it promises benefits for you and the people you pass it on to. Like genomes, chain letters undergo natural selection and sometimes parts even get transferred between coexisting "species."[1]

For present purposes, the most important part of this passage is the suggestion that chain letters may be subject to natural selection. As we know, natural selection creates entities that look like they were intelligently designed to perform certain tasks – entities, in other words, that have functions. What might the function of the chain letters be? It seems to me that their sole function is to pass themselves on. Like a gene or a virus, that's ultimately what they're designed to do. This is partly because that's what the people who wrote or modified the letters were aiming at. But it's also partly because any letters that failed to get themselves copied and passed on, despite their authors' best efforts, were automatically culled from the chain letter population. By the same token, any that were particularly persuasive, even if their authors had no idea why, multiplied at a faster rate. In short, the letters were shaped in part by blind, mindless selection. The successful ones weren't selected because they were good for the individuals sending or receiving them, or because they were good for society as a whole. They were selected because they were good for themselves, in the simple sense that they got themselves copied at a faster rate than less appealing variants.

Chain letters are not the only cultural artifacts that meet this description. Hoax computer virus warnings are another. These are emails advising people that a vicious computer virus is doing the rounds. A typical one might say something like: "If you receive an attachment titled Budweiser Frogs Screensaver, do not open it! If you do, it will steal your credit card details, wipe your hard drive, and blow up your computer. Forward this email to all your friends!" Needless to say, there's no such attachment and no such computer virus. The only virus is the warning itself, which spreads by exploiting people's gullibility and good nature. Hoax virus warnings and chain letters are "selfish" cultural elements, directly comparable to Richard Dawkins' selfish genes. The whole reason for their existence is to pass themselves on.

Initially, we might assume that these selfish cultural artifacts are very different from most of human culture, and especially from those aspects of culture that so puzzled the alien scientist: our mating and child-rearing patterns, our languages, our religions, our arts and entertainment. Most cultural products are much more useful to us than chain letters or hoax emails. They persist because they help us survive, make us happy, or in some way benefit the groups to which we belong. Is it possible, though, that the same forces that shape chain letters and hoax emails also shape culture in general? Is it possible, in other words, that the ultimate function of any cultural entity or institution, whether it benefits us or not, is to replicate itself?

This is certainly not an intuitive way to think about culture. But it's the central claim of a school of thought called *memetics*.[2] Memetics is based on the concept of the "meme," which Dawkins introduced in his 1976 bestseller, *The Selfish Gene*. Roughly speaking, a meme is an idea. Less roughly, it's a unit of culture. When people give examples of memes, they tend to fixate on quirky pieces of contagious culture: jokes, recipes, writing "clean me" in the dust on a dirty car, catchphrases, catchy tunes, ways of tying a knot, ways of tying *the* knot, viral Internet images, and so on. But the meme concept is much broader than that. It embraces anything and everything that can be passed on via social learning, from the trivial (facial expressions, mannerisms) to the historically momentous (agricultural techniques, political and religious ideologies). Just as genetics is the science of genes, memetics is the science of memes; it is, in essence, a collection of memes about memes (a collection of meta-memes, one might say). The core idea in memetics is that, like genes, memes are subject to natural selection, and that selection favors "selfish" memes – memes that, through accident or design, are good at getting themselves replicated and keeping themselves in circulation in the culture. This applies not only to chain letters and hoax virus warnings, but right across the board.

Memetics isn't the only approach to cultural evolution; there are many others.[3] Some stress the role of innate learning biases in determining which bits of culture people adopt. Some stress the role of cultural "attractors": ideas and practices that people naturally gravitate toward as a result of the native structure of the human mind. And some stress the idea that genes and culture coevolve – that new culture creates novel selection pressures, which leads to biological evolution, which in turn makes possible more new culture. What all these approaches have in common, and what distinguishes them from the meme approach, is that they focus almost entirely on how cultural products benefit the individual, the group, or both. Memetics, in contrast, focuses on how cultural products benefit the cultural products themselves.

In this final chapter, I'm going to argue for a memetic approach to cultural evolution. I'm *not* going to argue that memetics is the one true answer and that the other approaches are false. My argument instead will be that memetics can serve as a unifying framework for understanding cultural evolution: one that allows us to synthesize the insights of the other approaches, but which also goes beyond them. I should point out that, although the chapter is organized around Dawkins' concept of the selfish meme, most of the ideas I discuss come from other thinkers, such as Robert Boyd, Peter Richerson, and Joseph Henrich (who are not, incidentally, fans of the memetic approach). This is not to downplay Dawkins' contribution, however. It was Dawkins who first proposed that cultural elements are selected only in as much as that they benefit themselves, and this insight, I suggest, is the lynchpin that unites the various approaches to cultural evolution. Before we get to that, though, some readers may be skeptical of the notion that natural selection operates in the realm of culture. Thus, to kick off the discussion, let's look at some examples.

Selection beyond Biology

Certain things in the world look like they were intelligently designed. Most of these things – almost all of them, in fact – fall into just two categories: human artifacts and biological structures. Beer bottles and cactus stems look like they were designed to store liquid; computers and brains look like they were designed to process information. Where does this design come from? Opinions differ. According to Creationists, the design we see in the biological world comes from God or an Intelligent Designer, whereas the design we see in our artifacts comes from us. Darwinians disagree with the first part, holding instead that the design in life comes from the mindless process of natural selection. But they usually agree with the second. Many Darwinians like to point out that some things in the universe really are intelligently designed, exactly as the Creationists claim... just not living things. Human artifacts – flyswatters, microwaves, cheese sticks – are the only genuine products of intelligent design on the planet. The designer, though, is not a deity; it's us: an evolved animal. Ironically, then, intelligent design is ultimately a product of natural selection.

That's what many Darwinians say, but it's not quite correct. First of all, there is actually a smidgeon of intelligent design in the biological world. But again, this doesn't come from a deity; it comes from human beings selectively breeding new kinds of dogs, sweeter fruit, brighter flowers, fatter cows, and more productive hens. These organisms are, literally and to an important degree, human-made objects. Second, and more germane to the present topic, human artifacts and institutions are not pure products of intelligent design. Intelligence is clearly important – if it wasn't, it would never have evolved. But intelligence is just one source of the design found in the cultural arena. Blind selection plays a critical role as well, easing the workload placed on human intelligence and ingenuity. Here are six examples of selection beyond biology.

1. Breton Boats

The first example concerns the fishing boats used by Breton fisherman in the Île de Groix. Where did these boats come from? At first glance, it looks like a no-brainer: If anything's a product of intelligent design, it's a boat. On closer inspection, though, it turns out it really *is* a no-brainer... or at least a partial-brainer, in the sense that human brains played a more modest role in crafting the boats than we normally assume. This possibility was first mooted by the French philosopher Émile-Auguste Chartier (aka Alain), who in 1908, took a Darwinian hatchet to the common sense view. "Every boat," he observed,

> is copied from another boat ... Let's reason as follows in the manner of Darwin. It is clear that a very badly made boat will end up at the bottom after one or two voyages, and thus never be copied ... One could then say, with complete rigor, that it is the sea herself who fashions the boats, choosing those which function and destroying the others.[4]

If a boat returns, the boat makers may copy it. If it doesn't, they definitely won't. The boats that are most likely to be copied are therefore those that survive the longest. As Daniel Dennett points out, no one needs to know why these particular boats survive.[5] To make a good boat, you don't need to understand what makes a boat good; you only need to be able to copy another boat. How do you know you're copying a good boat? Well, you don't need to know, because the sea automatically culls the not-good ones from the boat population. Meanwhile, any especially good boats get copied at a faster rate. Over time, this process of culling and copying fashions more and more seaworthy boats.

Now *maybe* each and every step in the gradual evolution of the boat was a product of intelligent design: of a thousand forgotten boat makers figuring out a thousand different ways to make their boats more seaworthy. But maybe not. Maybe many steps along the path were simply fortuitous accidents, which were automatically preserved and propagated. To the extent that this is so, the design evident in Breton boats comes from blind, mindless selection, rather than the machinations of intelligent minds.

2. Conditioned Behavior

The idea that boats are only partly a product of intelligent design is surprising. Perhaps more surprising, though, is the idea that the same may be true of our behavior. According to the behaviorist psychologist B. F. Skinner, most of our behavior is learned, and learning is literally an evolutionary process. Skinner didn't mean learning in the everyday sense of the term, but rather learning via "operant conditioning." Operant conditioning happens when an animal's behavior is shaped by its immediate consequences. If your dog brings you your slippers and you give it some beef jerky, the dog will be more likely to bring you your slippers again in the future. In behaviorist lingo, the behavior has been *reinforced*. If, on the other hand, your dog chews up your favorite slippers and you shout at it, the dog will be less likely to chew up the next pair. The behavior has been *punished*. According to behaviorists, reinforcement and punishment are among the most important forces shaping the behavior of all animals, from rats to pigeons to people.

In his 1953 book *Science and Human Behavior*, Skinner discussed the deep parallels between operant conditioning and Darwinian evolution. "In certain respects," wrote Skinner, "operant reinforcement resembles the natural selection of evolutionary theory. Just as genetic characteristics which arise as mutations are selected or discarded by their consequences, so novel forms of behavior are selected or discarded through reinforcement."[6] Biological evolution involves selection by death or reproductive failure. Operant conditioning, in contrast, involves selection by behavioral consequences. Behavior that has positive consequences is selected: It increases in frequency within the organism's "population" of behaviors. Behavior that has negative consequences is selected against: It decreases in frequency and may ultimately go extinct. Through reinforcement and punishment, random behavioral "mutations" are slowly sculpted into complex patterns of adaptive behavior. Thus, just as large-scale evolutionary change is produced by the slow accretion of imperceptibly small modifications, so too is voluntary behavior.

As elsewhere, selection in the realm of behavior can produce an illusion of intelligent design. If your friends Jack and Jill have different senses of humor, you may end up cracking different kinds of jokes with Jack than the ones you crack with Jill. To an outside observer, it might appear that you're deliberately tailoring your jokes to your audience to get as many laughs as possible. And maybe you are... but you're not necessarily. Instead, your joke-telling habits may have been slowly sculpted by Jack and Jill's reactions to your wisecracks, without your awareness that this was even happening. In other words, your behavior may be a product of blind natural selection, rather than intelligent design. Of course, it'd be easy to push this point too far. Much of human behavior really is intelligently designed: We make and act on deliberate plans about what we're going to do. It's likely, however, that a larger fraction of our behavior is shaped by blind selection than we usually imagine is so.

Needless to say, the capacity for operant conditioning is itself a product of natural selection – we and other animals evolved to learn in this way. This has a fascinating implication, namely that natural selection set up a learning system that itself works according to the principle of natural selection. In a certain sense, natural selection discovered itself.

3. Language

Not only is individual behavior shaped by natural selection, so too are our cultural tools. Language is a prime example. Evolutionary psychologists have made a strong case that human beings are, by nature, a talking animal, as opposed to being merely a clever animal that invented talking just as we invented agriculture, the Internet, and reality TV.[7] The ability to learn language in the early years of life appears to be part of our biological endowment. The specific languages we learn, however, are obviously not. No child is born with a working knowledge of Mandarin, English, or Urdu, and if people are never exposed to these languages, they don't just spontaneously develop them. Where, then, do our languages come from? An alien anthropologist might initially wonder whether each language is the handiwork of an intelligent designer: a linguistic genius who sat down and invented the language for the rest of us. The alien would soon figure out, though, that with a handful of exceptions (such as Esperanto and Klingon), languages are not products of intelligent design at all. Like monkeys and meerkats, languages have no authors. They're products of an evolutionary process. Darwin made this very point in a remarkable passage in his book *The Descent of Man*:

> Dominant languages and dialects spread widely, and lead to the gradual extinction of other tongues … Distinct languages may be crossed or blended together. We see variability in every tongue, and new words are continually cropping up; but as there is a limit to the powers of the memory, single words, like whole languages, gradually become extinct. As Max Muller has well remarked: – "A struggle for life is constantly going on amongst the words and grammatical forms in each language. The better, the shorter, the easier forms

are constantly gaining the upper hand, and they owe their success to their own inherent virtue." To these more important causes of the survival of certain words, mere novelty and fashion may be added; for there is in the mind of man a strong love for slight changes in all things. The survival or preservation of certain favoured words in the struggle for existence is natural selection.[8]

In short, just as natural selection culls less seaworthy boats from the population of boats, and less rewarded behaviors from the individual's population of behaviors, so too selection culls less learnable, less useful words from each language's population of words. The inevitable result is that languages evolve to be more learnable and more useful over time.[9] And small changes in a language soon add up. In biology, a single species divided into isolated populations can evolve, first, into distinct breeds or subspecies, and ultimately into entirely new species. The same is true of languages: A single language spoken by two isolated populations can evolve, first, into distinct dialects (the linguistic equivalent of breeds or subspecies) and ultimately into new languages (the linguistic equivalent of species).[10] As Darwin astutely noted, the analogy between biological evolution and language evolution is surprisingly close.

4. Teddy Bears

Language evolution is at least as old as we are, but other arenas for cultural evolution have a much more recent pedigree. One of the most important is the capitalist marketplace.[11] Just as species compete for limited space in the local environment, so too products – from books to fizzy drinks to exercise equipment – "compete" for limited space on supermarket shelves and bestseller lists. This competition may foster the evolution of products exquisitely designed to suck money out of people's pockets and bank accounts – designed, in other words, to sell. Importantly, business people don't necessarily need to know why some products sell better than others. They only need to copy the ones that do. To the extent that that's what happens, the design we find in our products comes from blind selection rather than intelligent design.

An example concerns the cultural evolution of the teddy bear. The first teddy bears went on sale in the early twentieth century. In those days, they had long snouts and long, thin limbs. They were pretty ugly. As the century wore on, however, teddy bears became progressively cuter.[12] Their snouts receded, leaving them with cute flat faces. Their foreheads grew larger. Their limbs grew shorter and chubbier. In a word, they became more *neotenous* or baby-like. More and more they came to resemble the innate *Kindchenschema* we discussed in Chapter 4. Today's teddy bears are, in effect, the answer to the question: What do you get when you cross a human baby with a bear? And they raise a new question of their own: How do we explain the evolutionary trajectory of this enduringly popular children's toy?

Here's one possibility. Successful teddy bear makers were sensitive to market trends, and generally copied the designs that sold best last season. But they didn't copy them exactly. Some happened to push their designs a little further toward our evolved standards of cuteness; some happened to push them a little further away. The former sold better, and the better-selling bears became the baseline for the next season. Little by little, teddy bears drifted toward neoteny. Did successful bear makers know that increasing neoteny was the secret of their success? I doubt it. After all, if they did know, they could have just jumped straight to the most neotenous models. The trend toward neoteny is something that people only noticed after the fact. While it was happening, bear makers simply made more of whatever sold. In a sense, consumers redesigned the teddy bear with their aggregate preferences and purchasing decisions. If your parents bought you a teddy bear, they were contributing to the evolution of this beloved children's toy. Generalizing the point, any time you or anyone else buys anything, you're helping to guide the evolution of culture.

5. Businesses

It's not just products that are subject to natural selection in a market economy. So too are businesses. Roughly speaking, businesses that turn a profit survive, whereas those that don't, don't: They go down the plughole. As a result, businesses evolve into profit-making machines. The economist Milton Friedman took this view. According to Friedman, the typical business acts as if it's trying to maximize its profits, not only because that's what its directors are aiming at, but also because businesses that *don't* make a good profit are ruthlessly culled from the business population.[13] Obviously, this is a simplified account; it ignores government bailouts and other factors that, for better or worse, prevent the unimpeded action of natural selection in the economic realm. Nonetheless, to the extent that the market is free, and that profit is the company's motive, selection may create design in the structure of businesses, even without the intervention of a top-down, intelligent designer.[14]

My friend and colleague, Philip Tucker, came up with a good example of this. Organizational psychologists tell us that you can describe any organization in terms of a number of basic parameters. These include the extent to which workers have narrowly defined roles vs. broad roles with many responsibilities, and the extent to which power is centralized within the organization vs. distributed more diffusely.[15] The settings of these parameters are, in effect, the DNA of the organization. Crucially, different designs for organizations are more suitable for different industries. How do businesses end up with the right design for their economic niche? As Phil pointed out, there's no need to suppose that anyone worked out which parameters would work best for any company – no need, in other words, to imagine that companies are intelligently designed. Instead, we need only assume that any organizations whose specs were suboptimal were selected against, and that only those that were suitably constituted survived to propagate their kind.

And what's true of businesses may be true of institutions in general. As the anthropologist E. B. Tylor put it in 1871, "The institutions which can best hold their own in the world gradually supersede the less fit ones, and ... this incessant conflict determines the general resultant course of culture."[16] In this way, we're swept along by currents we barely understand, and over which we have little control.

6. Science

The marketplace isn't the only modern institution that functions as an arena for cultural selection. Another is the institution of science. The best-known voice on this topic is the Austrian-born philosopher, Karl Popper. According to Popper, the growth of scientific knowledge is literally an evolutionary process within the realm of ideas.[17] Science, he observed, involves the two key ingredients of Darwinian evolution: variation and selection. Scientists propose competing theories about the nature of the universe (variation), and then cull those theories that don't match what they see in the world and in the lab (selection). Thus, Aristotelian physics was replaced by Newtonian physics, which in turn was replaced by Einsteinian relativity. At each step, the earlier theory was displaced from the population of scientific ideas when a new, upstart theory matched our observations more closely. In effect, the scientific method establishes a struggle for existence among theories, which results ultimately in the survival of the fittest theories: those that best explain the facts. The end result is that our theories evolve – step by slow step – toward greater and greater accuracy. One minute, we're talking about God creating the world and all life in six days; the next, we're talking about the Big Bang and evolution by natural selection, and trying to figure out how to reconcile relativity theory with quantum mechanics. Just as a corporation functions as a giant profit-making machine, so too science, at its best, functions as a giant truth-producing machine – a distributed cognitive system designed to generate true statements about the world.

Hopefully, these examples have softened you up to the idea that natural selection can operate outside the realm of biology. It's important to stress, however, that in many ways, cultural evolution is very different from its biological cousin. For multicellular organisms like ourselves, genes are transmitted almost exclusively from parent to offspring. Unless you're a bacterium, you can't pick up genes from your friends. But you *can* pick up *memes* from your friends, or from your children, or from anybody else. Indeed, in a world of books, TV, and the Internet, you can even pick up memes from the dead. Another difference between biological evolution and cultural is that, although there's not much sharing of genes between different multicellular species, there's often a lot of sharing of *memes* across different "species" of culture. English, for instance, is classed as a Germanic language, but much of its vocabulary comes from Romance languages such as French and Latin. In a sense, modern English is a linguistic mongrel: a hybrid of Germanic and Romance languages. In the same sense, rock music is a hybrid of a variety of musical genres, including blues and country, gospel and jazz. And neither rock music nor English is exceptional. The sharing of code across species, though rare in multicellular organisms, is the norm in the cultural sphere.

But while the differences between biological and cultural evolution are significant, the similarities are often quite startling. Arguably, the most startling is the fact that, like plants and animals, cultural products can often be arrayed on a family tree.[18] Languages are the best example. Like biological species, languages descend from other languages, and – at least within the major language groupings – any two languages share a common ancestor if you trace it back far enough. Also like biological species, any two languages may be more or less closely related. By way of comparison, consider the great apes. Humans are more closely related to chimps than to orangutans, because the last common ancestor of humans and chimps lived more recently than the last common ancestor of humans and orangs. Similarly, French is more closely related to Spanish than to Hindi, because the last common ancestor of French and Spanish (i.e., Latin) existed more recently than the last common ancestor of French and Hindi (i.e., proto-Indo-European or PIE). Although language is the clearest example, family trees have been assembled for various other aspects of culture, including armor, Creationist tactics for thwarting the teaching of evolution in schools, fairy tales, food taboos, helmets, musical styles, myths, Native American projectile points, Neolithic pottery decorations, Persian rugs, skateboards, violins, world religions – and even plagiarized school work.[19] The fact that it's possible to place these cultural entities on family trees is important, because family trees are a telltale sign that the entities in question arose through a process of descent with modification... in other words, that they evolved.

The Ratchet Effect

Once upon a time, anthropologists thought that human beings were the only cultural animals. Culture, they argued, was our defining feature, the trait that walled us off from other creatures and insulated us from the sullying hand of natural selection. Since the middle of the twentieth century, however, animal researchers have slowly chipped away at this agreeable delusion. One of the biggest early blows came in the 1960s, when the primatologist Jane Goodall discovered that female chimpanzees sometimes use sticks to help them extract termites from termite mounds: a tasty snack.[20] The chimps have to learn this trick, but they don't each work it out for themselves. Instead, they learn it from watching others. Termite fishing is a cultural tradition.

Since this early breakthrough, the list of animal traditions has grown exponentially. In an important paper in the journal *Nature*, Andrew Whiten and a team of chimp experts identified thirty-nine unambiguous chimpanzee cultural practices – thirty-nine chimp memes, if you like.[21] These included using rocks to crack open nuts, using leaves as napkins, and greeting one another with a kind of overhead high five. Each of these behaviors is found in some groups but not others, even when the groups in question inhabit relevantly similar environments: when they all have access to both nuts and rocks, for instance. This tells us that the behaviors are not just products of independent, non-social learning. Furthermore, chimps from groups that don't normally perform the behaviors seem perfectly capable of learning them. This tells us that the behaviors are not just instincts found in some groups but not others. In short, the thirty-nine behaviors documented by Whiten and colleagues appear to be genuine, culturally transmitted traditions. Chimps are cultural animals.

And it's not just chimps. Carel van Schaik and colleagues conducted a similar investigation with orangutans, ultimately identifying twenty-four orangutan memes.[22] These included using leaves as gloves to protect their hands while foraging, blowing raspberries at each other while settling down to sleep for the night, and "snag riding": riding dead trees as they fall to the ground and then grabbing onto nearby branches at the last minute. Apparently, they engage in this last activity for the same reason we would – for fun. As with the thirty-nine chimp memes, these twenty-four orang memes seem to be bona fide cultural traditions.

But cultural traditions aren't just limited to apes. They're also found in some monkeys. Capuchins, for example, have a range of traditions, including cracking open nuts with rocks, putting other capuchin's fingers up their noses, and taking turns using one another's fingers to poke themselves in the eye (strange animals).[23] Cultural traditions aren't even limited to primates. As the biologist Hal Whitehead notes, millions of years before human culture appeared, there was culture in the ocean. The bearers of this underwater culture were whales. Different whale clans have their own dialects, social patterns, and foraging techniques. Again, these seem to be genuine traditions, rather than products of individual learning or genetic differences between groups.[24] And cultural traditions aren't even restricted to big-brained mammals. Some songbirds learn their song by listening to neighboring members of their own species, which leads to the emergence of local song "dialects." Similarly, birds in a number of species learn foraging techniques from one another, which leads to the emergence of local foraging traditions. The implication is unavoidable: Some nonhuman animals must be admitted to the culture club. We can no longer maintain a human-only membership.

Still, without disparaging other animals' cultural credentials, it has to be admitted that our nonhuman brethren simply aren't in the same league as us. There's a huge, gaping chasm between chimps fishing for termites and humans blasting themselves to the moon, sequencing their own genomes, and building hadron colliders to draw back the curtain on the fundamental constituents of reality. What bridges this great divide?

The standard answer is intelligence: Humans are smarter than other animals, and smartness can move mountains. As the futurist Eliezer Yudkowsky once put it, almost everything around us most of the time, other than the air we breathe, is the product of our big, clever brains. "The rise of human intelligence," he observed, "reshaped the Earth. The land sprouted skyscrapers and cities, planes flew through the skies, footprints appeared on the Moon."[25]

But intelligence only takes us so far. Although we're clearly smarter than chimps, we're nowhere near as much smarter as an alien scientist might surmise from comparing our cultural achievements (e.g., putting people on the moon) with theirs (e.g., using rocks to crack open nuts or sticks to fish for termites). As individuals, we're probably closer to the nut-cracking, termite-fishing end of the spectrum. If you doubt this, imagine being marooned in the jungle with no relevant knowledge. At a push, you might figure out how to obtain a little food. But you'd never figure out how to build a rocket ship and get to the moon. The rocket ship is just an extreme example of a much more general truth. Clever though we are, no individual could design even something as simple as a kayak from scratch, let alone a jetpack or a democratic nation-state.[26]

So where did these things come from? To answer that question, we need to look to something bigger than our brains – bigger, in fact, than any of us. A growing contingent of scholars argues that our "superpower" as a species is not our intelligence; it's our *collective* intelligence and capacity for *cumulative culture*: our ability to stockpile knowledge and pass it down from generation to generation, tinkering with it and improving it over time.[27] Biological evolution can give rise to the eye. But cumulative cultural evolution can give rise to entities every bit as complex as the eye: airplanes and smart phones, legal systems and the Internet.

What makes these cumulative cultural achievements possible for us when they're not possible for chimps or howler monkeys? No one knows for sure, but there are plenty of suggestions. These include language, theory of mind (that is, the ability to construe people as having thoughts, desires, and intentions), mental time travel, hypercooperativeness, practice, teaching, trade, shared attention, joint attention, imitation, true imitation, and overimitation – or, of course, some serendipitously fertile blend of these talents. Whatever it is, though, it makes all the difference in the world. It renders our cultural achievements utterly unique among the animals.

Certainly, some nonhuman animals may have some degree of cumulative culture. Consider nut cracking in chimpanzees. Chimps don't just hit their nuts with a rock. They use one rock as an anvil and a second as a hammer, and they strike the nut hard enough to crack the shell but not so hard that it crushes the kernel inside. It's difficult to imagine that a solitary chimp Einstein invented this entire procedure in a single saltational leap, after which it spread in its entirety from chimp brain to chimp brain, down through the chimp generations. As a number of experts have argued, it's more plausible that the practice emerged in cumulative, bite-size steps.[28]

Still, even if cumulative culture is not completely unique to our species, it's hard to deny that we take this trick a thousand times further than any other creature. And it's easy to see why it's useful. Cumulative culture is the ultimate time-saver. Because of cumulative culture, we don't need to reinvent the wheel each generation – quite literally. We don't each need to invent calculus because Newton and Leibniz did it for us. We don't each need to have our own Eureka moments to understand fluid dynamics; we don't each need to have an apple fall on our head to understand gravity; and we don't each need to dream of a snake eating its tail to understand the structure of the benzene molecule. All we need is a library card, a good teacher, or access to the Internet. We can then download into our brains some of the achieved knowledge of the species. This subsequently becomes the starting point for the next round of innovation.

The developmental psychologist Michael Tomasello dubbed this progression *the cultural ratchet*.[29] Extended across time, the cumulative effects of the ratcheting process are astonishing. I remember once watching a David Attenborough documentary in which an orangutan rowed a boat down a river. At first, it struck me as anomalous: Here was this animal skillfully piloting a vehicle it could never have invented itself. But then it occurred to me that human beings are in exactly the same boat, metaphorically speaking. In even the simplest human societies, people use tools they could never have invented themselves. And in our modern age, we routinely use technologies so complex that most people don't have the slightest clue how they work. It's as if we've appropriated the technology of an advanced race of aliens after they mysteriously vanished – except that the aliens were never really here. All of it came from us.

Cumulative culture doesn't just gift our species technology that none of us could have invented; it literally makes us smarter. The products of cumulative culture include not only our physical tools but also a well-stocked armory of cognitive tools: ideas and algorithms which we stamp into the gooey gray matter of our brains and which radically enhance our powers. These include, first and foremost, the words and phrases of the languages we speak. Each word and each phrase is a handy tool for thinking – a prosthetic aid to cognition, as Daniel Dennett puts it. Other life-altering cognitive tools include probability theory, cost–benefit analysis, time management, financial planning, and counting to ten when we're angry. These tools are a lot like smart phone apps. The more apps you put on your phone, the more your phone can do. Likewise, the more cultural tools you download into your brain, the more *you* can do. Unlike phone apps, however, most cultural apps have no identifiable author. Instead, they evolved, slowly but surely, as they passed through the minds of millions of people over hundreds or even thousands of years.[30]

And you don't have to be an incurable optimist to see that the tools get better with time. Consider the Roman number system. This cognitive apparatus is perfectly good for certain purposes, including measurement and record keeping. But as the biologist David Krakauer points out, it's not particularly good for calculation. There's no simple algorithm for dividing C by IV, for instance, or multiplying X by MCMLX. Europeans used the Roman system for 1,500 years. This meant that, for all of that time, they were unable to multiply or divide. They were physiologically capable of it, of course; they just hadn't installed the appropriate cultural software in their brains. These days, we use the Indian-Arabic system, which makes calculation much easier. It literally makes us smarter.[31]

Cumulative culture makes us smarter in another way as well: It frees us from the limitations imposed by the anatomy of our brains, furnishing us with knowledge far beyond the reach of any isolated individual. If you were to make a list of every person who's ever contributed in any way to the vast storehouse of our knowledge, and then add up every hour they devoted to making their contribution, you'd have a rough-and-ready estimate of the number of hours it would take for one individual to single-handedly assemble all the knowledge we now possess. What kind of time period are we looking at? Probably hundreds of thousands of years, and maybe even millions. This means that, by learning about science and getting a good education, we become as knowledgeable as a person who spent thousands or millions of years thinking and exploring the world. Bertrand Russell once quipped that "The average man's opinions are much less foolish than they would be if he thought for himself."[32] He had a point, but it doesn't just apply to the average person. It applies as well to the geniuses among us, all of whom build on the earlier cognitive achievements of the species. Newton is about as good an example of a genius as we might ever expect to find: a genius among geniuses, I'd argue. But even Newton was unable to comprehend the idea that matter warps space and slows time – not because of any constitutional incapacity, but just because he lived before Einstein. Einstein, for his part, couldn't have done what he did if he hadn't been able to build on the work of Newton and Newton's intellectual descendants. Matt Ridley captured the general point nicely in a discussion of the causes of economic growth: "I cannot hope to match [Adam] Smith's genius as an individual, but I have one great advantage over him – I can read his book."[33]

On our own, we're not particularly smart – certainly not smart enough to unravel the mysteries of the universe or put boot prints on the moon. We're smarter than chimps and bonobos, sure, but the gap between us and them isn't as great as we think. It's a river rather than a valley. Yet as a result of our ability to acquire knowledge distilled from thousands of years' worth of thinking, each of us can understand the universe to a degree unmatched by even our closest animal kin. As a result of cumulative culture, we have ideas in our heads that are orders of magnitude smarter than we are. As a result of cumulative culture, we have knowledge and technology it would take a single individual millions of years to create, if a single individual could create it at all. And as a result of cumulative culture, we're surrounded by machines and technology whose inner workings we don't understand and could never hope to understand. Humans are chimpanzees reciting Shakespeare – dunces with the technology of geniuses.

Often, though, these humbling facts are obscured from our vision. We routinely ascribe our species' cultural achievements to lone-wolf geniuses – super-bright freaks of nature who invented science and technology for the rest of us. This tendency is so pervasive it even has a name: *the Myth of the Heroic Inventor*. It's a myth because most ideas and most technologies come about not through the Eureka moments of solitary geniuses but through the hard slog of large armies of individuals, each making – at best – a tiny step or two forward.[34] As the historian of science Joseph Needham put it, "No single man was the father of the steam engine; no single civilization either."[35] In the same way, no single individual was the originator of evolutionary theory. People attribute the theory to Darwin, but the truth is that it's not really his. It's the product of the efforts of thousands of men and women working over several centuries. Nonetheless, friends of the theory and enemies alike want to attribute it to the great man.

Do I contradict myself by calling Darwin great? I don't think so. Some people – Darwin among them – plainly take larger steps forward than the rest of us. But even then, we need to remember that new ideas are rarely drawn from whole cloth. They come instead from the recombination of old ideas – from ideas having sex, as Matt Ridley puts it. The concept of natural selection, for example, involved combining Malthus's idea of the struggle in nature with the idea of selective breeding. This led Darwin to the insight that, because nature kills most of its children, it functions as a giant animal breeder. As important as this insight was, it's essentially just a remix. And so is most of culture. The birth of new technology, for instance, usually involves recombining existing elements in novel ways.[36] As L. T. C. Rolt observed, "The motor car was sired by the bicycle out of the horse carriage." Similarly, as Ridley reports in his book

The Rational Optimist, the Internet was born from the marriage of the computer and the phone, and the camera pill was born of a conversation between a gastroenterologist and a guided missile designer.[37] Perhaps the fairest summary of the situation is that most of our cultural achievements come not from super-bright freaks, but from cumulative culture, aided and abetted by some reasonably bright semi-freaks. In this way, our culture becomes smarter than we are.

Adaptive Culture

What is originality? Undetected plagiarism.
            —*William Ralph Inge (1929), p. 227*

If we accept, as I think we must, that natural selection operates on culture, and that cultural evolution is cumulative, then we should expect to find design without a designer in the realm of culture, just as we do in the biological realm. Our next question, then, is this: What are our cultural products designed to do? In Chapter 2, we ran through a series of hypotheses about what organisms are designed to do, ultimately concluding that they're gene machines – organic machines designed to pass on the genes giving rise to them. Let's now take the same hypothesis-testing approach into the cultural sphere.

A natural first thought for any Darwinian is that cultural products are designed to do exactly the same thing as biological products, namely to help their owners to pass on their genes. And it's easy to frame an argument to that effect. The capacity for culture is presumably a biological adaptation, just as thumbs and noses and sex organs are.[38] Natural selection could only have favored this capacity if it enhanced our ancestors' inclusive fitness (or more precisely, the replicative success of the genes giving rise to it). This implies that, generally speaking, our cultural products must enhance our inclusive fitness. Here, then, is our first hypothesis. Let's call it *the adaptive-culture hypothesis*.

> **Hypothesis #1:** Cultural evolution is all about the genetic fitness of cultural animals. Our cultural products – our memes – are tools and tactics for enhancing inclusive fitness. They're designed primarily to boost our survival and reproductive success, or to increase our ability to bestow benefits on our relatives. Culture is a genetic survival strategy of the human animal.

Based on this kind of view, cultural evolutionists have examined the psychological mechanisms that make culture possible, and the inclusive fitness advantages these mechanisms might have conferred. One of the most important mechanisms, it turns out, is the ability to copy each other. According to Rob Boyd and Pete Richerson, that's our species' Clever Trick: stealing ideas from one another's brains. Our Clever Trick is plagiarism! Elephants have their trunks; giraffes have their long necks – and humans have their ability to plagiarize. We do it compulsively. When a scientist couple raised a baby chimp alongside their own child, they soon discovered – much to their alarm – that the child aped the chimp a lot more than the chimp aped the child.[39] At the other end of the life cycle, people's last words on their deathbeds are apparently often quotations. Darwin wrote that "much of the intelligent work done by man is due to imitation and not to reason," and he was right.[40] When was the last time you invented a new knot, a new tool, or a new piece of furniture? Even if you *could* invent such things for yourself, it wouldn't be the best use of your time – after all, they're already there for the taking. And that's the primary advantage of imitation. We learn through imitation things it would take us forever to learn for ourselves. As such, we've evolved to do a little bit of innovating but a lot of mimicking.

Given the demonstrable benefits of imitation, it's a curious fact that people take such a dim view of it, especially in the individualistic West. Albert Schweitzer once observed that "When people are free to do as they please, they usually imitate each other." This was clearly not meant as a compliment. People quote Schweitzer's statement to thumb their noses at the way we copy each other – which, when you think about it, is rather ironic. We prize uniqueness and originality, but copy each other relentlessly. And we do it because it works. Although "copycat" is a derisive term, copying is a successful strategy. We're virtuoso imitators, who look down our noses at our own virtuosity.

But just as a virtuoso violinist doesn't play just any old tune, human beings don't imitate just anyone or anything. The biologists Charles Lumsden and Edward O. Wilson proposed that people have an evolved tendency to latch onto fitness-enhancing memes and to shun fitness-limiting ones.[41] In this way, as they famously remarked, our genes hold our culture on a leash. Other scholars, such as Rob Boyd, Pete Richerson, Joe Henrich, and Richard McElreath, have helped to flesh out exactly how our genes might do this.[42] In their view, humans have a suite of inherited learning biases, which strongly steer us toward adaptive memes. The biases fall into two main camps: *content biases* and *context biases*.

The content biases relate to the subject matter of our memes. People seem to be especially attentive to evolutionarily relevant stimuli, including food, fire, dangerous animals, violence, disease, kinship, sex, infidelity, babies, friendship, free riders, status, and group membership. Like our wariness of snakes and heights, our interest in these stimuli and circumstances is plausibly a product of natural selection. And it serves us well, facilitating the accumulation of knowledge on matters of evolutionary significance.

If it were just about the content of our memes, though, we'd be just as likely to copy the homeless person as the success story, the quack belief as the mainstream view. As a general rule, we're not. Alongside our content biases, humans have various *context* biases – that is, we have a tendency to acquire memes in some circumstances more readily than others. The context biases can be divided into several types. These include, to begin with, *model-based biases*: biases regarding *who* we copy, rather than what. The most important model-based bias is the *prestige bias*. This refers to the tendency to copy people who have prestige and status.[43] Because people's standing in the community and their success in life are partly determined by the memes they've acquired, copying prestigious people often means that we end up with success-promoting memes in our heads. The Protestant work ethic is one possible example. Individuals who acquire this meme increase their chances of accruing wealth and status, which in turn means that others are more likely to imitate them and the meme has a good chance of spreading. And in case you doubt whether the prestige bias is the sort of thing that could evolve, bear in mind that the tendency isn't just found in humans. Like us, chimpanzees imitate high-status individuals much more often than individuals further down the totem pole.[44]

Another model-based bias is the *similarity bias*. This refers to the fact that, rather than apples copying oranges or chalk copying cheese, we tend to copy birds of a feather: people like ourselves. For example, as the social psychologist Albert Bandura long ago observed, children most often copy individuals of the same sex and of a similar age to themselves.[45] This has a clear adaptive logic. If you're still a child, you're better off trying to copy your friends who hunt small animals than your dad who hunts buffalo. Copying your friends allows you to master skills that are realistically within your grasp, building up your skillset at a developmentally appropriate pace. Copying your dad, on the other hand, is likely to get you killed.

So much for the model-based biases. Another category of context biases is the *frequency-dependent biases*. The key here is not *who* holds a given meme but rather how rare or widespread the meme in question is. By far the most important frequency-dependent bias is the *conformist bias*. This refers to people's tendency to adopt the memes of the majority. Some of the most famous and well-replicated research in social psychology shows that human beings are a conforming animal.[46] Adults sometimes stereotype teenagers as sheep who follow the crowd. And they're right – but most adults are sheep as well. Peer pressure dogs us throughout our lives, and genuine trail blazers are few and far between. Even so-called non-conformists typically just conform to alternative, less widespread conventions (a phenomenon recently baptized the *hipster effect*).[47] Certainly, people do sometimes adopt memes precisely because they're *not* common, perhaps in a bid to stand out from the crowd. Henrich and McElreath call this the *rarity bias*.[48] But this tendency is the exception rather than the rule. The rule was best captured by Samuel Butler, who pointed out that "We think as we do, mainly because other people think so."

Although psychologists have done a lot of research on conformity, they've said relatively little about where our conformist streak might originally have come from. As with imitation, we often look down our noses at conformity, despite conforming constantly. For that reason, we might tend to assume that our sheep-like tendencies reflect a fault in our stars, rather than an adaptation. But conformity can be a successful strategy. It's often a good idea to adopt the practices and beliefs of the people around you. For one thing, the people around you aren't dead. If you do what they do – eat what they eat; avoid the dark alleys they avoid – *you* might continue not being dead as well. Mathematical simulations support the general thrust of this argument. Imagine you have to choose between two possible behaviors – two different tactics for catching a mammoth, say. Imagine also that everyone in your group has a 60 percent chance of ending up with the best tactic through individual learning alone. If you were instead to select ten individual learners at random, and adopt whichever tactic most of them had adopted, it turns out you'd have a *75 percent* chance of ending up with the best tactic: better than if you'd tried to work it out for yourself.[49] This means that your best bet would be to conform to the majority, rather than relying on individual learning.

Of course, like everything, conformity is a double-edged sword. When everyone copies the majority, it becomes difficult to dislodge majority memes and difficult for new ideas to get off the ground. As a result, in times of rapid social change, conformity can trap us in patterns of behavior that worked well in the past but are now well beyond their use-by date (a cultural analogue of evolutionary mismatch). In fact, even when there isn't much social change, conformity can trap us in inefficient or pointless practices and saddle us with false beliefs. Still, it's safe to assume that our conformist streak was generally adaptive for our ancestors, and that that's why we've got it today.

It's also safe to assume, I think, that the conformist bias, and all our learning biases, help to guide the evolution of culture. If these biases are biological adaptations (which seems likely), then presumably the memes of any culture will tend to enhance inclusive fitness. Sure enough, many cultural products – from helmets and seatbelts to vaccines and IVF – are indeed aimed at boosting the survival and reproductive success of the people who use them. And clearly it's working: The world's population has never been larger. But the adaptive power of culture is nothing new. Modern humans evolved and overran the Earth in less than 300,000 years, finding our way into every terrestrial nook and cranny, and colonizing every continent other than Antarctica. This was only possible because, whenever we ventured into new territories, our cultures evolved and adapted to fit our changing needs. Sure, *we* did some evolving as well; as we saw in Chapter 2, and as we'll discuss again soon, different human populations are partially adapted to different environments. But culture is the real secret of our success. A Kalahari Bushman could survive perfectly well in the Arctic with Inuit culture, but an Inuit *couldn't* survive in the Arctic without it.

The adaptive-culture hypothesis offers a promising new way to tackle the dilemmas raised in the Alien's Report. Consider morality, religion, and art. Some argue that these things are biological adaptations, and support the claim by pointing to the adaptive benefits of these practices. Others, however, argue that they're clearly not adaptations, but are highly variable cultural products. The adaptive-culture hypothesis suggests a resolution: Morality, religion, and art are cultural products, not adaptations, but they may tend to be adaptive because our culture-making minds are designed specifically to favor adaptive memes. In effect, these cultural phenomena are *acquired adaptations*, and the capacity for culture, at bottom, is the capacity to invent new adaptations for ourselves.

The adaptive-culture hypothesis is clearly onto something and deserves a place at the table. At the same time, though, it doesn't take us as far as we need to go. True, it's easy enough to reel off a list of cultural innovations that boost fitness. But we can't just score for a hit and not a miss. We also need to factor in culture that *doesn't* boost fitness. Examples are not hard to find. First, many aspects of culture have no impact on fitness, good or bad. There's nothing inherently fitter about driving on the right- rather than the left-hand side of the road, wearing trousers that flare out at the bottom rather than taper in, or letting your trousers hang off your hips rather than pulling them up to your belly button. These aspects of culture are neutral with respect to fitness, except in as much as that they're arbitrarily linked to local norms and fashions that people have to follow if they want to fit in or attract mates. Second, many aspects of culture actively hamper fitness. We discussed various examples in Chapter 2, including smoking, junk food, pornography, and contraception. All these cultural products are extremely common, but they're all also far from adaptive. (Remember we're using the word "adaptive" here not in the everyday sense of the term, but in the strict biological sense. In the everyday sense, most people these days would agree that contraception is adaptive. In the biological sense, however, it's not, because it shuts off people's fertility.)

Not only do we have lots of neutral and maladaptive memes, but our learning biases virtually guarantee that we will. Consider the prestige bias. When people copy a prestigious person, they don't just copy the things that made that person a success. They copy irrelevant things as well, such as clothes, mannerisms, political attitudes, and drug habits. One of my favorite animal anecdotes is about orangutans at a sanctuary in Borneo. One day, a group of orangs slipped into the kitchen and stole a pot. They made a big pile of rocks, placed the pot on the rocks, and then sat in a circle around it, waiting for the pot to give them soup. They were attempting to replicate what they'd seen humans do a hundred times before. This shows just how smart orangutans are – and also just how dumb! But how much smarter are we? Whenever we copy the style rather than the substance of a high-status person – whenever we wear the same shoes as a sporting hero, for instance, or try to emulate a rock star's debauched lifestyle – we're really not so different from the orangs around the pot. Advertisers exploit this chink in our armor when they pay celebrities to endorse their products. There's no good reason to think that skill on the football pitch goes hand-in-hand with skill in choosing the best brand of underwear or the best spray deodorant. But otherwise rational people act as if it does.

To be fair, this isn't especially harmful. However, the tendency to copy prestigious people clearly *can* be. In fact, it can be fatal. A stark example is the phenomenon of *copycat suicide*. When a suicide is splashed across the news, at-risk individuals sometimes copy the victim and take their own lives as well. There's some evidence that the more prestigious the original victim was, the more likely these copycat suicides become.[50] Thus, the prestige bias can sometimes lead to the acquisition of lethal memes.

How do we square neutral and maladaptive culture with the idea that the capacity for culture could only have evolved if it were adaptive? We do it in two easy steps. First, the fact that the capacity for culture was adaptive doesn't mean that every instance of culture must be. To say that *any* trait is adaptive is only ever shorthand for saying that it's adaptive *on average*. This loophole immediately opens up the possibility of maladaptive memes – or what Boyd and Richerson call *rogue cultural variants*. On top of that, as the hedgehog crossing the road in Chapter 2 reminds us, natural selection does not have foresight. As soon as the capacity for culture was in place, there was nothing to stop it spiraling off in non-adaptive or even somewhat maladaptive directions.

And that it seems to have done. To a significant extent, culture has come off the genetic leash and started evolving in its own right, quite independently of any fitness considerations. This helps to explain something that would otherwise be inexplicable: the fact that culture often clashes with our gut instincts. Many moral systems have rules against nepotism, which clash with our natural tendency to favor relatives over non-relatives. We tell each other to turn the other cheek, which clashes with our inclination to lash out at those who wrong us. And we tell each other that all lives have equal value, which clashes with the tendency to view some lives – including those of kin and members of our in-groups – as more equal than others. In these areas and elsewhere, the tension between our cultural inheritance and our gut-level inclinations hints that natural selection in the cultural sphere favors something very different from natural selection in the biological, some of the time at least. What might that something be?

Cultural Group Selection

One answer comes from a theory called *cultural group selection theory*.[51] The group selection we discussed in earlier chapters was genetic group selection. Cultural group selection is similar; it involves competition between groups, and the retention of traits that help those groups to out-compete rivals. The difference is that cultural group selection involves the retention of group-beneficial *memes*, rather than group-beneficial genes. Here's how it could work. Groups differ from each other culturally. They have different norms, different subsistence patterns, different ways of interacting with outsiders. As a result of these differences, some groups do better than others: They last longer; they grow faster; they bud off more daughter groups. Given sufficient time, the cultural practices of the better-performing groups come to predominate in the larger population.

Cultural group selection may sometimes involve direct competition between groups. One group might have superior weapons or military tactics, for instance, and use them to conquer its neighbors. In doing so, the weapons and tactics themselves are selected over less effective variants. But group competition needn't always be so direct. Sometimes, cultural group selection might just involve one group growing faster and crowding out others, a process known as *demographic swamping*. An example is the spread of farming. Throughout history, farming communities reliably expanded faster than forager tribes, and in doing so, slowly pushed the foragers into marginal territories. As this happened, farming itself bubbled up into prominence.[52]

The distinguishing claim of cultural group selection theory – the claim that sets it apart from other theories of cultural evolution – is this: Memes that are good for the group can be selected *even when they're not good for the individuals possessing them*. To take a hypothetical example, cultural group selection could theoretically favor a meme such as "Risk your life to advance the interests of the group." Such a meme would not be good for the individual. However, if it were sufficiently good for the individual's *group*, it could potentially persist nonetheless. From a strict gene's-eye perspective, the meme would be a maladaptive side effect of the flexible learning mechanisms that make culture possible. But that's not to say that it would merely be a random error. Beliefs and institutions shaped by cultural group selection have an evolved function, just as biological adaptations do; it's just not the same one. Their function is to benefit the group. With that idea under our belts, we're ready to formulate our second hypothesis.

Hypothesis #2: Cultural evolution is sometimes about individual fitness, but it's not always. In some cases, it's about the good of the group. Group competition may favor memes that benefit the group, even when they take a toll on the individual meme holder. As such, many cultural products are designed, to a greater or lesser extent, to enhance the survival and success of the groups in which they're found.

To make this concrete, let's consider some possible examples of cultural group selection. The first concerns the institution of monogamous marriage. This is something that an alien scientist would find utterly mysterious. Monogamy itself is no mystery; as we've seen, left to our own devices, the pair bond is our most common mating arrangement. The mystery is that, in some societies, high-status men are forbidden from taking a second wife (or a third or a fourth), even if everyone involved is happy with the arrangement.

Historically, this is a peculiar situation for human beings to be in. As discussed in Chapter 4, in the vast majority of societies, polygyny has been permitted and not especially rare.[53] And that's exactly what evolutionary principles would predict. In many species, polygyny can greatly increase males' reproductive success, so selection favors a male push toward polygyny. At the same time, polygyny tends not to be hugely disadvantageous to females, so there's not much of a female push against it. Moreover, according to the *polygyny threshold model*, polygyny can sometimes even *increase* females' reproductive success.[54] It all depends on how much the males differ from one another in what they can offer the females. If they don't differ much – if most males are fairly healthy, for instance, or if most have good access to resources – then it's probably better for females to have a male's undivided attention than to have to share him with other females. But if males vary a lot, then it can often be better to be the second mate of a healthy or resource-rich male than to be the only mate of an unhealthy or resource-poor one. The prediction, therefore, is that the more that males differ in what they can offer females, the more common polygyny will be.

This turns out to be true of a wide range of species, and until recently, it was true of us as well. In the early civilizations of the world, the gap between the richest men and the poorest was vast, and the minority of men at the top took dozens or even hundreds of mates (recall Ismail the Bloodthirsty). In forager societies, on the other hand, the wealth gap was nowhere near as large, and polygyny was nowhere near as common.[55] When we look back at our past, in other words, the polygyny threshold model describes our species well. But here's the strange thing. In modern Western societies, the gap between the richest and poorest is larger than ever. Admittedly, the wealth is less concentrated in male hands. Nonetheless, our alien scientist would probably predict that modern Western societies would be the most polygynous on Earth. Billionaires like Bill Gates and Jeff Bezos would take thousands of wives – more, even, than the despotic leaders of old. But they don't. Why not?

Well, it's not that human nature is radically different in modern Western societies and that people are never inclined to polygyny – if that were the case, we wouldn't need laws against it. And of course therein lies the answer to our question. The main reason that Westerners never marry polygynously is that we have laws and social norms that forbid polygynous marriage, insisting instead on monogamy. The real issue, then, is how these laws and norms came about. How did the "monogamy-only" meme come to prosper in the West, despite the fact that, to some extent, it clashes with human nature?

Henrich, Boyd, and Richerson chalk it up to cultural group selection.[56] In their estimation, societies that insist on monogamous marriage have a number of advantages over those that permit polygyny. These trace back to the fact that polygynous marriage skews the sex ratio of potential mates. When one man has two mates, another man must have none... and when King Solomon has a *thousand* mates, 999 men must have none. This means that, in societies that permit polygyny, there's always a surplus of young, unmarried men. As we know from Chapter 3, and from life, young unmarried men have a nasty habit of getting up to mischief (or to put it more fairly, a high proportion of society's mischief-makers come from the ranks of young unmarried men, even if most are fairly well behaved). The net effect is that polygynous societies are plagued to a greater extent than monogamous ones by the kinds of social problems associated with this demographic. They have higher crime rates and more violence. They have more child abuse, more fraud, more kidnapping, robbery, and murder. And they're also more likely to go to war. All of this hands a significant advantage to monogamous societies. Because these societies are less besieged by such problems, they tend to become wealthier and more productive. They grow faster, persist for longer, and are more often copied by others. As this happens, the monogamy-only meme becomes more and more widespread. It flourishes because it's good for the groups that adopt it.

How did some groups come to institute norms of monogamous marriage in the first place? One possibility is that the norms were intelligently designed – that far-seeing leaders figured out that banning polygyny would eventually reduce social problems, making their societies more prosperous and peaceful. But you wouldn't want to bet your life on it. Scientists have only recently noticed the link between polygyny and the social problems under discussion. It seems unlikely, therefore, that moral teachers and lawmakers in earlier, less-informed centuries would have been so perspicacious. A more probable scenario is that the norms and laws are products of blind cultural selection. Groups that happened to lean toward monogamous norms, even for reasons unrelated to the reduction of social ills, automatically started doing better and overtaking the rest. Any random stumble in that direction was captured and magnified. In this way, cultural group selection may explain how norms of monogamous marriage established themselves in the West without any mastermind pulling the strings, and without anyone even noticing the benefits.

Cultural group selection may also explain social norms related to the next step on the conveyer belt of human life: children. Specifically, it may shed light on norms regarding how many children one ought to have. In the moral systems of the world's early civilizations – those that survived, at any rate – the prevailing view was that people should have as many children as humanly possible. Of the 641 commands issued in the Hebrew Bible, the very first is "Be fruitful and multiply."[57] Similarly, in the Hadith, Muhammad advised a man to "Marry women who are loving and very prolific, for I shall outnumber the peoples by you."[58] Why did ideas like these go viral?

The answer suggested by cultural group selection theory is that groups adopting norms of the be-fruitful-and-multiply variety tended to outperform those that were more reproductively restrained. The death rate after the turn to agriculture was extremely high, and thus the mere survival of any agricultural group was dependent on maintaining a relatively high birth rate. Furthermore, groups that pumped out children like bees in a hive grew much more rapidly, which gave them three key advantages in competition with other groups. First, larger groups are harder to plunder because they have more eyes to keep a lookout and more fists to mount a defense. Second, larger groups are more likely to prevail in any group-on-group altercation, even when they face better fighters. And third, larger groups are better able to plunder or swallow up smaller, more vulnerable groups. For all these reasons, we'd expect cultural group selection to favor norms of high fertility.[59]

In any case, that's what we'd expect in the early agricultural societies of the world. In other times and places, different fertility norms may come to the fore. Among nomadic hunter-gatherers, having too *many* offspring is just as big a threat to the group as having too few. If a woman gets pregnant immediately after the birth of her last child, she could end up with two fully dependent offspring at once. This is hard for her, of course, but it's also a drag on the group. In such circumstances, any norms or beliefs that helped cut the birth rate down to more manageable levels – norms, for instance, about how soon one should resume sex after giving birth – would stand a good chance of being selected. Notice that it wouldn't matter what those beliefs were, or how closely tethered they were to reality. All that would matter is how the beliefs affected their hosts' behavior. Consider the U'wa people of South America. According to U'wa mythology, twins attract bad luck. That being the case, if a woman gives birth to twins, she has to kill one or both of them immediately.[60] Needless to say, this is pure superstition. It's possible, however, that the superstition was selected because it helped the U'wa to optimize their fertility levels. As far as cultural group selection is concerned, the fact that the practice is rooted in a false view of the world is irrelevant. Indeed, an accurate view on this matter – a belief, in other words, that twins do *not* portend bad luck – might actually work against the interests of the group, as it would undermine their rationale for keeping the birth rate low. In short, it's the behavioral effects of a belief that matter in cultural evolution, not the belief's correspondence with reality.[61]

This insight helps open up an explanation for another major domain of culture: religion. Various scholars, including David Sloan Wilson and Ara Norenzayan, have argued that religions are shaped in large measure by cultural group selection.[62] Just as the function of sharp teeth is to tear apart prey, the function of religions is to knit together collections of individuals into socially cohesive groups. This might help explain some central features of the world's religions. Among the most important, in Norenzayan's view, is the widespread belief in *Big Gods*. Big Gods are powerful supernatural beings which keep track of what we do and punish us if we step out of line. The belief in such beings is not a human universal; it's found only in large-scale societies. And that's because it's only in large-scale societies that we need them. As the evolutionary psychologist Robin Dunbar has argued, for groups of one-to-two-hundred people, our social instincts are sufficient to keep society running smoothly.[63] We know everyone in the group, we can keep tabs on what everyone's up to, and we can generally keep people in line with everyday social tools such as approval and opprobrium. But the moment groups get much bigger than this, people start to encounter strangers and near-strangers at an unnaturally high frequency, and the usual social tools no longer do the job. Social cohesion starts to erode and groups start to break down... unless, that is, new institutions step into the breach and foster group cohesion where our social instincts fall short.

According to Norenzayan, that's where Big Gods come in. Belief in these beings helps to foster the unnaturally high levels of cooperation found in our unnaturally large societies. This works on a simple principle, namely that *watched people are nice people*. Just as we slow down to the speed limit when we spot a police car, so too we tend to follow the social rules when we believe that an all-powerful, all-knowing, and moralistic god is watching our every move. Big Gods aren't the *only* way to keep people on the straight-and-narrow. Some traditional societies opt instead for the doctrine of karma and reincarnation, and modern secular societies opt for police, law courts, and CCTV cameras. Nonetheless, Big Gods are a historically important solution to the problem of taming our more antisocial inclinations – a solution that may have been molded by cultural group selection.

Because people often assume that "for the good of the group" is synonymous with "good," it's worth stressing that the products of cultural group selection are not always pretty. Certainly, cultural group selection can favor cooperation and loyalty within groups. But it can also favor inhumanity and hostility toward other groups. In fact, even within groups, cultural group selection doesn't guarantee brotherly or sisterly love. That's one way to foster group success, but it's not the only one. If lies, tyranny, and violence hold groups together and help them outperform their enemies, then lies, tyranny, and violence will be favored by cultural group selection. The moral of the story is clear: The fact that something is good for the group doesn't imply that it's good for the individuals in that group, or good in the moral sense.

Like genetic group selection, cultural group selection is controversial. There's a lot more theory than evidence in the area, and not everyone's convinced that the idea is sound. Still, the theory offers us another potential lifeline in our quest to answer the alien's questions. Why do our moral systems and religions so often push us to submerge our interests to those of the group, jarring against human nature in the process? Why are people more altruistic and cooperative than we'd expect based on a pure gene-machine model of behavior? And why are people so willing to help members of their own groups, while ganging up against everyone else? The answer to all of these questions may be that, over long ages, cultural group selection favored memes that promoted the interests of the group, even at a cost to the individuals harboring those very memes.[64]

With cultural group selection added to the toolkit, we've now got the resources we need to explain elements of culture that are good for the individual and elements that are good for the group. We already know, though, that this leaves much of culture unexplained. As chain letters and hoax virus warnings remind us, some facets of culture are not good for the individual *or* the group; they're only good for themselves. To incorporate this insight into our conception of culture, we're going to need a fundamental shift in perspective. We're going to need the cultural equivalent of the gene's-eye view of evolution.

The Selfish Meme

> A remarkable parallel, which to my mind has never been noticed, obtains between the facts of social evolution and the mental growth of the race, on the one hand, and of zoological evolution, as expounded by Mr Darwin, on the other.
>
> —*William James (1880), p. 441*

Natural selection is one of the great ideas in science, on a par with universal gravitation and plate tectonics. But how far does the theory extend? Where, in other words, is it applicable? Universal gravitation applies to all matter everywhere in the universe; plate tectonics applies only to the Earth's outer crust. What about natural selection? For many, the answer is obvious: Natural selection applies to hereditary material, and on Earth, that means genes. According to Richard Dawkins, however, this is a mistake. Properly understood, natural selection goes much deeper and has far greater generality. In *The Selfish Gene* and later works, Dawkins outlined a view he calls *universal Darwinism*. The core idea is that natural selection operates not just on genes, but on any *replicator*. As discussed in Chapter 2, a replicator is anything in the world – indeed, anything in the universe – that gets itself copied. As long as the copying process is less than perfect, errors inevitably creep in, giving rise to new variants. Purely by chance, some of those variants will have properties that cause them to be copied at a faster rate than the rest. Barring some disaster, these variants will eventually come to dominate the replicator population. For most of the last four billion years, there has been only one replicator on this planet: the gene. But within the relatively recent past, argues Dawkins, a new replicator has emerged and is sitting right under our noses. Indeed, it's sitting right *behind* our noses, in our brains. Dawkins, a gifted term-coiner, coined the term *meme* for this new replicator.[65]

Memes, as we've seen, are units of culture: ideas, beliefs, practices, and anything else that can be passed on via social learning. In Dawkins' view, memes jump from mind to mind, and are subject to natural selection in much the same way that genes are. To give an example, the concept of the meme is itself a meme, one that's currently passing from this book into your brain. If you like it, you may pass it on to someone else (positive selection).

If you don't, you won't (negative selection). So far, the meme meme has done well in the competition for survival in the marketplace of ideas (or *meme pool*). Boyd and Richerson proposed an alternative label for the units of culture – *cultural variant* – but this never really took off. And as for Wilson and Lumsden's offering – *culturgen* – that sunk like a lead balloon.[66] The meme meme proved to be a much catchier meme than any of its rivals. In fact, it's done so well that it's made it into the *Oxford English Dictionary*.

What made the idea so successful? I suspect that if Dawkins had intended the term simply as a label for the building blocks of culture, the meme meme might long since have perished. But there's much more to memes than that. The central and distinguishing claim of memetics is that, just as natural selection in the biological realm favors selfish genes, so too natural selection in the realm of culture favors selfish *memes*: memes that, through their effects on the people who hold them, act as if their one goal in life is to survive and propagate themselves in the culture. To paraphrase William Hamilton's 1963 statement on selfish genes, the ultimate criterion which determines whether a meme will spread is not whether it benefits us or our groups, but whether it benefits the meme itself.

Two examples will illustrate the point. The first is apple pie and ice cream. The apple-pie-and-ice-cream meme has prospered in human societies because it powerfully activates the brain's pleasure centers – more powerfully, in fact, than anything in our natural environment. Eating too much of the stuff isn't good for us, but that's irrelevant. The meme proliferates, not because it's good for *us* but purely because it's good for itself – purely, that is, because it's good at proliferating. To be clear, it doesn't *want* to proliferate or *know* what's good for it, any more than genes do. It's apple pie! The idea is simply that if we want to understand which memes come to predominate in a culture, then rather than looking at how memes affect our fitness or the fitness of our groups, we need to look at how they affect their own chances of getting passed on.

A second example concerns language. Most words in any given language are useful; that's why we use them and why they survive in the culture. But as Daniel Dennett points out, not *every* word is useful. Some survive despite being entirely useless to us, just because they're good at surviving. Such words are the linguistic equivalent of junk DNA. And one of the secrets of their success is that we often don't even know that they're there. Dennett once outlined this idea to a student, and the student asked for an example. Dennett replied, "Well, like, there might be, like, a catchphrase or, like, a verbal tic that was, like, a bad but infectious habit that could, like, spread through a subpopulation and, like, even go to fixation without, like, providing any communicative benefit at all." To which the student responded, "I got the point; I want, like, an example."[67]

In sum, memes are not selected because they're adaptive, either in the technical sense used by evolutionary scientists (i.e., because they enhance gene transmission), or in the everyday sense used by civilians (i.e., because they promote happiness or wellbeing). They're not selected because they're good for us: The smoking-tobacco meme is successful despite the fact that it kills its bearers. They're not selected because we like them: Earworms are successful despite the fact that they're often as irritating as a fly buzzing round your head. And they're not selected because they're true: Freud's memes were successful despite the fact that most of his best-known claims are almost certainly false. Ultimately, memes are selected for one reason and one reason only: because they have properties that keep them in circulation in the meme pool.

From this modest starting point, some rather profound consequences follow. First, through cultural competition, memes get better and better at surviving and spreading. Food gets more appetizing; music gets catchier; stories get more appealing. As Paul Graham put it, the world gets more addictive.[68] Second, given enough time, memes start falling into mutually supportive clusters, known in the trade as *memeplexes*. Memeplexes include everything from calculus to political ideologies. Like genes in a genome, the memes of a memeplex may coevolve and coordinate with one another. Thus, the belief that "You should die for your religion" may coevolve with the belief that "Life continues after death." Each may reinforce the other, and each may flourish more reliably in the other's presence. Third, just as selection in the biological realm creates an illusion of intelligent design, so too may selection in the realm of culture. Memes and memeplexes may come to look like they've been intelligently designed to protect themselves, promote themselves, and propagate themselves in our minds. Of course, many memes really are intelligently designed for one purpose or another. But blind, mindless selection may create hidden layers of design within our institutions and belief systems. Like biological adaptations, our memes and memeplexes may be designed to achieve ends that none of us are aware of, in ways that none of us need understand.

The meme approach to culture is often seen as an alternative to the approaches we've looked at already: the adaptive-culture approach and cultural group selection. In my view, though, that's not the right way to think about it. Instead, the meme approach provides an overarching framework that incorporates the other approaches, but goes beyond them as well. Because memes live inside us and depend on us for their continued existence, the best way they can help themselves is often to help us. As such, the memes that do best in a culture are often those that boost our fitness or aid the survival of our groups: boats that don't sink; gods that keep us in line when no one else is watching. This isn't always the case, however, and when it's not, the meme approach picks up the slack. It explains things the other theories cannot. This includes chain letters, earworms, and smoking. As we'll see later, it may also include more consequential aspects of culture.

By incorporating but also going beyond the insights of the other approaches to culture, memetics is very like the gene's-eye view of evolution. As we saw in earlier chapters, genes are often selected because they help their owners, boosting their survival or reproductive success. As we also saw, though, there are circumstances in which a gene can be selected even when it doesn't help its owner – when it helps its owner's kin, for example. The general rule encompassing both scenarios is that *genes are selected to the extent that they help themselves*. It's the same with memes. Often memes are selected because they help their hosts, boosting the host's fitness or the fitness of the group. But there are circumstances in which a meme can be selected even when it doesn't help its hosts. The general rule encompassing both scenarios is that *memes, like genes, are selected to the extent that they help themselves*. This brings us to our third hypothesis.

> **Hypothesis #3:** Cultural evolution isn't about the survival of the fittest genes, the fittest individuals, or the fittest groups; it's ultimately about the survival of the fittest memes. Through cultural competition, memes and memeplexes come to look as if they were intelligently designed to propagate themselves in our minds and cultures. Often, they do this by helping the people who hold them or the groups to which those people belong. Sometimes, however, they do it without helping us at all, or even while actively harming us.

The meme approach to culture might initially sound somewhat outlandish. The next step, then, is to look at these ideas in more detail, and see whether they hold any water.

The Meme-ing of Life

Memes are everywhere. Every time we talk with friends, we're exposed to new memes. Every time we switch on the TV or log onto the Internet or walk down the street, we're exposed to new memes. We're exposed to more memes every day than we could ever hold in our brains: more jokes, more factoids, more opinions. Which ones live long and prosper and which burn out or fade away? As Susan Blackmore put it in her book *The Meme Machine*, "imagine a world full of hosts for memes (e.g. brains) and far more memes than can possibly find homes. Now ask – which memes are more likely to find a safe home and get passed on again?"[69]

As with genetic evolution, it's partly a matter of luck, or what we might call *memetic drift*.[70] So, for example, if a cause or conspiracy theory happens to find its way into the head of a celebrity, or any other thought leader, it can suddenly spread like wildfire, even if it wouldn't have made it far on its own steam. But memetic success is more than just a throw of the dice. Some memes have a better chance of making it than others. As Blackmore observes, there are two main things that memes must do to increase their "market share." First, they must find their way into people's minds. This creates a cultural selection pressure for memes that grab our attention and which we can't stop thinking about. Second, memes must influence our behavior in ways that cause them to spread. This creates a cultural selection pressure for memes that motivate us to talk about them, to pass them on, or to impose them on other people.[71] Earworms tick both boxes: Once you've got one, it's hard to get it out of your head (Criterion 1), and you often end up humming or whistling it, thereby passing it on to your neighbor, who now hates you (Criterion 2). From the meme's-eye perspective, this is emblematic of culture in general. Culture is a big bag of earworms.

Let's look more closely at how each of these selection pressures operates. First, how do memes grab our attention and take up residence in our heads? At the most basic level, memes spread best when they mesh well with the native structure of the human mind. My favorite example concerns color terms.[72] Different languages have different numbers of words for the basic colors. The smallest number is two; the largest is eleven. Thus, there's a great deal of cultural variability in the ways people talk about color. But there's also order amidst the chaos. If a culture has only two color terms, they're always black and white. If a culture has *three* terms, they're always black, white, and – can you guess the third? Take a guess before reading on...

(This is a filler sentence, so you don't accidentally read ahead before guessing.)

The answer is red. Interestingly, people usually guess correctly. It's as if we can peer into our own minds and introspectively discern that red is the most salient color. And the fact that it's so salient reflects something about the basic design of our visual systems – something that's left its mark on the color terms of the world's languages.

It doesn't stop there. For cultures with four or five basic color terms, the next ones to come in after black, white, and red are usually blue and yellow, in no particular order. For cultures with six or seven terms, the next to come in are usually green followed by brown. Finally, for cultures that have more than seven terms, the next to come in are orange, gray, pink, and purple, again in no particular order. The upshot is that, although there's a lot of variability in the color terms of the world's cultures, there are also some general trends, and these trends come from the species-typical design of the human visual system. People are more likely to come up with color concepts that gel well with the basic structure of the mind, and these concepts are more likely to go viral.

The principle applies not only to low-level perceptual categories, but to more complex social thought. Humans appear to have an inbuilt, almost irresistible tendency to divide the world into "us" and "them," good guys and bad, in-groups and out-groups.[73] We see this everywhere: in politics, in religion, in sports and national pride. For Jesus, the poor and downtrodden were the in-group and the Pharisees and the rich the out-groups; for Marxists (much as for Jesus), the working class was the in-group and the ruling class the out-group; for the Nazis, the Aryan race was the ingroup and Jews the main out-group; and for some radical feminists and "lesbian separatists," women are the in-group and men the out-group. On the surface, these philosophies look very different. But they all share a deep structural commonality: the division of humanity into friends and foes. Because this way of slicing up the social world comes so naturally to people, memes that play on this theme slot like well-placed Tetris pieces into many people's minds. This boosts their chances of spreading.

Other cases fit the same pattern. Humans, as we saw earlier, are naturally attentive to information about sex, relationships, kinship, status, and threats. As a result, memes related to these evolutionarily relevant topics are more likely to spread than those related to, say, ball bearings. Even within these subject domains, some memes are more likely to spread than others. People have a natural aversion to incest, which means that incest taboos are more likely to spread than pro-incest propaganda. We have a natural desire for sex and intimacy, which means that love spells and relationship advice are more likely to spread than celibacy. And we have a natural dread of death, which means that comforting afterlife memes are more likely to spread than the belief that death is the end.[74] The general rule is that memes that mesh well with the contours of human nature have a leg-up in the fight for human brain space. Conversely, memes that don't pay homage to our evolved nature tend to be smashed on the rocks of biological reality.

Of course, not all minds are alike, and different memes flourish in different kinds of minds. The memes of quantum mechanics, for instance, take root only in highly intelligent minds, whereas extreme political views and fundamentalist religious beliefs seem to thrive more reliably in less intelligent ones.[75] Personality matters as well. Novel memes do better with people high in the trait of openness, whereas misanthropic memes do better with people low in the trait of agreeableness. Note that, like virtually all psychological traits, intelligence and personality are partially heritable – in other words, differences between people in these attributes are due in part to differences in their genes.[76] This leads to an interesting conclusion, namely that our genes help determine which memes we end up holding.

But grabbing our attention and sticking in our minds is only half the battle. Memes also need to find a way out again and into other minds. They can do this in many ways. One simple strategy is to increase the inclusive fitness of their bearers. To start with, any meme that increases its bearer's longevity – any food preparation technique that eliminates toxins, for instance – stands a very good chance of spreading, if for no other reason than that it has more time to get copied by others. The flipside of this coin is that any meme that shortens its bearer's life simultaneously curtails its chances of spreading, because people don't generally copy the dead (or didn't, at any rate, before the invention of TV).[77] Here's an example. During the heyday of European colonialism, a number of indigenous groups came to the view that, if they had faith, the Europeans' bullets couldn't harm them. Needless to say, this meme had disastrous consequences. One group infected with the meme – the Mahdists of Sudan – lost 11,000 men in a single battle to the bullets of Kitchener's army. This was bad for them obviously, but it was also bad for the meme. In effect, the meme removed itself from the meme pool through its effects on its hosts' behavior.

So, one way a meme can spread itself in the culture is by boosting its owners' longevity. Another is by boosting its owners' reproductive success. Any meme that encourages its bearers to have lots of children inevitably upgrades its chances of spreading.[78] This is because children tend to inherit the belief systems of their parents and the parents' community, and therefore fertility-promoting memes automatically build lots of new homes for themselves. This can be seen most clearly in the realm of religion. Members of conservative religious denominations typically have more offspring than moderates or the non-religious. This enhances the genetic fitness of conservative religious parents. But it also enhances the *memetic* fitness of conservative religious memes, because this fertility pattern creates lots of receptive children for those memes to infest. Some argue that, by boosting their hosts' fertility, conservative religions are currently gaining ground round the world.[79] Compare this to the case of the Shakers: an eighteenth-century American sect. The Shakers forbade reproduction altogether and insisted on strict celibacy – not just for the priesthood but for everyone. This, as you can imagine, was about as good a recipe for memetic success as the belief that bullets can't hurt you. In Dennett's words, Shakerism was a sterilizing parasite. Unsurprisingly, Shaker memes are now all but extinct, preserved only in the history books.

In the preceding examples, the fitness of the memes coincides with the fitness of the people who hold them. As we already know, though, memes don't necessarily need to boost their owners' fitness in order to boost their own. Other roads lead to Rome. For example, in addition to encouraging their hosts to have more children, memeplexes can propagate themselves by increasing the reliability with which they're transmitted to the children their hosts already have. In his book *Thought Contagion*, Aaron Lynch gives the example of religious traditions that insist that children are schooled only with members of the same tradition.[80] Sometimes called *cocooning*, this practice insulates children from competing memes (or did before the invention of the Internet). Any religion that does this gives itself a substantial head start in the competition for people's souls (so to speak). The tactic facilitates the spread of the memeplex, without necessarily raising the fitness of its bearers.

An important difference between genes and memes is that memes aren't passed only from parent to child. As such, memeplexes that include instructions to "spread the word," not just to one's offspring but to anyone and everyone, can boost their own chances of spreading. One reason for the success of Christianity is that its teachings include instructions to convert other people to the faith. In Matthew 28:19–20, for instance, Jesus exhorts the faithful to "go and make disciples of all nations, baptizing them in the name of the Father and of the Son and of the Holy Spirit, and teaching them to obey everything I have commanded you." It's no accident that Christianity and other proselytizing religions have many more adherents than religions such as Jainism or Judaism, which don't proselytize nearly as much.[81] Proselytizing doesn't immediately benefit either the proselytizers or the proselytizees. But it does benefit the religion that mandates it in a clear and obvious way. Just as sex evolved to spread genes, proselytizing evolved to spread memes.

In evolution, success is always relative: If you and I are running a marathon and you fall behind, my relative rank improves even if I don't run any faster. Thus, another way that a memeplex can get ahead in life is by cutting down the competition. Many memeplexes, including religions and political ideologies, come fully equipped with ideas about why competing views are false or evil. These mudslinging memeplexes often have an advantage over their live-and-let-live rivals – even when the mud being slung is grossly unfair or inaccurate. Admittedly, life is never easy for memeplexes adopting this strategy. Mudslinging often begets counter-mudslinging, initiating spiraling mudslinging arms races (consider, for instance, the sorry state of political discourse in many modern nations). Nevertheless, in spite of the risks, mudslinging and name-calling often give a memeplex an advantage.

(I don't mean to imply, by the way, that attacking rival memes is always a bad thing. Often, it's a very good thing. At the societal level, criticizing ideas is the ultimate source of our moral and intellectual progress. And at the individual level, critical thinking functions as a memetic firewall, designed to attack and neutralize certain sorts of toxic memes: those backed up with poor evidence and logically dubious arguments.)

It's important to keep in mind that the fate of any meme is not determined solely by its inherent stickiness or spreadability. Other factors come into play. One is the authority of the meme vector (that is, the person spouting the meme). For example, in principle if not in practice, Catholics automatically accept memes emitted by the pope. Indeed, the Catholic Church traditionally claimed that, on special occasions, the pope emits infallible memes. As a consequence, memes that find their way into the pope's head may do better in the meme pool than they would otherwise have done. Clearly, then, the success of any meme is dependent on factors external to it. Still, given a particular set of external factors, selection will tend to favor those memes that grab our attention, that stick in our minds, and that influence our behavior in ways that cause them to spread.

The Human Mind as the Ecology of Memes

Critics of memetics sometimes charge that the field ignores the nature of the mind, treating it as a blank slate that robotically replicates any meme it happens to encounter. I hope it's clear from what I've said already that this is a mistake. A meme's success is shaped overwhelmingly by the nature of the minds it inhabits. How, though, should we conceptualize that fact? In this section, I want to explore two possibilities. The first is that memes adapt to our minds in the same way that organisms adapt to their environments. The second is that memes *parasitize* our minds, reshaping them for their own ends. (For a discussion of other common criticisms of memetics, see Appendix B.)

Let's start with the idea that memes adapt to our minds. One way to frame this would be to say human nature creates selection pressures on the memes of a culture. Although in principle memes could evolve in any direction they please, in practice, they don't. They mold themselves to the peaks and valleys of the minds that contain them – minds shaped in part by personal experience and in part by natural selection. In effect, the human mind is the environment to which our memes adapt.

This angle on cultural evolution suggests a new way to answer the alien's questions. Take, for example, the question of God. Why does belief in a divine being come so naturally to us? Why does it so stubbornly sink its claws into our gray matter? It's a bit odd when you think about it. No one has any trouble shaking off their belief in Santa Claus or the tooth fairy. The God meme, on the other hand, is almost impossible to dislodge from some minds, regardless of evidence or arguments. It's a little like an earworm, except that, while earworms are often irritating, most believers are extremely fond of the God meme, and are more worried about losing their belief in God than about keeping it. Some argue that humans evolved specifically to believe in God – in other words, that theistic belief is an adaptation.[82] But this seems unlikely. If we look at uncontroversial adaptations, such as arms and the basic emotions, we find that every normal human being possesses them, and that they're fairly similar from person to person and from culture to culture. In contrast, millions of human beings lack a belief in God, and among those who do believe, their conceptions of God vary so greatly that it's debatable that they actually believe in the same thing.[83] None of this sits easily with the idea that theistic belief is an adaptation. And yet there *is* a puzzle. The various conceptions of God all find a comfortable home in the human mind, unlike, say, Heisenberg's uncertainty principle. If we didn't evolve to believe in God, why does God fit our minds like a glove?

The meme's-eye view suggests a solution – a solution that flips the adaptationist explanation on its head. Rather than the human brain evolving for God, God evolved for the human brain. The God meme is compelling, not because we evolved to find it so, but because it's the grizzled survivor of many hundreds of generations of cultural competition for our attention and allegiance. It wins our allegiance in a number of ways. As Dawkins wrote,

> The survival value of the god meme in the meme pool results from its great psychological appeal. It provides a superficially plausible answer to deep and troubling questions about existence. It suggests that injustices in this world may be rectified in the next.[84]

To Dawkins' suggestions, I would add that the God meme plays on some very natural human fears: the fear of death, of misfortune, and of everlasting torment should we fail to believe. In these and other ways, the God meme meshes well with human nature and human motivations. Just as our lungs evolved biologically to fit the ecology of the world above water, so the God meme evolved culturally to fit the ecology of the human mind.

It's not just God; music may have a similar backstory. People everywhere take to music like a duck takes to water. From our earliest years, most of us "get" music, are emotionally moved by it, and develop particular affections for certain musical genres. (Interestingly, most people also have musical pet hates.) How can we explain our duck-to-water relationship with this most abstract of art forms? As with God, many evolutionary psychologists argue that music is an adaptation. Some suggest that it's a way to boost group morale, others that it's a way to attract mates – a kind of auditory peacock's tail.[85] Again, though, the adaptationist view is less than compelling. Music is extremely varied in its forms and effects, and many people do little in the way of music-making. This hardly screams adaptation. Yet we're left with the awkward fact that music comes so naturally to us. How can we explain the fit between music and the human mind?

Once again, the meme's-eye view comes to the rescue. Of all the billions of musical "experiments" that people have conducted throughout the ages, the only ones that persisted were the minority that happened to push our emotional buttons in an especially provocative way – the ones, in other words, that were best able to make a home in the human brain.

Our brains didn't evolve for music; music evolved for our brains.[86] A cultural animal with a different brain would evolve a different system of music, perhaps entirely unrecognizable to us.

When it comes to God and music, cultural evolution plausibly did most of the heavy lifting, while biological evolution sat on the sidelines. In other areas, however, biological and cultural evolution may have shared the workload more evenly. Language is a plausible example. Spoken language comes extremely easily to young children – much more easily than music or God. Before children can tie their own shoelaces or find their own way home, they effortlessly develop language abilities that far exceed those of the best-trained signing apes. How do they do it? A big part of the answer seems to be that children have a built-in sensitive period for acquiring language. As Steven Pinker and Paul Bloom have argued, children's brains contain specialized neural machinery designed to suck up language like a vacuum cleaner in the early years of life.[87] But this machinery probably isn't the only reason that kids learn language so easily. Cultural evolution may play a starring role as well. For as long as children have been learning languages, the words and grammatical rules they found easiest to learn persisted for longer in the culture. In effect, the minds of each new generation functioned as a selective sieve through which the elements of language had to pass. In the course of millennia of sieving, languages evolved to be more and more easily learnable. As the anthropologist Terrance Deacon once argued, this may help to explain why children find language so easy to learn.[88] Not only did children's brains evolve to acquire language, but human languages evolved to infest children's brains. Biological and cultural evolution met in the middle.

The idea that memes evolve to inhabit human brains is compelling. There is, however, another, more unsettling way to look at it. Rather than simply adapting to human brains, memes may sometimes *parasitize* our brains. In biology, a parasite is an entity or organism that uses another organism to survive without immediately killing it, and which either has no impact on the host's fitness or actively harms it. Many memes, much as we might love them, meet this definition. Apple pie and ice cream, for instance, survives in human societies by stimulating our pleasure circuits, but isn't particularly good for us. We might say that it evolved culturally to parasitize those brain circuits – that is, to take advantage of our weakness for sweet and fatty foods. Likewise, we might say that teddy bears evolved to parasitize our parental instincts, and that chain letters evolved to parasitize our greed and attentiveness to potential misfortune. To be clear, in suggesting that some memes are parasites, I'm not saying they're necessarily bad. We could do without the chain letters, certainly, but if apple pie is a parasite, it's a welcome one as far as I'm concerned. Nonetheless, it's an interesting fact about the human condition that many of our memes meet the technical definition of a parasite.

Chain letters and apple pie are relatively inconsequential examples of parasitic memes. Other examples might be more significant. Various thinkers have argued that religions are parasitic mind viruses.[89] This is probably overstating the case; religious beliefs and practices are presumably sometimes good for us, at least in some ways. But they're not always. Sometimes they sterilize the people who hold them (as when people take a vow of celibacy and devote themselves to spreading the religion). And sometimes they even *kill* the people who hold them (as when people refuse life-saving medicine or blow themselves up for their faith). Even leaving aside such catastrophic outcomes, there's often a price to be paid for religion. Most religions involve complex and time-consuming rituals and practices. Time spent praying, proselytizing, or worrying how many angels can dance on the head of a pin is time that could have been spent looking for food or mates, or keeping an eye out for predators. Religious memes burn up precious fuel without any obvious countervailing advantage. To be more precise, they burn up precious fuel without any obvious advantage *to us*. There may, however, be an advantage to the religious memes themselves. People infested with these informational parasites may end up acting for the good of the parasites, rather than for their own good.

Does the same apply to music? In other words, is music a parasite that evolved to survive in the ecological niche of the human brain? Maybe so. At first glance, music appears to be evolutionarily useless: It monopolizes time and resources without any clear Darwinian payoff. This is most obvious when we think about earworms. Most of us have these irritating musical papercuts playing on the inner jukebox for a significant chunk of each day. This doesn't come for free. It's like leaving the radio on all day; it uses up energy. Seeing earworms in this light makes it hard to resist the idea that they're informational parasites, siphoning off some of the energy budget of the brain to pointlessly propagate themselves over and over again, against the will of their hosts. Could this be true, though, not just of earworms, but of music in general? Our immediate reaction is to say no. After all, most of the time we choose what music we listen to and we listen to it because we like it. But of course, that's the main tactic that music has evolved to propagate itself: It appeals to us. In a certain sense, then, music parasitizes the brain. Again, this isn't to say it's a bad thing. For most people, music is one of the joys of life and there's no good reason we should give it up. The point, though, is that the human brain evolved to propagate genes, but through the vagaries of cultural evolution, it now spends a great deal of time hosting and broadcasting musical memes.

One reason to take the memes-as-parasite view seriously is that, rather than simply molding themselves to the human brain, in some cases memes literally take over and reshape the human brain. The neuroscientist Stanislaw Dehaene calls this *neuronal recycling*.[90] It happens when areas of the brain that evolved for one purpose are co-opted and restructured for new, culturally constructed purposes. Dehaene's main example is reading. When people learn to read, this recalibrates a number of brain regions involved in visual perception, turning them to the new task of recognizing and making sense of written words. In the process, it also makes them worse at performing their original function. From a memetic perspective, we might say that the reading meme evolved to parasitize these evolutionarily ancient areas of the brain, retraining them for its own purposes and thereby constructing for itself a suitable home in an initially inhospitable environment. And what's true of reading may be true more broadly. From religion to music to advanced mathematics, our memes evolved culturally to colonize the brain, just as our remote ancestors evolved biologically to colonize the land. In doing so, however, they transformed the very landscape they were colonizing.

Memetic Adaptations and Meme Machines

As I mentioned at the start of the chapter, when people think of memes, they tend to think of contagious scraps of culture: buzzwords, catchphrases, earworms, fairy tales, urban legends, viral YouTube videos – the usual suspects. But to focus on these cases is to radically underestimate the power of the meme's-eye view. The real power of Dawkins' idea is revealed only when we consider the effects of cumulative cultural evolution over hundreds or even thousands of years. Rather than simply promoting some memes over others, cultural selection may, given sufficient time, build up complex memetic "structures" – informational edifices with some of the properties of organisms.

To start with, selection may create large-scale memeplexes: religions, political ideologies, and other collections of co-adapted memes. As we saw in Chapter 2, the ultimate function of any organism is to pass on its genes. The ultimate function of any *memeplex*, in contrast, may be to pass *itself* on and keep itself afloat in the culture. To this end, selection may equip memeplexes with the memetic equivalent of adaptations.[91] In order to pass on their genes, organisms have adaptations designed to accomplish a diverse range of tasks, including finding food, maintaining a suitable body temperature, and choosing suitable mates. Similarly, in order to pass *themselves* on, memeplexes may have adaptations designed to accomplish an equivalent range of tasks, including gaining and retaining adherents, spreading themselves from mind to mind, and protecting themselves from rival memeplexes, criticisms, and attacks.

The notion of memetic adaptations has been most thoroughly explored in regard to religion. One interesting example comes from the psychologist Darrel Ray.[92] Why, asks Ray, do many religions proscribe common sexual activities such as masturbation, sex before marriage, and sex for pleasure rather than procreation? These activities are hard for most people to resist, and for the most part are relatively harmless. The proscriptions seem odd... until, that is, we look at them from a memetic perspective. Consider what happens when people slip up and break the rules, as they usually do from time to time. Most often, they're wracked with remorse and redouble their religious efforts: They crawl back to their priest or preacher, confess their sins, or devote themselves anew to their holy books. And that, says Ray, is the whole point of the proscriptions. It's not to stop people doing these things; it's to make them feel guilty about things they were probably going to do anyway. This binds the faithful ever-more tightly to the religion.

If this is right, then the fact that these religious rules are hard to follow is not a bug; it's a feature. Like the mammalian eye, however, it may be an instance of design without a designer. We needn't assume that some conniving religious leader woke up one morning and thought: "What can I do to keep people from deserting the religion? I know! I'll invent a bunch of hard-to-follow moral rules that will make people feel guilty about their natural cravings and keep them crawling back for forgiveness!" Instead, the design may have come from the blind, mindless process of natural selection. Religious groups presumably varied in how strict they were, and groups that happened to have hard-to-follow rules were more likely to retain their members. These groups outlived and outlasted their more liberal compatriots, and thus it's these groups that we see around us in the world today. The rules can therefore be seen as an adaptation – not an adaptation of religious individuals but an adaptation of the religion itself. People don't need to understand the ultimate function of the adaptation in order for it to work, any more than a rose needs to understand the ultimate function of its thorns.

Retaining followers is a crucial adaptive task for any religion. Another crucial task, especially since the Enlightenment, is neutralizing skeptical questions and deflecting rational criticism. Just as turtles evolve shells and trees evolve bark, religions evolve an assortment of defenses against this memetic threat. Among the most important is the concept of faith: the notion that one ought to accept the tenets of one's religion even in the absence of evidence and even if they don't make sense. This is a common feature of many of the world's religions. Many Christians hold, for instance, that it's hugely important to believe in God, and that if one feels one's faith slipping, one should do whatever it takes to keep it alive. When you think about it, this is rather strange. As Richard Dawkins asks in his book *The God Delusion*, why is it so important to the all-powerful, all-knowing, and all-good creator of the universe that we believe in him on insufficient evidence? Wouldn't a benevolent God think it more important that we're good to each other? Conversely, if it's really so important that we get our metaphysical beliefs straight, wouldn't a benevolent creator furnish us with more persuasive evidence and arguments for his existence?

Once again, it doesn't make a lot of sense – until we look at it from the meme's-eye perspective. As Dawkins points out, if you wanted to find a way to insulate a memeplex from rational criticism, you couldn't do much better than the idea that accepting that memeplex on blind faith is the highest virtue, and doubting it a terrible sin.[93] Did anyone actually sit down and invent this idea for that purpose, though? Quite possibly not. The idea might just be a product of blind selection, which bubbled up to the top of the memetic food chain simply because it happened to work – not for us, but for the religions of which it's a part. Here are some other possible religious adaptations, designed – perhaps by no one – to neutralize skeptical memes. The list is based largely on the work of Dawkins and Dennett.[94]

- God moves in mysterious ways.
- God planted false evidence to test our faith.
- The devil planted false evidence to tempt us away from the faith.
- It is disrespectful and dangerous to question the doctrine.
- Unbelievers will burn in hell.
- Unbelievers must be expelled from the community.
- Unbelievers must be killed.

Of course, just as it's possible to overextend adaptationist explanations in the biological realm, it's possible to overextend them in the cultural. Still, as soon as we grant just two premises – first, that natural selection favors selfish memes, and second, that selection acting over long periods of time creates an illusion of intelligent design – it's difficult to dodge the conclusion that memeplexes will come to possess sophisticated adaptations designed to help them to spread.

And once we've gone that far, the next step seems almost inevitable. As we saw in Chapter 2, natural selection operating on genes gives rise to gene machines: organisms designed to perpetuate their genetic material. Is there any parallel in the memosphere? Here we get into speculative territory, but territory worth exploring. Memes can't exist independently of their human hosts, so cultural evolution couldn't give rise to the memetic equivalent of organisms – that is to say, independently existing "meme machines." What it could do, though, is give rise to social institutions and belief systems that, in effect, convert *us* into meme machines. Just as viruses evolve to take over cells and use them to spread themselves, so memes may evolve to take over our brains and use them to spread *them*selves. To a greater or lesser extent, they may transform an ape designed to pass on its genes into an ape designed to pass on its religion, its political philosophy, or even just its hobbies and interests. More generally, the evolution of selfish memes over the course of human history may have turned us all into publicists for whatever memes we happen to have caught or contracted.[95]

If there's any truth to this, the implications are enormous. For one thing, it puts an entirely new spin on the reasons I wrote this book: I've been infected with memes and they've turned me into their mouthpiece. Among other things, I've been infected with the meme meme, and right now I'm working to infect you, too. Every advocate of the meme perspective is a walking, talking advertisement for the power of contagious memes. And that's just the beginning. With our memetic spectacles on, everything looks strikingly, frighteningly different. Universities suddenly appear as a great conspiracy of memes, designed to get infected people (professors) to pass them on to uninfected people (students). Similarly, the Internet suddenly appears as an immense and rapidly evolving vortex of information, designed to keep us glued to our computers, slavishly spreading memes.

Perhaps most striking of all, though, human history suddenly appears as a vast conflict between competing "species" of memes. Leftist memes and conservative memes wrestle each other for control of the democratic state. The memes of Islamic fundamentalism and those of Enlightenment liberalism clash with each other in various ways around the world, as do various families of memes within each of these broad traditions. Much of the history of the twentieth century consisted of Marxist memes, fascist memes, and the memes of capitalist democracy slugging it out through our bodies, our voices, and our communication technologies. These historical skirmishes were, in essence, battles of ideas, and the people caught up in them merely vessels through which these ideas fought for supremacy. Ninety-nine times out of a hundred, if they'd been born in a different part of the world, or even just a different part of town, they'd have been fighting for the other side. Thus, it's almost as if the ideas themselves were fighting each other through whichever humans they happened to infest. The writer Jonnie Hughes made this point well in his cleverly titled book *On the Origin of Tepees*, when he wrote: "Throughout history, we humans have prided ourselves on our capacity to have ideas, but perhaps this pride is misplaced. Perhaps ideas have us."[96]

Lest this sounds like too alarming a picture of our predicament, remember that today, more than ever, we have some degree of control over memetic evolution. I'll give you two examples. First, during the course of human history, we've developed various ways to put memes we consider important, but which might not survive unaided, on "life support." Dennett gives the example of calculus.[97] The memes of calculus have survived in human societies for hundreds of years. But their survival is not a result of the fact that calculus is a catchy memeplex. No one ever complains that "Calculus is so darned catchy; I just can't get it out of my head!" Actually, scratch that; some people might: geniuses and autistic savants (and note that there's some overlap between those categories). Most of us, though, are not particularly susceptible to the memes of calculus. Like domesticated sheep, calculus memes survive in our culture only through the efforts of dedicated stewards, including teachers, textbook writers, and curriculum developers. The job these stewards do is not an easy one. Calculus was invented by minds very unlike the everyday, run-of-the-mill mind. We might say, in fact, that the difficulty that most of us have in learning calculus is a result of memetic mismatch: mismatch between the environment that calculus memes initially evolved to inhabit (nerd minds) and the environment afforded by the typical human mind. In short, teaching calculus is a little like growing a crop in foreign lands: It takes a lot of time, effort, and energy. But it is possible! And this tells us that we can keep valuable but non-catchy memes alive in our culture, whenever we deem it worthwhile.

So, that's one reason we're not entirely at the mercy of catchy, self-interested memes. A second concerns the jewel in the crown of our cultural achievements: science. As we've seen, memes are not necessarily selected because they're true. On the contrary, false-but-catchy memes can often trump true-but-less-catchy ones. Again, though, this isn't inevitable. Over the last several centuries, we've slowly pieced together cultural mechanisms that reliably favor truth over catchiness. These include critical thinking, careful observation, peer review, open discussion, independent replication, and the rejection of authority, tradition, and revelation as reliable sources of knowledge. Taken together, these habits and tactics constitute the scientific method. In a sense, science is a system of selectively breeding accurate memes. In the "wild," there's no guarantee that accurate memes will do better than inaccurate ones. But in the carefully controlled memetic ecosystem of science, we can and do breed memes for closer and closer correspondence to truth. In this way and many others, memetic evolution is no longer entirely beyond our control. Our genes and our memes have created a creature that, to some extent, can seize control of its memetic destiny.

Gene–Meme Coevolution I: Sculpting the Body

I've argued that the history of the human animal consists, to an important degree, of different species of meme battling it out for control of our minds and our bodies. Based on what I've said so far, we might imagine that this takes place against the backdrop of a static biology. That would be a mistake, though. The gene pool and the meme pool are two vast oceans of information, but these oceans are not hermeutically sealed off from one another. Evolutionary change in one pool can catalyze evolutionary change in the other. New memes, from stone tools to cooking to agriculture, create new selection pressures on our species. This leads to biological change... which in turn makes possible new memes, which then create new selection pressures, and on and on. The dynamic interplay between genes and memes is the focal point of a rapidly growing area of research known as *gene–culture coevolutionary theory*.[98] Kevin Laland and Gillian Brown describe this area as "a hybrid cross between memetics and evolutionary psychology, with a little mathematical rigor thrown into the pot."[99] With gene–culture coevolution in hand, we're now ready to put together the pieces and frame our final hypothesis.

> **Hypothesis #4:** Cultural evolution is about the survival of the fittest memes. Memes are selected to the extent that they're good for themselves, regardless of whether they're good for their hosts or their hosts' groups (although often they are). Through cultural competition, memes and memeplexes evolve to inhabit and exploit human minds. In doing so, however, they create new selection pressures on their hosts. As a result, long-standing memes may bring about evolutionary change in the minds and the bodies they reside in.

Let's look at some examples. By far the best-studied example is the evolution of lactose tolerance.[100] Lactose is the sugar found in milk. Some adults are *lactose intolerant*: Drinking milk gives them stomach cramps, flatulence, diarrhea, and other unpleasant side effects. Milk, for these people, is a mild toxin. In the West, we often view lactose intolerance as a deficit, a disorder, a deviation away from normality. But this is misreading the data. Virtually all mammals are lactose intolerant after weaning. In infancy, they produce an enzyme in their gut called *lactase*, which breaks down the lactose in their mother's milk. For that brief window of time, mammals are lactose tolerant. Once they're weaned, however, their milk-drinking days are over and they stop producing the enzyme. It's simple Darwinian cost-cutting: Why produce the enzyme when it's no longer needed? After weaning, then, the vast majority of mammals join the ranks of the lactose intolerant. This means that lactose tolerance, rather than intolerance, is the real deviation away from normality. It represents a profound break with mammalian tradition.

And not only are most nonhuman mammals lactose intolerant after weaning, so are most human beings! Only roughly one in three continues to produce lactase throughout the lifespan. Lactose tolerance is concentrated in certain hotspots around the globe. It's especially common in Northern Europe, and gets progressively less common as you move south to the Mediterranean or east to India. It's also found in certain parts of East Africa and the Arabian Peninsula. More precisely, lactose tolerance is found in people who trace their ancestry to these parts of the world, wherever they happen to live today. These people are the exceptions, though. Almost everyone else, including most Africans, Australian Aborigines, Asians, and Native Americans, have the typical mammalian phenotype: They can only digest lactose efficiently in infancy and early childhood. The fact that people of European ancestry tend to assume that most humans are lactose tolerant reflects a kind of ethnocentrism: a biological ethnocentrism.

It also reflects historical myopia. Until recently, *no one* could digest lactose post-weaning. But then, in the last several thousand years, various groups began herding animals such as cows, goats, and camels. Suddenly, there was a new supply of protein available to them: the milk of these nonhuman mammals. It's not clear how the habit of milk drinking initially got off the ground. Vegans sometimes try to turn people off the practice by emphasizing how biologically weird it is. No one, they observe, would get up in the morning and start sucking on a cow's udder. At some point, though, someone must have done something very like that to get the milk-drinking ball rolling. Some early adopter must have had the bright idea of stealing milk from their cows, goats, or camels. Maybe they were starving; maybe they were caring for an infant whose mother had died before the infant was weaned. But whatever the reason, others must have followed suit. The meme must have caught on. At first, milk drinkers were mismatched with their new diet. However, the milk-drinking meme created a powerful new selection pressure for the ability to digest lactose throughout the lifespan – and once that selection pressure was in place, it was only matter of time before lactose tolerance evolved.

In fact, it evolved several times, independently, in various parts of the world. The target of selection in each case was a gene called the LCT regulatory gene. Normally, this gene shuts down the ability to digest lactose after weaning. But in populations with a long history of dairy farming, the gene is often broken. That's what the lactose-tolerance gene is: a broken lactose-intolerant-after-weaning gene. The oldest version of the gene is the Eurasian one. Around eight thousand years ago, a baby was born with a broken LCT gene. This variant was then selected: It found itself copied into more bodies than working versions of the gene. A *lot* more bodies: People with the broken gene had an average of 10 percent more offspring than those without it, one of the strongest selection differentials known to science. Today, this variant is found in millions of people across Eurasia, from Ireland to India, and every last copy traces back to that one mutant baby. Other versions of the gene, including the various African and Arabian versions, spread in much the same manner, but more recently. Importantly, these versions are broken in different ways than the Eurasian variant. That's how we know that lactose tolerance evolved independently on several occasions.[101]

Every time it evolved, though, it enabled a way of life that, by mammalian standards, can only be described as bizarre. A fully grown tiger would never drink milk from an adult female, especially not one belonging to another species. But people in dairying populations do essentially that. In effect, they're breastfed for life by cows and other dairy animals. Just as mosquitoes are parasites on human blood, and earworms parasites on human brain power, people who have milk on their cereal or cheese on their crackers are, quite literally, parasites on cow's milk (or goat's milk or camel's milk). They are, as Gregory Cochran and Henry Harpending put it, "mampires."[102]

What's the selective advantage of mampirism? Stomach cramps and flatulence aren't pleasant, but how do they stop people from having babies or helping their relatives to have babies? Harpending put it well. We can think of the lactose-tolerance gene as a magic pill: If you take the pill, you get 40 percent more energy from your milk. Anyone who *doesn't* take it is therefore throwing nearly a third of the available calories down the drain. In a resource-poor environment, that could easily be the difference between life and death. But even during the good times, the lactose-tolerance gene could boost inclusive fitness. The amount of milk you'd need to feed three lactose-intolerant children would be sufficient to feed *four* lactose-tolerant ones. This means that, in premodern conditions, parents with the lactose-tolerance gene could easily produce more surviving offspring than those without.[103]

I've mentioned that the vast majority of mammals steer clear of lactose after weaning, but at least two mammals buck this trend: mice (which eat cheese) and cats (which drink milk). How do we explain these anomalies? Let's start with the easier case: Why do mice like cheese? The answer is... mice *don't* like cheese. The idea that they do is a catchy-but-false meme. They'll eat it if there's nothing else on offer, but given a choice, most mice will shun cheese in favor of fruit, grains, or seeds. That said, there are reasons to think that mice, and indeed most mammals, would do better with cheese than with milk. Fermentation during the cheese-making process eliminates much of the lactose in milk, making it a lot easier to digest. The same is true of yogurt and sour milk. The fact remains, however, that mice are not the cheese fans they're sometimes made out to be.

Now, the harder question: Why do cats drink milk? Cats seem to be a genuine exception to the rule that mammals only drink milk in infancy. But not all cats. Cats in certain parts of the world are more tolerant than those in others. And it turns out that the geographical distribution of lactose tolerance in cats closely matches the geographical distribution of lactose tolerance in humans.[104] European cats, like European humans, are more likely to be lactose tolerant. Apparently, dairy-farming peoples have been sharing their milk with cats for thousands of years. This cultural habit caused cats to evolve into (slightly) different animals – just as it did for dairy-farming people themselves.

Lactose tolerance is the best-known example of gene–meme coevolution, but it's not the only one. A second is the ability to efficiently digest starchy foods: foods such as tubers, rice, and wheat. People everywhere eat such foods, and some scientists argue that the extra energy they provide (along with the extra energy provided by meat and cooking) is what allowed our species to evolve its big, clever brain.[105] But although all humans eat starchy foods, some eat more than others. Agriculturalists eat a lot, as do foragers living in deserts. Everyone else eats significantly less (or did, at least, until recently). For populations with a starch-heavy diet, the ability to rapidly break down starches into sugars is another magic pill, enabling people to squeeze more energy out of their food. The source of the magic pill in this case is an enzyme called *amylase*. Amylase is found in saliva, and helps to turn starches into fuel for the brain and body. People with a long history of feasting on starchy foods, including Europeans, East Asians, and Hadza hunter-gatherers, have more amylase in their saliva than do other populations. This isn't because they have a new version of the salivary amylase gene; instead, it's because they have extra copies of the gene. Like lactose tolerance, this is a biological response to a cultural innovation: eating lots of starchy foods.[106]

But it's not just high-starch people who have extra copies of the amylase gene. So do their dogs! Dogs, as you probably know, evolved from wolves. To be more precise, dogs *are* wolves: domesticated wolves that evolved to live with humans. But dogs are much better than their wild cousins at digesting starches. This is because the ancestors of today's dogs scavenged from the scrapheaps of the early human farmers. Their scavenging ways created a selection pressure for the ability to efficiently digest the farmers' starchy leftovers. Better digesters had more puppies, and thus this ability was soon the norm.[107]

Gene–meme coevolution doesn't just adapt people and their pets to local foods and beverages, and it doesn't just explain differences between human populations. It also helps to explain certain traits that all human beings share, and which distinguish us from all other animals. Anthropologists have various examples. One is the fact that human hands are so much more dexterous than most nonhuman hands. A chimpanzee could never learn to play the guitar, use a pen, or build a model airplane, even in the most enriched learning environment imaginable. In contrast, most people are capable of acquiring these talents or equivalently demanding ones.

Where did our digital dexterity come from? Most likely, it's a product of gene–culture coevolution – or rather, gene–*technology* coevolution. New tools ratcheted up the selection pressure for dexterous hands, which made possible the cultural evolution of newer, more sophisticated tools. These new tools then further ratcheted up the selection pressure for dexterous hands... and so on. Put simply, our hands coevolved with the tools we so dexterously manipulate. At the start of this process, we had simple stone tools; at the end, we had needles and threads and pianos and looms – all courtesy of gene–culture coevolution.[108]

In addition to our dexterity advantage, humans greatly outperform our fellow apes when it comes to throwing. A chimp gliding through the trees is poetry in motion; a chimp trying to throw a projectile is comical. Humans, on the other hand – especially male humans – are so good at throwing, and this talent is so rare among the animals, that some anthropologists call us the *throwing ape*. Where did our knack for throwing come from? The most plausible suggestion is that it coevolved with the projectiles we used for hunting. As we got better and better at throwing, we were able to wield more sophisticated projectiles; as our projectiles got more sophisticated, we got better and better at throwing. The thrower and the thrown coevolved.[109]

Hurling projectiles was integral to the survival and success of our pre-agricultural ancestors. Arguably, though, our most important tool back then was fire. The ability to make fire is clearly a cultural innovation. If our alien scientist were to suck you into its spaceship, dump you in a forest, and then order you to make a fire, you probably couldn't do it – and if you could, it's because someone taught you how. Humans don't have a fire-making instinct; fire-making is a product of culture. But it's a product of culture that's had a big impact on our biological makeup. As the anthropologist Richard Wrangham argues, fire allowed us to cook our food, and this simple innovation had profound downstream consequences.[110] For one thing, it resized our teeth. An important part of the job description of teeth is to tenderize food so it's easier to digest. However, one of the main things that cooking does to our food is tenderize it. In effect, fire took over some of the work that was formerly done by our teeth. As a result, we no longer needed such fearsome knives in our mouths, and natural selection downsized them: Our teeth and our mouths got smaller. Natural selection took a similar tack with our guts. For various reasons, cooking our food meant that we got more energy from each mouthful we ate; the practice is another magic pill we can add to our list. Because we needed fewer mouthfuls of food to fuel our activities, selection was able to reduce the size of the human gut. That's why, unless we're overweight, our torsos taper in at the waist, rather than flaring out as they do in other apes.

More important than any of this, though, is the fact that cooking our food meant that we suddenly had a surplus of energy – a surplus that could be devoted to growing and running an oversized brain. In other words, cooking helped make possible the evolution of human intelligence. And with that, we enter a whole new arena of gene–meme coevolution.

Gene–Meme Coevolution II: Sculpting the Mind

Gene–culture coevolution is one of the most important tools in our Darwinian arsenal. From our heads to our toes, our bodies bear the indelible stamp of this powerful selective force. But natural selection doesn't just magically lose its power at the outskirts of the skull – and there's no reason to think that selection due to culture is an exception to that rule. If culture can reshape the body, there's every reason to think that it can also reshape the mind.

Which aspects of the mind owe their existence to gene–meme coevolution? Perhaps the best example is one of our most unique and important attributes: our intelligence. According to the *cultural intelligence hypothesis*, human intelligence evolved in lockstep with our evolving culture.[111] The argument is straightforward and persuasive. Human beings have an inbuilt capacity for culture. This implies that, throughout most of our evolutionary history, culture was adaptive: Our tools and traditions generally enhanced our fitness. Because culture was adaptive, those of our ancestors who were more adept at acquiring it had a distinct advantage over those who were less adept – they passed on their genes at a greater rate of knots. Thus, culture created a selection pressure for bigger, smarter brains. As our brains got bigger and smarter, we became capable of more sophisticated culture. This in turn ratcheted up the selection pressure for bigger, smarter brains, which made us capable of yet more sophisticated culture, which ratcheted up the selection pressure for... well, you get the idea. In this way, our clever brains and our clever cultures coevolved. One minute, we had simple tools and fire and little else besides, the next we were cooking our food and using sophisticated tactics for tracking prey, and the next after that, we were building shelters and making clothes. Crucially, with each further step down the path, we got a little smarter, and a little smarter, and a little smarter. Humans invented culture, but culture invented us as well.[112]

The cultural intelligence hypothesis focuses on the *usefulness* of culture in driving the evolution of intelligence, but there may be more to the story. Culture isn't just useful, after all; it's also often dangerous. The same tools and techniques that boost our fitness if we use them wisely – fire, canoes, bows and arrows – can kill us if we don't. This means that as new culture appears in the world, so do new dangers. And unlike snakes, heights, and predators, these are dangers that our species hasn't evolved specifically to deal with. According to IQ researcher Linda Gottfredson, lower IQ individuals are probably just as good as anyone else at dealing with ancestral threats, but they're much worse at dealing with evolutionarily novel ones. People with higher IQs are more likely to spot novel hazards in advance and better able to take precautionary measures. Even today, in our bubble-wrapped modern world, smarter individuals are less likely to drown, less likely to die in car or motorbike accidents, and less likely to injure themselves with guns or explosives. Their greater adeptness at skirting disaster gives them a big advantage in the competition to pass on their genes – an advantage that other animals don't need because other animals don't have such powerful, dangerous culture. According to Gottfredson, then, intelligence evolved partly in response to the dangers it created for itself in the form of our tools and technology.[113] This is the dark flipside to the sunny optimism of the cultural intelligence hypothesis: the yin to the latter's yang. If ideas like these are on the right track, then intelligence belongs in the same conceptual basket as lactose tolerance. Both are products of gene–culture coevolution.

Another candidate for that basket is language. Earlier I mentioned that humans have an innate capacity to learn language, but that the specific languages we learn evolve culturally to mesh with our language-hungry brains. There may be a twist in the plot, however. It's possible that our languages themselves helped to wire a language-learning instinct into the human genome.[114] Here's what might have happened. It all began with the cultural evolution of a rudimentary proto-language: a system of grunts and gestures not too far removed from what we see today in wild chimps. We didn't have a dedicated language faculty at that time, so we acquired this proto-language via general learning mechanisms. (This is presumably what captive apes do when they learn to communicate with signs.) The proto-language wasn't nearly as useful as our modern ones. But as anyone who's visited a foreign-speaking country knows, even a little language is better than none. As such, any ancient human who acquired the proto-language more easily, mastered it at a younger age, or used it more adroitly, would have had an advantage over her more linguistically ham-fisted contemporaries. And what an advantage! Language is useful in virtually every sphere of human life: communicating needs and wants, organizing hunts and other cooperative ventures, entertaining mates, conveying useful information to offspring, finding out who to trust and who not to.[115] In these areas and others, better language-users would have had a definite edge. Given the evolutionary importance of these activities, such individuals would plausibly have had more offspring, and their linguistic advantage would have spread through the population. As humans became more verbally adept, this would have triggered the cultural evolution of a more complex proto-language. That in turn would have created a selection pressure for even greater linguistic giftedness, which would have spurred the cultural evolution of an even more complex language – the usual story. In short, the cultural evolution of language resulted in the biological evolution of a language-learning instinct, and vice versa.

If so, this has an interesting implication. We saw earlier that genes for lactose tolerance were a *consequence* of milk drinking, rather than a cause. The same may be true of language. We tend to assume that genes for language came first, thereby making language possible. It's equally likely, however, that it's the other way around: that language came first and then created a selection pressure for genes promoting the rapid acquisition of language. In other words, our gift of the gab may have started with a cultural mutation, rather than a genetic one.[116]

In my view, the gene–culture coevolutionary explanations for language and intelligence are both fairly convincing. Before wrapping up, though, we should once again dip our toes into more speculative waters. Let's start with religion. My working assumption till now has been that religion is *not* a biological adaptation but rather a product of culture. However, gene–meme coevolution may complicate the picture.[117] Religion has been around for a long, long time: 50,000 years, according to some estimates. When it first appeared, it was presumably purely a product of culture, just like the first proto-language. But as religion got a foothold in the human population, it may have brought new selection pressures to bear on our species. People who were congenitally suited to life in a religious community may have fit in better and thrived to a greater extent than their more atheistic counterparts. As such, humans may have evolved over time to be more and more open to religious beliefs and practices. Religion, in other words, may have started life as a pure product of learning, but have become somewhat less learned and more innate over time. As Peter

Frost put it, "Man has made religion in his own image, but religion has returned the favor."[118]

Would this mean that religion is an adaptation after all? It's a tricky question. We often think about these issues in black-and-white terms: either a trait is an adaptation or it's not. It may, however, be more accurate to imagine that there's a continuum between full-blown adaptations on the one hand and unambiguous by-products on the other, with various indeterminate stages in between. If that's right, then while it might not be true to say that religion is an adaptation, it might not be entirely false, either. Religion may sit somewhere in the no-man's land between adaptation and by-product. Clearly, babies marooned on an island wouldn't spontaneously start praising the lord or reciting prayers in the way that birds spontaneously start building nests.[119] But perhaps they'd have a better chance of inventing beliefs and practices along those lines than would people who lived prior to the appearance of religion. And perhaps when they did invent them, these beliefs and practices would have a better chance of going viral.

One last example. As the alien mentioned in its report, humans engage in a broad range of activities that seem to have no utilitarian function but instead are done purely for fun. Among these activities are art, music, and humor. Some evolutionary psychologists argue that these phenomena are adaptations, others that they're merely by-products.[120] As we've just seen, though, there's another possibility. Art, music, and humor may occupy the gray area between by-product and adaptation. By now, you know the routine. Step one: Art, music, and humor emerged as pure products of culture. Step two: Through cumulative cultural evolution, these cultural products became more and more contagious, and more and more widespread. Step three: As these contagious memes staked out their territory in human culture, there were fitness benefits for anyone who was particularly adept at producing them. Maybe they were more desirable as mates, maybe as friends or allies. Whatever the reason, though, if artistic, musical, or humorous people ended up having more offspring, the species would inevitably have evolved to be more artistic, more musical, and more humorous. This wouldn't imply that art, music, or humor are now full-blown adaptations. But it might imply that the average level of artistry, musicality, and hilarity in our species is higher than it would have been otherwise. Natural selection may have taken the general-purpose cognitive mechanisms underlying art, music, and humor and started fine-tuning them for these initially unforeseen purposes. In this way, selection may have made the human brain a somewhat better home for artistic, musical, and humorous memes. I have no idea if this is true. But if it is, then gene–meme coevolution is partially responsible for some of the most puzzling, most treasured aspects of human culture.

Survival of the Weakest?

In all the examples we've look at so far, culture created new selection pressures, and in doing so, shaped new adaptations (or in some cases, proto-adaptations). But culture can affect our evolution in another, very different way as well: It can remove pre-existing selection pressures, thereby triggering the collapse of adaptations we already have. Selection pressures, you see, don't just *build* adaptations; they keep them in place once they're there – a process biologists call *stabilizing selection*. If the selection pressures stabilizing an adaptation are weakened, natural selection can no longer filter out harmful new mutations in the genes that shape that adaptation. As a result, mutations start to drift unopposed into the gene pool, like Hitler into Austria, ultimately causing the adaptation to disintegrate.

This is not an uncommon occurrence. When our distant primate ancestors started munching on vitamin C-rich fruit, they no longer needed to synthesize vitamin C in their own bodies, and they soon lost that ability. Likewise, when our more recent hominin ancestors started eating meat, they no longer needed to synthesize vitamin A or vitamin B12, and they lost the ability to do that. (This is one line of evidence, incidentally, that humans are natural-born meat eaters.) The broader lesson is that, when it comes to adaptations, the use-it-or-lose-it principle applies.

Could human culture and technology cause adaptations to deteriorate? Consider an animal example. Among baboons, females with larger rumps get more matings. There's therefore a selection pressure for bigger bottoms in female baboons. In the wild, there are non-negotiable limits on how big their butts can get. If females can't move around easily or escape from hungry predators, they tend not to pass on their big-butt genes. In captivity, though, this selection pressure is relaxed and the size limits are removed. One consequence is that, in some zoos, female baboons are evolving butts that are too big. This isn't an aesthetic judgment; their butts are getting so big that they could never return to the wild. As Vicky Melfi of Paignton Zoo remarked, "it's very important for us as a zoo to understand the implications if we are breeding animals that don't have this environmental cut off. Their bottoms shouldn't be allowed to just keep getting bigger."[121]

What about humans? Have our bottoms been getting bigger, metaphorically speaking? In other words, has culture caused any human adaptations to erode, rendering us unfit for life in the state of nature? The answer is probably yes. As the archaeologist Timothy Taylor notes, for a long time now, humans have lived in the "rain shadow" of technology, insulated from many of the selection pressures that traditionally assailed us.[122] This has fostered what Taylor calls the "survival of the weakest." So, for instance, once we began building artificial shelters and devising clever tools, we had less need for big, calorie-hungry muscles. Consequently, we evolved to be much punier specimens of apehood. Compared to our cousins the chimps, even the strongest of strong men are giant weaklings – the least weak of a weak bunch.

Another, more recent casualty of culture is our eyesight. In hunter-gatherer times, individuals with severe myopia were less likely to spot a lion or notice the edge of a cliff, and thus less likely to pass on their myopia genes. Nowadays, glasses and other technologies mean that the short-sighted can see just as well as our most sharp-eyed ancestors. Indeed, with modern technology, even people with severe myopia can see everything from tiny germs to remote galaxies. Myopia is partly due to novel environmental factors, such as too much close-up work and not enough sunlight.[123] Like virtually everything, though, it's also partly a product of genes. That means that by enabling the short-sighted to survive and reproduce, our technology could be slowly eroding the genetic quality of human eyesight. Of course, there's nothing we can do about this. We can't deprive short-sighted people of glasses, so that they'll wander into traffic and therefore fail to pass on their short-sighted genes. We simply have to recognize that the longer we live with technology, the more dependent on it we'll inevitably become – not just habitually but biologically as well.

The Future is Unwritten

> As a result of a thousand million years of evolution, the universe is becoming conscious of itself, able to understand something of its past history and its possible future. This cosmic self-awareness is being realized in one tiny fragment of the universe – in a few of us human beings. Perhaps it has been realized elsewhere too, through the evolution of conscious living creatures on the planets of other stars. But on this our planet, it has never happened before.
>
> —*Julien Huxley (1957), p. 13*

We started our journey by looking at our species through the eyes of an outsider: an intelligent being from another world, for whom all that we take for granted about ourselves and our lives is a puzzling novelty. In the intervening chapters, we've seen that many of the mysteries this perspective raises can be solved with a simple set of Darwinian principles. Human beings are animals, and like all animals, we have a nature, crafted and honed by natural selection over countless generations. Human nature is, at bottom, a complex set of strategies for passing on the genes giving rise to it. This simple insight explains many of the most central and deepseated features of our species. It explains the fact that men are more interested than women in casual sex and more inclined toward violence. It explains the fact that women are choosier than men about their mates and generally more parental. It explains the fact that we fall in love and get jealous. It explains the fact that we love our children more deeply than we love anyone or anything else in the world. And it explains the fact that we tend to favor relatives over non-relatives, while also managing to cooperate extensively with non-relatives most of the time.

Perhaps most significantly, though, an evolutionary perspective explains the fact that human beings are cultural animals: animals designed to import into our heads ideas and behaviors from the world and the people around us. Like every aspect of human nature, our knack for culture evolved initially as a gene-copying strategy: just another way to make a living in a harsh Darwinian world. But unlike any other gene-copying strategy – unlike trunks or feathers or lust or fear – our talent for culture opened up an entirely new arena for evolution by natural selection. It brought into existence a new replicator: the meme. And memes had a very different agenda than the genes that made them possible. As memetic evolution picked up steam, humans were transformed. No longer were we devices designed solely to pass on our genes. Suddenly, we became hybrid creatures, torn between passing on our genes and passing on our memes. This vision of our species helps to explain much of what most puzzled the alien scientist: our moral systems, our religions, our art and music and science. Cultural evolution is the key to unravelling the deepest mysteries of the human animal.

If this is right, then our alien scientist would have just one last question: What's next? What does the future hold for the gene–meme hybrids we call human beings? Will we escape the Earth and colonize other worlds, or will we drive ourselves to extinction? Will we engineer ourselves into a species of Einsteins, or will our intellectual faculties deteriorate, like our ability to make vitamin A? Will we cast off our superstitions by exposing them to rational scrutiny, or will our superstitions evolve into more virulent forms, like bacteria in response to antibiotics? Will we tame our inner demons – our tendency to scapegoat, our proneness to moral panics – or will we just keep on swapping one fashionable prejudice and mass delusion for another until the very end of time?

The future is notoriously difficult to predict, even for hyperintelligent alien scientists. As such, a wise alien would refrain from taking too confident a guess. One thing is certain, however. With each passing moment, the answer to all these questions falls more and more squarely into our own hands. In our earliest days, as we slipped into existence from the womb of our apish ancestors, we were entirely at the mercy of the nonhuman world: the weather, predators, famine, and disease. Today, however, after thousands of years of accelerating cultural evolution, the tables are rapidly turning. We're no longer at the mercy of the nonhuman world; the nonhuman world is largely at our mercy. The evolution of culture has been the ultimate game-changer for our species. It has enabled us to understand ourselves and the world to a degree far beyond what a neutral observer could reasonably expect of an ape. It has allowed us to start reshaping the world in accordance with our wishes and whims. And it has begun to entrust us with the power to direct not only our own evolution but the evolution of all other life on this planet. This is an awesome responsibility, and one we may or may not be fit to carry. Whether we like it or not, though, our evolving culture is pushing our species ever-more firmly into the driver's seat of planet Earth as a whole. For better or for worse – perhaps for better *and* for worse – this appears to be the destiny of the strangest animal in the world: the ape that understood the universe.

Appendix A

How to Win an Argument with a Blank Slater

Boy, this shit ticks me off!

—*Jonathan Marks on evolutionary psychology, cited in*
Hagen (2005), p. 145

Evolutionary psychology has a unique ability to divide opinion. Some people love it; it captivates them and excites them in a way that academic fields rarely do. Others, though, hate it. They don't just disagree with some of its conclusions. They *hate* it. Arguably, some of the blame for this lies with evolutionary psychologists themselves. Some can be overzealous about the field and dismissive of other approaches, and a handful take a "shock jock" approach to science, delighting in causing offense. But this isn't the only reason for the antipathy. Some critics of evolutionary psychology seem to suffer from a malady which Martin Daly and Margo Wilson dubbed *biophobia*. Biophobia is an allergy to evolutionary and genetic explanations for human psychology – an allergy that seems to be particularly prevalent among social scientists. People afflicted with this debilitating condition often agree that "*Obviously*, humans evolved, and obviously nature and nurture are both involved in shaping the human mind!" Having made such declarations, however, they then fight tooth and nail any specific claims about evolved sex differences, genetic contributions to intelligence, or anything else in that vein. Some critics seem to think that, if we entertain such ideas, it will inevitably promote sexism and racism, dissolve responsibility, and maybe even lead to eugenics. No wonder they're in denial about genes and evolution! As we'll soon see, though, such worries are groundless. Certainly, there are reasonable criticisms of evolutionary psychology. But there are also many weak and fatuous criticisms, which cling to the field like flies. In the following pages, we'll dissect some of the most common.

1. *"Evolutionary psychology aims to justify traditional sex roles and societal inequalities. It is right-wing propaganda dressed up in the guise of science, carefully designed to prop up the status quo."*

This first criticism is more a conspiracy theory than a sound academic critique. First, leaving aside the automatic assumption that "right-wing" equals "bad," most evolutionary psychologists (like most academics in general) lean to the Left politically.[1] Second, and more importantly, the criticism is premised on an unjustified, unjustifiable assumption, namely that if something has an evolutionary origin, it must therefore be acceptable, legitimate, or good. This assumption probably explains why the politically correct view on issues such as traditional sex roles and violence is that they're entirely due to nurture, whereas the politically correct view on homosexuality is that it's entirely due to nature. But the whole assumption is misguided. It's an example of what philosophers call *the naturalistic fallacy*: the fallacy of concluding that what is natural must be good.[2] The reason this is a fallacy is that, in a Darwinian universe, there's no reason to think that naturalness and goodness will necessarily coincide. This applies as much to adaptations as to anything else. The fact that a trait is a product of natural selection means that it passed the test of survivability. How would that imply that it also passed the test of moral permissibility or goodness? To assume that it would is like assuming that, because a person won a gold medal for swimming, that person must also have won a gold medal for golf: It simply doesn't follow. Whether a trait is natural or unnatural is irrelevant to the question of whether it's good. Moral worth should be judged not in terms of the naturalness of a trait, but rather in terms of how that trait impacts the wellbeing of everyone affected by it. Thus, violence is natural but bad, medicine unnatural but good.

2. *"Even if evolutionary psychology doesn't imply that bad things are actually good, it does still provide an excuse for bad behavior – in particular, for the kinds of bad behavior that men specialize in, such as uncommitted sex, infidelity, sexual coercion, and rape. Evolutionary explanations provide a biological version of the Nuremberg defense: 'I was just obeying orders – the orders of my genes.'"*

The first thing to say here is that, if this is a problem for evolutionary explanations, it's also a problem for sociocultural ones. Why would one type of causal explanation dissolve people's responsibility for their behavior but the other leave it intact? The second thing to say is that, even if evolutionary explanations *did* dissolve responsibility in a way that other explanations did not, this wouldn't imply they were false – not unless there were some mechanism in nature that ensured that the way things are is the way that we'd like them to be. And if that were the case, you could get rich just by digging up your backyard, because nature would presumably have made sure that pirates had buried their treasure there.

Both these points are moot, however, because there's no good reason to think that evolutionary explanations dissolve responsibility. To see why, we need to think about the reasons we hold people responsible in the first place: why we reward people for good behavior and punish them for bad. What function does this practice serve? The short answer is that it incentivizes good behavior, deters bad behavior, and protects the rest of society. But it can do all those things whether the intentions and the motivations that drive people's behavior come from the environment, from genes, or (as is most probable) from a mixture of both. Regardless of the precise causes of human behavior, explaining our crimes and misdemeanors is not the same as excusing them.[3]

3. *"Even if evolutionary psychology doesn't justify bad behavior, and even if it doesn't imply that we can't hold people responsible for their actions, it does imply that certain undesirable phenomena – old-fashioned sex roles, xenophobia, war and aggression – are ultimately ineradicable. And that's a terrible message."*[4]

First, it's worth noting again that, even if this were true, it would not imply that evolutionary psychology is false. Wishful thinking is a bad argument. Second, it's not true. The fact that something has an evolutionary origin doesn't imply that it's necessarily fixed or immutable. As far as I'm aware, no evolutionary psychologist holds that it does. Most explicitly hold the reverse. Steven Pinker, for instance, wrote a long book arguing that violence and war have declined over the decades, centuries, and millennia, despite the fact that aggression is part of human nature.[5] The decline happened because aggression isn't the *only* part of human nature. Empathy, the moral sense, reason, and self-control are integral to our nature as well, and given the right circumstances, these "better angels" can trump our baser instincts. "Evolved" does not mean "inevitable."

On reflection, this shouldn't be too surprising. Anyone who's been on a diet, or stayed up all night to watch the sunrise, knows that people can resist acting on evolved motivations. Admittedly, it's not always easy; going against human nature is often like swimming against the tide. But that's not evolutionary psychologists' fault! And if we've got any chance of improving our situation, it's presumably not going to come from deluding ourselves about the causes of our behavior. As policy wonks like to say: wrong diagnosis, wrong cure.

4. *"Evolutionary psychology is the latest incarnation of a disturbing trend in Western thought: genetic determinism. Practitioners of this dangerous pseudoscience assume that genes – often single, solitary genes – contain the blueprints for entire traits. But this is a naive and simplistic conception of human development, and one which overlooks the crucial role of the environment in shaping who we are."*

How should evolutionary psychologists plead to the charge of genetic determinism? It depends. If genetic determinism is taken to mean that genes and genes alone control the development of any trait, then the plea should be "not guilty." No one holds that view. Anyone who did would presumably think that identical twins are literally identical, as opposed to just surprisingly similar – but I challenge the critic to find a single evolutionary psychologist who thinks that. Genetic determinists of that ilk are mythical creatures, like griffins, centaurs, and the Freudian unconscious. Evolutionary psychologists hold an entirely uncontroversial view of development, according to which every trait develops through a complex interaction between genes and non-genetic factors. They realize of course that most traits are influenced by many hundreds or thousands of genes. And they realize as well that, rather than containing blueprints for traits, genes simply nudge the developmental process in this direction or that, making some outcomes likelier than others. This view of development is standard among evolutionary psychologists, and consistent with all their theories.[6] Somehow, however, many critics mishear the modest claim that "genes are involved" as the ludicrous claim that "genes are *all* that's involved."

So, if genetic determinism is taken to mean that genes alone shape who we are, then evolutionary psychologists must plead "not guilty" to the charge of genetic determinism. But what if it's taken to mean simply that genes play a role – a large role – in shaping the mind and behavior? Well in that case, evolutionary psychologists would have to plead "guilty as charged." However, given that definition, *everyone* should have to plead guilty as charged to the crime of genetic determinism. There's now overwhelming evidence that genes have a profound impact on many features of human psychology previously attributed to learning and culture alone. This includes many sex differences, the basic emotions, and even some complex emotional states, such as love and romantic jealousy (see Chapters 2 to 5). If the critic wants to attribute these things solely to the environment, rather than to genes and environment jointly, then it's the critic who holds the extreme view, not the evolutionary psychologist.

Why, though, do people think that, according to evolutionary psychologists, everything is down to genes? Part of the problem is the widespread assumption that, if a trait is an adaptation, it must therefore be a product of nature rather than nurture, genes rather than learning. This assumption is seductive but false. In many cases, learning is part of the developmental program that *installs* the adaptation. A good example concerns play-fighting in kittens. Play-fighting looks like fun. It *is* fun. But it's also a serious business, because it helps the animal to develop its species-typical hunting and fighting skills. It's easy to imagine genes that bias feline development in such a way that kittens are naturally inclined to jump around and play. These genes were presumably selected because jumping and playing stimulated the development and fine-tuning of the animals' hunting and fighting skills. Thus, hunting and fighting are adaptations – they're there because selection favored them – but they're also partly learned. Learning plays a crucial role in many human adaptations as well, including kin altruism, incest avoidance, and language.

5. *"Hypotheses in evolutionary psychology fail the most basic test of scientific validity: They're unfalsifiable. There's no way to prove them wrong, even in principle. After all, we don't have a time machine and behavior doesn't fossilize!"*

First point: If you argue that evolutionary explanations are unfalsifiable, you can't also argue that they're false. Many critics do, but the two claims contradict each other. Second point: If evolutionary explanations are unfalsifiable, then competing sociocultural explanations are unproven. After all, if we'd shown that the sociocultural explanations are true, we would simultaneously have shown that the evolutionary ones are false, which would imply that they're falsifiable. If the evolutionary explanations are *not* falsifiable, as the critics assert, then we can't have shown that they're false and thus we can't have shown that any competing theories are true. In short, anyone who uses the unfalsifiability argument cannot then claim that evolutionary explanations are false or that sociocultural explanations are true. That's a little awkward.

But is it actually the case that evolutionary explanations are unfalsifiable? This claim is itself falsifiable – and has been repeatedly, resoundingly falsified. *Many* hypotheses in evolutionary psychology have been discarded after empirical scrutiny. The most famous is E. O. Wilson's adaptationist explanation for homosexuality.[7] According to Wilson, homosexuality is a product of kin selection: People with a same-sex sexual orientation enhance their inclusive fitness by helping care for siblings, nieces, and nephews. In other words, gay men and lesbians are functionally equivalent to non-reproductive worker insects. This hypothesis was taken seriously for many years. As the data began to roll in, though, it became apparent that Wilson was wrong and that homosexuality is not an adaptation at all. Wilson's theory was falsified.[8]

Another example comes from David Buss, one of the founders of evolutionary psychology. In his famous cross-cultural study of human mate preferences, Buss predicted that, in every culture, men would have a stronger desire than women to marry someone with no prior sexual experience: a virgin. Sure enough, in every culture where there was a difference, it was in the predicted direction. However, in nearly half the cultures Buss surveyed, there was no sex difference either way. Furthermore, there were large cross-cultural differences in the extent to which virginity mattered to anyone: In some cultures, both sexes cared a lot about virginity; in others, neither did. Once again, the hypothesis had to be abandoned.[9] These cases, and others like them, falsify the hypothesis that hypotheses in evolutionary psychology are unfalsifiable.

Perhaps a better way for the critic to frame the criticism would be to say that it's *difficult* for evolutionary psychologists to prove that their explanations are definitely correct and to rule out other, non-evolutionary explanations. But this is just a special case of a more general truth, namely that it's difficult for anyone to prove that *any* explanation is definitely correct and to rule out other explanations of any kind. There's no magic-bullet study that can single-handedly establish that a given trait is an adaptation – or that it's not. The best we can ever do is to look at the preponderance of evidence and see which way it points.

6. *"Evolutionary psychologists sit around spinning* just-so stories *about traits we already know we have, and then conclude that their stories are true based solely on surface plausibility. This is a recipe for scientific*

disaster, because you can make up a plausible-sounding adaptationist story about anything."[10]

It's probably fair to say that evolutionary psychologists sometimes accept adaptationist hypotheses on insufficient evidence. I'll say more about this soon. However, to suggest that the entire field consists of nothing more than just-so storytelling is unfair. Evolutionary psychologists *start* with just-so stories – also known as *hypotheses*. But they then go out and test their hypotheses, and publish the findings in peer-reviewed journals. In doing so, they convert the just-so stories into tentatively accepted or rejected scientific claims. This isn't an occasional exception to the just-so story rule; there are literally thousands of published empirical studies in evolutionary psychology. It's also worth pointing out that *all* scientific hypotheses are just-so stories before they're tested. Why do people only level the criticism at *evolutionary* hypotheses? In most cases, it's purely because they've heard other people doing it. They're parroting a criticism that's been bouncing around since the sociobiology controversy first ignited in the 1970s. Back then, in the early days of the field, the criticism had some merit. Today, though, it's simply outdated. Certainly, you can make up adaptationist explanations for pretty much anything. But you can make up *sociocultural* explanations for pretty much anything too. Why should the sociocultural explanations win by default?

7. *"Evolutionary psychologists present a highly ethnocentric vision of the nature of the human mind. From sex differences to family arrangements, they mistake the transient idiosyncrasies of Western culture for unchanging facts about human nature. And it's little wonder. Evolutionary psychologists usually only test their theories on undergraduate students from WEIRD nations – that is, nations which are Western, Educated, Industrialized, Rich, and Democratic."*[11]

There is always a risk of mistaking elements of one's own culture for aspects of human nature, and we should clearly be alert to this pitfall. But there's also a risk of making the opposite blunder: of mistaking aspects of human nature for elements of one's own culture. It's odd to focus on one possible error while steadfastly ignoring the other. It's a little like watching someone teetering precariously on a tightrope and shouting "Be careful not to fall to the left! Whatever you do, don't fall to the left!" It's especially odd given that, since at least the mid-1900s, social scientists have more often "fallen to the right" – that is, they've more often mistaken human nature for local culture than vice versa. Generations of Western social scientists, misled by a blank slate view of the mind, denied non-Westerners even basic facets of our shared humanity, such as romantic love, the facial expressions of emotion, and grief in response to the death of an infant.[12] Shunning evolutionary explanations is no guarantee you'll avoid error.

As for the claim that evolutionary psychologists too often test their theories only on undergraduates or WEIRD populations, I'm willing to concede this point, up to a point. WEIRD populations are psychologically atypical in a number of ways, and the field would no doubt be improved if more evolutionary psychologists tested their theories on a wider range of people.[13] To be fair, though, this is a problem for all of psychology, not just evolutionary psychology. And evolutionary psychologists have a better track record than most in this regard. Some of the best-known studies in evolutionary psychology are also some of the largest cross-cultural studies in psychology as a whole. They include David Buss's research on sex differences in mate preferences and David Schmitt's research on sex differences in sexual inclinations.[14] On top of that, evolutionary psychology draws heavily on research in anthropology and animal behavior, and is grounded in deep sociobiological principles that apply to everything from butterflies to buffalo. This helps evolutionary psychologists to set aside their cultural biases, and pay proper attention to things that WEIRD psychology often overlooks, such as the importance of kinship and genetic relatedness in all known human societies.[15] Needless to say, there's always room for improvement. But it would be easy to overstate how WEIRD evolutionary psychology is. It's reasonably normal.

8. *"Evolutionary psychologists claim that everything is an adaptation. They wantonly apply adaptationist explanations to every nook and cranny of human life. They're guilty, in other words, of what's sometimes called* panadaptationism."

This criticism came originally from the paleontologist Stephen Jay Gould.[16] When I first went into evolutionary psychology, I shrugged it off. After all, there are plenty of things that evolutionary psychologists fully acknowledge are not adaptations. Just for a start, virtually all would agree that agriculture, reading and writing, complex mathematics, and formal logic are not adaptations but cultural tools.[17] And evolutionary psychologists are hardly shy about deploying non-adaptationist explanations. The mismatch explanation for obesity, for instance, is a non-adaptationist explanation, but it's also one of the most famous ideas in evolutionary psychology's oeuvre. It is true that evolutionary psychologists have attempted to apply the adaptationist framework to a very wide range of phenomena. But wherever you find an adaptationist hypothesis, you almost always find a non-adaptationist hypothesis too. Thus, while some evolutionary psychologists argue that female orgasm, homicide, music, rape, and religion are adaptations, others argue that these things are instead by-products of other traits.[18] In short, it's simply not true that evolutionary psychologists think that everything is an adaptation, and that they never consider other possibilities.

So, initially, I rejected the charge of panadaptationism. As time went on, however, I began to worry that maybe Gould was onto something. (For an evolutionary psychologist, this is like confessing a murder!) Certainly, no one argues that *all* traits are adaptations. But evolutionary psychologists may too often *overextend* the adaptationist framework, and accept adaptationist explanations on relatively flimsy grounds. Here are some hypotheses that I suspect push the adaptationist envelope too far. (I acknowledge that I could be wrong in any particular case.)

> Yawning is an adaptation for cooling the brain. • Dreaming is an adaptation for practicing important behaviors "offline." • Premenstrual syndrome is an adaptation for dumping one's current partner if he fails to get you pregnant. • Male masturbation is an adaptation for offloading sperm before it gets too old and decrepit. • The flared head of the human penis is an adaptation for removing other men's sperm from the woman's reproductive tract. • Autism is an adaptation for solitary foraging.[19]

All of these hypotheses set off my panadaptation alarm. Of course, the fact that someone, somewhere, has proposed a dodgy adaptationist explanation doesn't mean that all or even most evolutionary psychologists accept it. For myself, I like to think that my training in evolutionary psychology means that, rather than buying any old adaptationist hypothesis, I'm a more discerning consumer of these hypotheses. And although some evolutionary psychologists accept weak adaptationist explanations, many critics of the field reject strong ones. Still, it's all-too-easy to go crazy with adaptationist reasoning. If we're not careful, we'll be guilty of what Gould accused us of all those years ago: overextending the adaptationist mode of explanation.

I hope I've made it clear that I don't think that evolutionary psychology is perfect. It's not. But many of the criticisms leveled at the field are simply misguided. Critics who rail against status quo bolstering, genetic determinism, and just-so storytelling are like the crazy person in the bus shelter, fighting with a sparring partner who isn't really there: They've invented their own evolutionary psychology and are arguing loudly with that. Unfortunately, onlookers might not always realize that the crazy person is fighting an imaginary foe, and might mistake the critics' version of evolutionary psychology for the real deal. That's why it's important to tackle these criticisms. It has to be said, though, that it's a rather frustrating task. No matter how often the criticisms are refuted, they keep coming back from the dead, like zombies in a B-movie. Here's how the evolutionary psychologist Robert Kurzban summed up the situation:

> Critics assert that evolutionary psychologists are wrong in believing behavior is genetically determined, that every aspect of the organism is an adaptation, and that discovering what is informs what [should] be. Evolutionary psychologists reply that they never made any of these claims, and document places where they claim precisely the reverse. The critics then reply that evolutionary psychologists are wrong in believing behavior is genetically determined, that every aspect of the organism is an adaptation, and that discovering what is informs what should be.[20]

Fighting the evolutionary psychologists' corner is like weeding a garden, or cutting the head off a hydra. It's like a Nietzschean eternal recurrence, or pushing Sisyphus's rock up the hill, again and again, forever. And it's also a pain in the butt. But it is worth the effort. Why? Simple: Because the evolutionary psychologists are right! More precisely, they've taken a big step in the right direction. And that's why, in the end, they're going to win the argument.

Appendix B

How to Win an Argument with an Anti-Memeticist

[I]f you contribute to the world's culture, if you have a good idea, compose a tune, invent a sparking plug, write a poem, it may live on, intact, long after your genes have dissolved into the common pool ... The meme-complexes of Socrates, Leonardo, Copernicus and Marconi are still going strong.

—Richard Dawkins (1976/2016), p. 259

Memetics has never been a punching bag in quite the way that socio-biology and evolutionary psychology have. But there's been plenty of discussion of the field and plenty of criticism too. While some think that memetics is the answer to the riddle of human life, others think it's junk – and dangerous junk, at that. Luis Benitez-Bribiesca, for instance, warned that memetics is "a dangerous idea" and a "pseudoscientific dogma," which "poses a threat to the serious study of consciousness and cultural evolution."[1] In this appendix, we'll consider some of the main objections to the field, and see whether such dire prognostications are warranted.

1. *"What exactly* is *a meme? The whole concept is poorly defined – and we can't do good science with poorly defined concepts."*

This is one of the most common criticisms of memetics. Let me start my response to it with a concession: Memes are indeed difficult to define. It's hard to say where one meme ends and the next begins. As Susan Blackmore once asked, are the first four notes of Beethoven's Fifth a meme? Or is the entire composition a meme, and the first four notes just a fragment of that meme? Are they both memes?[2] You say tomayto; I say tomahto – are these two memes or only one? Is a wheel a meme, or is the *idea* of a wheel a meme? There are no definite answers to any of these questions.

So, at one level, the critics are right: It is hard to define memes. The question is, however, whether this justifies writing off memetics altogether. And the first thing to say about this is that, if it does, then it also justifies writing off a whole lot more than memetics. There's a large literature in the social sciences looking at the *diffusion of ideas* – that is, at how ideas and technologies spread. We'd have to throw all of that away for a start. After all, what exactly is an idea? The whole concept is poorly defined! And when you think about it, so are all the concepts we use to describe culture: belief, custom, norm, ritual – even the concept of culture itself (and the concept of concept). If we're going to reject memetics because memes are fuzzy around the edges, then we're also going to have to reject most theories of human culture and most of the social sciences. And nor would the purge end there. There's no perfect, incontestable definition of genes, so we'd have to throw away genetics. Similarly, there's no non-arbitrary dividing line between hills and mountains, or between different types of clouds, so we'd have to throw away geography and meteorology. In short, if we took the "what exactly is a meme?" criticism seriously, and if we were consistent, most of science would be consigned to the garbage heap.

So, this is our situation: Either the criticism is legitimate and we need to throw away large swathes of science, or there's something wrong with the criticism and we need to throw that away instead. The success of science stands as a towering monument to the latter option. From chemistry to cosmology to consciousness studies, many areas of science have made startling progress without perfectly precise definitions. Meanwhile, one of the great meta-discoveries of philosophy is that it's extremely difficult to define anything. Philosophers have yet to settle on definitions for even such basic concepts as *knowledge* or *truth*. (Google "Gettier problems" if you dare.) We need to avoid what I call *the philosopher's trap*: getting so hung up on definitions that we never leave the starting gate and begin making new discoveries about the world. As Laland and Brown observe, we should stop worrying about finding a perfect definition for the word "meme," and just see where we get with the research.[3]

2. *"Memetics adds nothing to the stockpile of human knowledge. All it does is replace the word 'idea' with the word 'meme,' and then declare itself a science. Its alleged scientific insights are merely the pretentious codification of common sense. As the neurophilosopher Paul Churchland*

observed, memetics boils down to the trivial observation that 'good ideas spread.' "

Let's start with the suggestion that the meme concept is merely a scientific-sounding replacement for the everyday word "idea." For several reasons, this criticism misses the mark. First, memes *aren't* just ideas. They're anything that can be passed on socially, including mannerisms, rituals, and practices. Second, any time the word "meme" is used, it references the hypothesis that the elements of culture don't necessarily flourish because they're advantageous to us or our groups. They flourish – *if* they flourish – because they're advantageous to themselves. This hypothesis is not implicit in the word *idea*, and it's also far from obvious. Furthermore, it may or may not be true. Like hypotheses in evolutionary psychology, it's potentially falsifiable.

What about the suggestion that memetics simply re-describes agreed-upon facts? The skeptic and math popularizer Martin Gardner took this view when he wrote that "memetics is no more than a cumbersome terminology for saying what everybody knows."[4] But was he right? The first point to make is that anyone who suggests that memetics tells us nothing we don't already know is tacitly agreeing that memetics is accurate, in general if not always. To be consistent, these people can't then turn around and argue that memetics is false. The second point – a rather important point, one might think – is that it's *not actually true* that memetics simply re-describes agreed-upon facts. Consider some of the traditional theories of religion.[5] According to the wish fulfilment theory, religious beliefs spread because they're good for us: They comfort us and help us through the hard times. According to the social glue theory, religious beliefs spread because they're good for societies: They bind individuals together into socially cohesive groups. According to the social control theory, religious beliefs spread because they're good for leaders: They help keep the rabble in line. And according to adaptationist theories, religious beliefs spread because they're good for our genes: They enhance the transmission of the genes inclining us toward religion. The memetic theory of religion, on the other hand, claims something else entirely. It claims that religious beliefs spread purely because they're good *for themselves*. Thus, anyone who argues that memetics simply re-states the obvious must think that it's just obvious that the traditional theories of religion – many of which have held scholars' attention for more than a century – are false, and that religions are obviously just designed to perpetuate themselves. If this is so obvious, though, why are only the memeticists making the claim?

As for Paul Churchland's assertion that memetics boils down to the banal observation that "good ideas spread," there's really only one problem with this: It's not what memetics claims! On the contrary, it's central to the memetic perspective that bad ideas sometimes spread as well – think earworms and smoking tobacco. Of course, critics could respond that, when they talk about "good ideas," they don't mean ideas that are good for *us*. Instead, they mean ideas that are good at spreading. Taking this tack, the criticism would be that memetics boils down to the trivial-sounding claim that "ideas that are good at spreading tend to spread." *Does* it boil down to this claim? In a sense, yes – but in exactly the same sense that evolutionary theory boils down to the trivial-sounding claim that "*genes* that are good at spreading tend to spread." In both cases, the simplistic summary hides a world of interesting and unexpected implications. For memetics, this includes such ideas as that memes evolve to propagate themselves in the culture, that religions and other institutions possess adaptations designed to help them survive, and that clusters of mutually supportive memes evolve to grab our attention and convert us into their mouthpieces (see Chapter 6). Critics of memetics should focus their efforts on contesting claims like these, rather than denying that memetics makes any contestable claims in the first place.

3. *"The hallmark of a good scientific theory is that it generates research: It makes novel predictions about the world, which lead scientists to make otherwise unexpected discoveries. Memetics, however, has been woefully unsuccessful on this front. Indeed, the field's flagship journal,* The Journal of Memetics, *had to close its doors because it didn't get enough submissions."*

This is the "if you're so smart, why aren't you rich?" criticism. Of all the criticisms on offer, it's probably the one that worries me most. In the end, though, I think it fails. It is certainly true that memetics has yet to deliver much in the way of new research. It's also true that many specific meme-based explanations have yet to be adequately tested. However, when it comes to evaluating a theory or explanation, what we ultimately want to know is not how many publications it's generated, or how many surprising discoveries, but rather something far more basic: whether or not the theory is true. That, in the final analysis, is what science is all about. And despite the current research shortfall, there's good reason to believe that, at least in its general outline, memetics is indeed a true and accurate theory.

My main argument for this conclusion is as follows. The key difference between memetic and non-memetic approaches to cultural evolution is in their answer to the question: What are cultural entities designed to do? For non-memetic approaches, the answer is that they're designed to benefit the individual or the group. For memetics, on the other hand, the answer is that they're designed to benefit themselves. Often they do this by benefiting the individuals or the groups in which they reside, consistent with the other theories. Sometimes, however, they benefit themselves *without* benefiting the individual or the group, or even while actively harming them. We saw various examples of this in Chapter 6, including chain letters, hoax virus warnings, earworms, cigarettes, and verbal tics such as saying "like" in every sentence. For non-memetic approaches, these are anomalies – awkward exceptions to the rule that memes persist when they're good for individuals or groups. For memetics, in contrast, they're confirmation of the broader rule that memes persist when they're good for themselves, regardless of whether they're good for us or our groups. In short, memetics provides a covering rule that incorporates the other approaches but also goes beyond them. Its greater explanatory reach suggests that the memetic approach more accurately depicts the nature of cultural evolution.

This isn't to suggest that memeticists don't need to provide new evidence for their theory; they do. However, the fact that the theory does a good job of explaining the evidence we already have gives us a strong reason to think that it's broadly accurate. And that in turn suggests that the existing research deficit needn't be a permanent affliction.

4. *"Evolution can only happen when you have high-fidelity copying of replicators. If genes weren't copied perfectly most of the time, natural selection could never promote particular gene variants, because they'd degrade before it had a chance. Here's the problem for memetics. Although genes are generally copied perfectly, memes rarely are. If you read a story to a room full of children, each will recall a slightly different version of it. How could selection get a fix on any meme when memetic transmission is so imprecise?"*[6]

This criticism sounds menacing in the abstract, but shatters the moment we drill down into the details. Clearly, memes are not copied as faithfully as genes. Equally clearly, though, culture is not just a giant game of Chinese whispers (or what Americans call the telephone game). Memetic transmission is accurate enough that scholars can track the spread of ideas across history, and children can chat with their grandparents, despite learning the language half-a-century later.[7] Tackling the issue from the other direction, if memetic transmission *weren't* generally accurate, cumulative cultural evolution would be impossible: Good ideas would quickly degrade as they passed from person to person, preventing us from building on each other's insights and innovations. Gene–culture coevolution would also be ruled out. As the geneticists Luigi Cavalli-Sforza and Marcus Feldman note, the evolution of lactose tolerance relied on high-fidelity copying of the milk-drinking meme. Lactose-tolerant parents must have passed on not only their lactose-tolerant genes to their offspring, but also the habit of drinking milk. Thus, at least some memes are copied faithfully enough to cause historically significant evolutionary change in the species. Like bumblebee flight and Wikipedia, the meme concept might not work in theory, but it works just fine in practice.

5. *"All right, you dodged that bullet. But there are other important differences between genetic and cultural evolution – differences that sink the ship of memetics. One is that, whereas genes are copied from organism to organism, memes are not copied from mind to mind. As Dan Sperber points out, they're reconstructed in each new mind that adopts them. People are exposed to the barest bones of an idea, and then fill in the gaps based on their background knowledge and innate psychological biases. As such, the fact that certain ideas are widespread doesn't necessarily mean that they were selected over competing variants. It may just reflect the fact that the minds containing these ideas are structured in similar ways."*[8]

First things first, what exactly does it mean to say that ideas are reconstructed rather than copied? A good illustration comes from Pascal Boyer's work on the concept of ghosts.[9] The ghost concept is found in cultures all over the world, and wherever it's found, people construe these supernatural entities in remarkably similar ways. Specifically, ghosts are construed as beings that don't have solid bodies, but which are psychologically a lot like us: They can see and hear; they have beliefs; they have desires; they have intentions. How do we explain the fact that people in diverse cultures converge on this conception? According to Boyer, the explanation is *not* that they acquire each element of the ghost concept through a process of social learning. Social learning is involved, sure. But for the most part, each individual builds up the concept from scratch. As soon as people get the idea that ghosts are sentient beings, rather than inanimate objects, they automatically infer that ghosts have sensory experiences, beliefs, desires, and intentions. And this, argues Boyer, is because human beings have an evolved tendency to construe sentient beings in this way. The ghost concept is an example of what Dan Sperber calls a "cognitive attractor": an idea that people everywhere naturally gravitate toward, as a result of the native structure of the human mind.

There's certainly some merit to this approach. If nothing else, it reminds us that we can't assume that every widespread feature of our memes is a product of memetic selection. However, rather than being inconsistent with memetics, Sperber and Boyer's ideas can profitably be integrated with the meme's-eye view. We saw in Chapter 6 that a complete science of memetics would have to include a rigorous understanding of human psychology. Innate biases and attractors would be part of that understanding. They help to explain why people come up with the memes they do, and why some memes sit more easily in the human mind than others. Biases and attractors would only undermine memetics if they left no room at all for social transmission or competition among socially transmitted ideas. But that's not realistic. Indeed, beyond explaining broad trends in the structure of our memes, it's not clear how much work attractors actually do. In some cultures, people believe that the creator of the universe took the form of a bipedal primate, was killed by other primates of the same kind, and then came back to life three days later without the aid of medical science. The fact that we have a handful of unlearned cognitive biases doesn't begin to explain why people in some cultures believe this whereas people in others don't. The only way to explain that is in terms of the social transmission of ideas. And once we've got social transmission, we've got competition among ideas to be the ones that get transmitted.

Clearly, memes are not copied directly from mind to mind in the way that genes are copied from cell to cell; they're reconstructed in each new mind they encounter. But this is only a problem for memetics if reconstructed memes are not allowed to count as memes. And why shouldn't they? Reconstructed or not, memes are subject to a winnowing process, and this process inevitably favors selfish memes.

6. "*We still haven't discussed the weakest part of the analogy between biological and cultural evolution. New genes are products of random mutations: A letter in the genetic code is incorrectly transcribed, or a chunk of DNA is inserted, inverted, deleted, or duplicated. But new* memes *are* not *products of random mutations; we deliberately invent them for particular purposes. Thus, cultural evolution is very different from biological evolution, and memetics gives a misleading account of the process.*"

Here's how Steven Pinker articulated this criticism in his book *How the Mind Works*:

Taken literally, [the meme theory] predicts that cultural evolution works like this. A meme impels its bearer to broadcast it, and it mutates in some recipient: a sound, a word, or a phrase is randomly altered. Perhaps, as in Monty Python's *Life of Brian*, the audience of the Sermon on the Mount mishears "Blessed are the peacemakers" as "Blessed are the cheesemakers." The new version is more memorable and comes to predominate in the majority of minds. It too is mangled by typos and speakos and hearos, and the most spreadable ones accumulate, gradually transforming the sequence of sounds. Eventually they spell out, "That's one small step for a man, one giant leap for mankind."[10]

This is not, of course, how new culture comes about. New music doesn't arise when someone accidentally whistles a tune incorrectly, and the new "mutant" melody happens to be catchier. Einstein didn't come up with $e = mc^2$ because he misunderstood Newton. Good ideas don't come from mishearings or misrememberings, followed by the selective retention of useful accidents. All these things, argues Pinker, are the deliberate inventions of creative minds.

No doubt, there's a lot of truth in this: Simple mistakes and copying errors are not nearly as important in cultural evolution as they are in biological evolution. But for several reasons, this is far from fatal to memetics. To begin with, new memes *do* sometimes come from simple copying errors. Here are some examples of historically significant memes that came about that way.

- "Mary was a virgin when she conceived Jesus." This was based on a mistranslation of Isaiah. The original version described Mary not as a virgin but as an unmarried woman.
- "It is harder for a rich man to get to heaven than for a camel to get through the eye of a needle." Probably it was a rope, not a camel.
- "Martyrs will be rewarded with seventy-two virgins when they enter paradise." Apparently it was seventy-two grapes or raisins – a reward that, while not entirely unwelcome, might be somewhat disappointing if you were expecting the virgins.[11]

So, random errors do play a creative role in culture. Some even argue that genuine creativity *must* involve some randomness. As the anthropologist Gregory Bateson put it:

Creative thought must always contain a random component. The exploratory processes – the endless *trial and error* of mental progress – can achieve the *new* only by embarking upon pathways randomly presented, some of which when tried are somehow selected for something like survival.[12]

It's true that many, probably most, new cultural variants don't emerge completely at random; people have explicit goals in mind and they work studiously toward them. It's also probably true that this reduces the workload placed on natural selection, at least to some degree. But random or non-random, selection does still play a role – a large role, an indispensable one. As the great William James observed, natural selection in the realm of ideas operates at two main levels: first, within the individual's mind, and second, within the wider society.[13] Regarding the first level, many have argued that creativity involves a process of variation and selective retention. The physicist and philosopher Ernst Mach noted, for example, that "Newton, Mozart, Richard Wagner, and others ... say that thoughts, melodies, and harmonies had poured in upon them, and that they had simply retained the right ones."[14] Even the best composers and songwriters come up with bland tunes and dud lyrics every now and then; we just never get to hear them. In the words of Linus Pauling, "The best way to have a good idea is to have lots of ideas."

A second selection process takes place in the wider world. As with biological evolution, there are always many more variants than can possibly survive in the culture. In rap music, for instance, there are "too many MCs, not enough mics," to quote the Darwinian rapper, Baba Brinkman. As a result, there's always a struggle for existence among the available variants. The fact that these variants aren't entirely random doesn't eliminate the struggle or negate the role of selection. On the contrary: *Most* new variants are not adopted by the culture. Every songwriter wants to write a hit song, but most songs are not hits. People can't just think "I'm going to invent the best possible cultural variant now," and then simply sit down and do it. And even if we could, it's not as if all culture is deliberately invented. Sure, some people do sit down with the aim of creating hit songs, books, or movies. But with rare exceptions, no one sits down with the aim of creating hit moral systems, religions, or languages. These cultural products almost always emerge spontaneously in the course of our interactions, without any central planner or overarching goal. Nonetheless, as we saw in Chapter 6, they often show signs of design. Much of this comes from blind cultural selection, rather than conscious, deliberate planning.

One final point: We hear a lot about "random" genetic mutations in biology, but it's important to be clear what this means. To say that a mutation is random is only to say that it's random *in relation to the organism's fitness.* But of course mutations have causes – radiation, chemicals, viruses – and they're *not* random in relation to those. (The only exception would be mutations resulting from uncaused quantum perturbations, which would be truly and irreducibly random.) The same applies to memes. Like genetic mutations, new memes are not random in relation to their causes, which are the minds and intentions of the people who create them. Most of the time, however, they *are* random in relation to their own memetic fitness. People rarely engage in creative efforts with *no other goal* than to create a meme that will survive and promote itself in the culture. They have different goals in mind: creating a song that captures a mood, a book that captures an idea, a trap that captures a bear. Thus, our creative outpourings, though not random in relation to our goals, *are* at least somewhat random in relation to the selection pressures operating on memes. The analogy between biological evolution and cultural evolution is closer than we might have suspected.

7. *"According to memetics, memes are viruses that invade our minds, rob us of free will, and act against our reproductive interests. If that were true, though, vast numbers of people would be suicide bombers or celibate. Instead, harmful memes are rare – as rare as apes with tails. Checkmate!"*[15]

Not so fast! To begin with, memetics doesn't claim that memes are *always* bad for us. It claims that, in order to be selected, they just have to be *good for themselves.* Often memes are good for themselves precisely *because* they're good for us. Memes live inside us, after all, and thus their interests and ours often overlap. Admittedly, memeticists do sometimes describe memes as mind viruses, which seems to imply they're usually harmful.[16] But that's just loose talk. A more apt analogy would be with bacteria. Some bacteria are bad for us, but plenty of bacteria are good: We couldn't survive without them. Ditto memes.

That said, it is worth emphasizing that bad memes are not quite as rare as the critic suggests. The alien scientist spotted several widespread memes that work against our reproductive interests, including condoms and the pill. We can add to that list any cultural innovation that satisfies evolved desires without us having to engage in the behavior that the desires were designed to prompt – think porn, recreational drugs, and keeping pets as surrogate children. These cultural habits, though common, are adaptively neutral at best and at worst actively maladaptive (in the technical sense of the term; I'm not saying they're bad, or good, or anything in between).

It is true that *highly* virulent, fitness-diminishing memes, such as suicide bombing and celibacy, are rare. It's also unsurprising: Memes that kill off or sterilize their hosts will inevitably burn themselves out before long. But that's not an argument against memetics. *Viruses* that kill off or sterilize their hosts inevitably burn *them*selves out before long, but that's not an argument against virology. On the contrary, it explains an important fact about the prevalence of different diseases, namely that the most common diseases are usually rather mild.[17] The same applies to maladaptive memes. The more virulent a meme is, the less common that meme will be. Thus, the martyrdom meme and the celibacy meme will always be less common than, say, the smoking meme or the apple-pie-and-ice-cream meme. The fact that we see relatively few highly virulent memes is not inconsistent with memetics. On the contrary, memetics makes good sense of the pattern.

Permissions

Extract from *Has Man a Future?* by Bertrand Russell (1961), reprinted by permission of the Bertrand Russell Peace Foundation.

Extract from *Sociobiology: The New Synthesis* by Edward O. Wilson (1975), reprinted by permission of Harvard University Press.

Extract from "The psychological foundations of culture," by John Tooby and Leda Cosmides (1992), in Jerome H. Barkow, Leda Cosmides, and John Tooby (Eds.), *The Adapted Mind: Evolutionary Psychology and the Generation of Culture*, pp. 96–97, reprinted by permission of Oxford University Press.

Extract from "Sexual strategies: A journey into controversy," by David M. Buss, (2003), *Psychological Inquiry, 14*, p. 225, reprinted by permission of Taylor & Francis.

Extract from *The Science of Intimate Relationships* by Garth J. O. Fletcher, Jeffry A. Simpson, Lorne Campbell, and Nickola C. Overall (2013), p. 43, reprinted by permission of John Wiley and Sons Inc.

Extract from *The Science of Good and Evil* by Michael Shermer (2004), p. 84, reprinted by permission of Michael Shermer.

Extract from *New Bottles for New Wine* by Julian Huxley (1957), reprinted by permission of Peters Fraser & Dunlop (www.petersfraserdunlop.com) on behalf of the Estate of Julian Huxley.

Extract from "Controversial issues in evolutionary psychology," by Edward H. Hagen (2005), in David M. Buss (Ed.), *The Handbook of Evolutionary Psychology*, p. 145, reprinted by permission of John Wiley and Sons Inc.

Extract from *The Selfish Gene* by Richard Dawkins (1976/2016), p. 259, reprinted by permission of Oxford University Press.

Notes

Chapter 1: The Alien's Challenge

1 The phrase comes from "The Eve of War," from the 1978 album *Jeff Wayne's Musical Version of The War of the Worlds*.
2 Hawking (2008).
3 The otherworldologist in question is actually an Earthling, the astronomer Sir Martin Rees. This paragraph draws on ideas from Rees's 2004 book, *Our Final Century*.
4 Cf. Kelly (2010); McLuhan (1964).
5 Cf. Pollan (1990).
6 Shubin (2008).
7 Barkow, Cosmides, and Tooby (1992); Buss (2014); Cronin (1991); Daly and Wilson (1988); Dawkins (1976/2016); Hamilton (1996); Pinker (1997); Trivers (2002); Workman and Reader (2014); Wright (1994).
8 Blackmore (1999); Boyd and Richerson (1985); Cavalli-Sforza and Feldman (1981); Dawkins (1976/2016); Dennett (1995); Durham (1991); Henrich (2016); Laland (2017); Lumsden and Wilson (1981); Mesoudi (2011); Richerson and Boyd (2005).
9 Alexander (1987), p. 3.
10 Dawkins (1976/2016).
11 Blackmore (1999).

Chapter 2: Darwin Comes to Mind

1 Darwin (1859); Stewart-Williams (2010).
2 For an excellent overview of evolutionary theory, and the evidence that shows that it's true, see Coyne (2009).
3 Dawkins (1986).
4 Goodall (1986).
5 Darwin (1871).

6 Hamilton and Zuk (1982); Miller (2000b); Thornhill and Gangestad (1993).
7 Zahavi (1975).
8 Andersson (1994); Darwin (1871).
9 Miller (2000b).
10 Stewart-Williams and Thomas (2013a, 2013b).
11 Hamilton (1964).
12 J. L. Brown (1975).
13 Hamilton (1963), p. 354.
14 See, e.g., Hamilton (1964); Trivers (1985, 2002); Williams (1966).
15 Dawkins (1976/2016, 1982).
16 Dawkins (1982).
17 Burt and Trivers (2006); Dawkins (1998).
18 Dawkins (1976/2016), p. 26.
19 Darwin (1871); Wynne-Edwards (1962).
20 Williams (1966); see also Dawkins (1976/2016).
21 Nowak, Tarnita, and Wilson (2010); Sober and Wilson (1998); D. S. Wilson and Sober (1994); D. S. Wilson and Wilson (2007).
22 West, Griffin, and Gardner (2007a).
23 For a recent confident proclamation regarding the demise of neo-Darwinism, along with a rejoinder, see Laland *et al.* (2014).
24 Nesse (1990).
25 Curtis, de Barra, and Aunger (2011); Rozin, Haidt, and McCauley (1993).
26 Marks and Nesse (1994); Seligman and Hager (1990).
27 Öhman and Mineka (2001).
28 Cook and Mineka (1987); Darwin (1872).
29 See, e.g., Conley (2011).
30 Pinker (2002), pp. 53–54.
31 Montagu (1973), p. 9.
32 In case you think the blank slate and the SSSM might be straw men, see the numerous flesh-and-blood examples given in Pinker (2002) and Tooby and Cosmides (1992).
33 Pinker (1997), pp. 21–22.
34 Trivers (1972).
35 Buss *et al.* (1992); Daly, Wilson, and Weghorst (1982); Symons (1979).
36 Daly and Wilson (1988).
37 D. Lieberman, Tooby, and Cosmides (2007); Westermarck (1891).
38 On the kinds of evidence that suggest that a trait may be an adaptation, see Schmitt and Pilcher (2004).
39 Miller (2000b), p. 134.
40 Polderman *et al.* (2015); Turkheimer (2000).
41 Miller (2013); Stewart-Williams and Thomas (2013b).
42 Relatively egalitarian: E. A. Smith *et al.* (2010). Lived with a mix of relatives and non-relatives: Hill *et al.* (2011). Cooked their food: Wrangham (2009). Division of labor by sex: Murdock and Provost (1973). Mothers and others cared for offspring: Hrdy (2000, 2009); Sear and Mace (2008). Violence: Pinker (2011). Language: Pinker (1994); Pinker and Bloom (1992). Camping trip: Cosmides and Tooby (1997).

43 Eaton, Konner, and Shostak (1988), p. 739; Wright (1994), p. 191.
44 Tooby and Cosmides (1992).
45 Hawks *et al.* (2007).
46 Cochran and Harpending (2009).
47 Gould (2000b).
48 Jablonski (2006).
49 Cosmides and Tooby (1997).
50 Zuk (2013).
51 Pinker (1997), p. 42.
52 Power and Schulkin (2009), p. 110.
53 Walpole *et al.* (2012).
54 Nesse and Williams (1994), p. 213.
55 Seligman and Hager (1990).
56 Bekoff and Byers (1998); Burghardt (2005).
57 Geary (1995).
58 Nesse and Williams (1994).
59 Strassmann (1997).
60 D. E. Lieberman (2013).
61 Pinker (2011, 2018); Ridley (2010).
62 Miller (2009).
63 Hahn-Holbrook and Haselton (2014).
64 Principe (2012).
65 R. A. Friedman (2014).
66 Sommers (2013b).
67 Gould and Lewontin (1979).
68 It used to be thought that the default sex for mammals was female. We now know, however, that the development of both sexes is an active, gene-governed process, and thus that there's no default sex. See, e.g., Jordan *et al.* (2001).
69 Symons (1979).
70 King and Belsky (2012); Miller (2000b); Puts, Dawood, and Welling (2012).
71 See, e.g., Fausto-Sterling (2000); Lloyd (2006).
72 Kenrick *et al.* (1995).
73 Glocker *et al.* (2009); Lorenz (1943).
74 Pinker (1997).
75 Art and music: Pinker (1997). Morality: Haidt (2012); Stewart-Williams (2015a). Religion: Atran (2002); Boyer (2001); Stewart-Williams (2015b).

Chapter 3: The SeXX/XY Animal

1 Geary (2010); Halpern (2012); Maccoby and Jacklin (1974).
2 Jussim (2012).
3 For a good discussion of this view, see Eagly (1995).
4 Bernard (1974), p. 13; Money (1987), cited in Halpern (2012), p. 8.
5 Le Bon (1879), cited in Gould (1981), p. 104.
6 Geary (2010); Halpern (2012); Maccoby and Jacklin (1974).
7 Buss and Schmitt (1993); Symons (1979); Trivers (1972).

8 See, e.g., Chasin (1977). A good account of the history of the sociobiology controversy can be found in Segerstråle (2000).

9 For overviews, see Andersson (1994); Janicke *et al.* (2016).

10 Gwynne and Rentz (1983); Haddad *et al.* (2015); Moeliker (2001); Pelé, *et al.* (2017).

11 Thornhill (1976).

12 Tidière *et al.* (2015).

13 Janicke *et al.* (2016).

14 Clutton-Brock and Vincent (1991); Trivers (1972); Williams (1966).

15 Oberzaucher and Grammer (2014). Note that some argue that Ismail might not be the real record holder, and that Genghis Khan might have had more offspring.

16 Trivers (1972).

17 Stewart-Williams (in press-b).

18 Campbell (2002).

19 Stewart-Williams and Thomas (2013a, 2013b).

20 Trivers (1985).

21 Emlen and Wrege (2004).

22 Paczolt and Jones (2010). Seahorses are commonly given as an example of a sex-role-reversed species, because the males carry the eggs in a specialized pouch before they hatch. Overall, however, males and females invest about equally in offspring, and thus seahorses are only partially sex-role reversed (Eens & Pinxten, 2000).

23 Lack (1968).

24 Stewart-Williams and Thomas (2013b).

25 Lassek and Gaulin (2009).

26 Kruger and Nesse (2006).

27 Wood and Eagly (2002).

28 Stewart-Williams (in press-b); Stewart-Williams and Thomas (2013a, 2013b). The strongest rejoinder to the claim that humans have a relatively low level of overall dimorphism comes from Marco Del Giudice and colleagues (2012), who argue that, although the sex difference in any given trait may be modest, when we consider multiple, related traits simultaneously, the resulting multivariate differences are considerably larger. For a response to this argument, see Stewart-Williams and Thomas (2013b), pp. 167–168, and for a response to our response, see Del Giudice (2013). Note that, whether or not Del Giudice's multivariate approach is useful (and we agree that it sometimes is), it remains the case that humans are *less* dimorphic than most other mammals – including our immediate pre-human ancestors.

29 Buss (1988); Campbell (2002).

30 Miller (2000b).

31 Feingold and Mazzella (1998); Ford and Beach (1951).

32 Marlowe (2000); Stewart-Williams and Thomas (2013b).

33 Kaplan (1994).

34 Hrdy (2000, 2009); Sear and Mace (2008).

35 Marlowe (2000).

36 Miller (2000b).

37 Stewart-Williams and Thomas (2013a, 2013b).

38 Anthropological evidence: Betzig (2012). Genetic evidence: Labuda *et al.* (2010).

39 Sex drive: Baumeister, Catanese, and Vohs (2002). Sexual fantasies: Ellis and Symons (1990). Sexual regret: Kennair, Bendixen, and Buss (2016).

40 Sex outside a committed relationship: Simpson and Gangestad (1991). Desire for sexual variety: Buss and Schmitt (1993). Infidelity: Schmitt (2014).

41 Bendixen *et al.* (2017); Lippa (2009); Schmitt (2005); Schmitt and 118 Members of the International Sexuality Description Project (2003).

42 Clark and Hatfield (1989).

43 Summarized in Schmitt (2014).

44 Friday (1973).

45 Conley (2011).

46 Note that this difference isn't *just* a product of the sex difference in interest in casual sex. Other factors come into play, including the fact that consenting to be alone with an other-sex stranger is riskier for women than for men. See Stewart-Williams and Thomas (2013b), pp. 154–155.

47 Bell and Weinberg (1978).

48 Symons (1979), p. 300.

49 McClintock (2011).

50 Betzig (1986).

51 Stewart-Williams and Thomas (2013b).

52 Ellis and Symons (1990), p. 543.

53 Salmon (2012).

54 Baumeister and Vohs (2004); Symons (1979).

55 Buss (2016).

56 Saad and Gill (2003).

57 van den Berghe (1979), pp. 60–61.

58 Morris (2016).

59 Cited in Pinker (1997), p. 474.

60 Buss and Schmitt (1993); Gangestad and Simpson (2000); Stewart-Williams and Thomas (2013b).

61 Buss and Schmitt (1993); Gangestad and Simpson (2000).

62 Buss and Schmitt (1993); Kenrick *et al.* (1993).

63 C. R. Harris (2013).

64 Allison and Risman (2013); Stewart-Williams, Butler, and Thomas (2017).

65 The quote is from Ken Wishnia's novel *Soft Money* (2015), p. 187.

66 Note that the exceptions make good sense in evolutionary terms as well; see, e.g., Hrdy (1977).

67 For evidence that the ceiling number of offspring has been greater for men than women throughout human evolutionary history, see Betzig (2012); Labuda *et al.* (2010).

68 Buss and Schmitt (1993); Kenrick *et al.* (1993).

69 Stewart-Williams and Thomas (2013b).

70 Buss (1989).

71 Buss (1989); Feingold (1990); N. P. Li, Bailey, and Kenrick (2002).

72 Summarized in Buss (2014).

73 Kenrick and Keefe (1992).

74 Sohn (2017).

75 Sohn (2016).

76 Symons (1979).

77 Ogas and Gaddam (2011).

78 See Stewart-Williams (in press-b) for qualifications to this claim.

79 Muller, Thompson, and Wrangham (2006).

80 Thornhill and Gangestad (1994).

81 Ogas and Gaddam (2011).

82 Millward (2013).

83 Buss (2016).

84 Buss (1989).

85 To be precise, this is men's preference with respect to physical appearance. Importantly, though, physical appearance isn't all that matters in choosing a mate; people also want someone who's similar to them and who they can relate to. As such, the ideal partner age for men tends to fall somewhere between the age of peak fertility and their own age (Kenrick & Keefe, 1992).

86 Kenrick and Keefe (1992).

87 Cited in Symons (1989), p. 34.

88 Gottschall *et al.* (2004).

89 Kenrick *et al.* (1995).

90 Buss (1989); Feingold (1992).

91 Barrett, Dunbar, and Lycett (2002).

92 Betzig (1989).

93 Buss (2014).

94 F. Harris (1975), p. 58.

95 Buss (1989).

96 Buss, Shackelford, Kirkpatrick, and Larsen (2001).

97 Gottschall *et al.* (2004).

98 Eagly and Wood (1999).

99 Eagly and Wood (1999).

100 Buss (2014).

101 On greater female attractiveness, see Feingold and Mazzella (1998); Ford and Beach (1951).

102 Lubinski, Benbow, and Kell (2014).

103 Ridley (1993).

104 Sommers (2013b), p. 204.

105 Archer (2004); M. Wilson and Daly (1985).

106 Wall punching: Collins (2015). Homicidal fantasies: Kenrick and Sheets (1993). Homicide: Daly and Wilson (1988).

107 Archer (2002).

108 Daly and Wilson (1988).

109 M. L. Wilson *et al.* (2014).

110 LeBoeuf (1974).

111 Stewart-Williams and Thomas (2013a, 2013b).

112 Stewart-Williams and Thomas (2013b), pp. 160–162.

113 Lassek and Gaulin (2009); Puts (2010).

114 Puts (2010).
115 Deaner and Smith (2013).
116 Daly and Wilson (1988).
117 For a critique of Eagly and Wood's Social Role Theory, see Stewart-Williams and Thomas (2013a), pp. 258–263.
118 Stewart-Williams (2002).
119 Maccoby and Jacklin (1974).
120 Starr (2015).
121 Low (1989).
122 Maccoby and Jacklin (1974).
123 Lonsdorf (2017); Pasterski *et al.* (2007); Wallen (2005).
124 Campbell (2005).
125 Archer (2009).
126 M. Wilson and Daly (1985).
127 Lykken (1995), p. 93.
128 Clutton-Brock (1991); Trivers (1972).
129 Schärer, Rowe, and Arnqvist (2012).
130 Wood and Eagly (2002).
131 Hewlett (1991).
132 Buss (2014).
133 Møller and Birkhead (1993).
134 Badinter (2010/2011).
135 Stewart-Williams and Thomas (2013a).
136 Eagly and Wood (1999); Wood and Eagly (2012).
137 Janicke *et al.* (2016).
138 Tiger and Shepher (1975).
139 Lubinski *et al.* (2014); K. Parker and Wang (2013); Susan Pinker (2008).
140 Buss (2014).
141 Bekoff and Byers (1998); Burghardt (2005).
142 Lamminmäki *et al.* (2012); G. Li, Kung, and Hines (2017); Pasterski *et al.* (2005).
143 Lonsdorf (2017).
144 Kahlenberg and Wrangham (2010).
145 Wood and Eagly (2012).
146 Gray, McHale, and Carré (2017).
147 See, e.g., Chasin (1977); Fausto-Sterling (2000); Fine (2010, 2017).
148 Stewart-Williams (2017).
149 Sommers (2013a, 2013c); Stewart-Williams (2014, 2017).

Chapter 4: The Dating, Mating, Baby-Making Animal

1 Schopenhauer (1818/1966), p. 534.
2 N. Wolf (1991).
3 Darwin (1871), pp. 341, 582.
4 Darwin (1871), p. 353.
5 Cloud and Perilloux (2014).
6 Cunningham *et al.* (1995).
7 Langlois *et al.* (1987); Slater *et al.* (2000).

8 This raises the challenging question of why some people have an exclusively same-sex sexual orientation. Space forbids me from going into this in any detail, but I will say just two things. First, the simple evolutionary prediction regarding sexual orientation – that is, that people will be primarily attracted to the other sex – gets it right more than 95 percent of the time, which is a far better success rate than most theories in psychology. Second, whatever the explanation for same-sex sexual orientation turns out to be, it's probably not an adaptation (Bobrow & Bailey, 2001; Camperio-Ciani, Corna, & Capiluppi, 2004; LeVay, 2011; see also Appendix A).

9 Summarized in Perrett (2010).

10 Gangestad and Simpson (2000); although see Kordsmeyer and Penke (2017).

11 López, Muñoz, and Martín (2002); Møller (1993); Swaddle and Cuthill (1994); Thornhill (1992); Waitt and Little (2006).

12 Nedelec and Beaver (2014); Shackelford and Larsen (1997); Stephen *et al.* (2017); Van Dongen and Gangestad (2011).

13 See, e.g., Pound *et al.* (2014).

14 Møller (1997).

15 Johnston and Franklin (1993); Perrett (2010); although see Jones *et al.* (2017).

16 Singh (1993). Some suggest that a better generalization would be that the preferred WHR is always within the feminine range, or that it's always lower than the local average.

17 B. J. Dixson *et al.* (2010); King (2013); Singh (1993); Singh *et al.* (2010); Singh and Luis (1995).

18 Thornhill and Gangestad (1993); although see Scott *et al.* (2013).

19 Body shape and athleticism: S. M. Hughes, Dispenza, and Gallup (2003); Sell, Lukazsweski, and Townsley (2017); Singh (1995). Voice pitch: Apicella, Feinberg, and Marlowe (2007); Feinberg *et al.* (2005).

20 Perrett (2010).

21 B. J. Dixson and Vasey (2012); Puts (2010).

22 Cited in Jahr (1976).

23 Based on D. Lieberman *et al.* (2007).

24 H. E. Fisher (2009).

25 Kokko and Ots (2006).

26 Dawkins (1979).

27 Bixler (1992); Joshi *et al.* (2015).

28 Boulet, Charpentier, and Drea (2009); Pusey (2004).

29 Shepher (1971).

30 A. P. Wolf (1966, 1970).

31 Weinberg (1963).

32 DeBruine (2005).

33 D. Lieberman *et al.* (2007).

34 Pinker (1997), p. 455.

35 Stewart-Williams (2015a).

36 Scheidel (2004).

37 Bixler (1982).

38 See, e.g., Archie *et al.* (2007); Hoffman *et al.* (2007); Lihoreau, Zimmer, and Rivault (2007); Muniz *et al.* (2006).
39 Penn and Potts (1998).
40 McClure (2004).
41 Tallis (2005).
42 Gaiman (1996).
43 H. E. Fisher (2006, 2008, 2016); H. E. Fisher *et al.* (2002); Tennov (1979).
44 Cited in Tennov (1979), p. 241.
45 Shostak (1981), p. 268.
46 Sternberg (1986).
47 de Bernières (1994), pp. 344–345.
48 Maugham (1938/2001), p. 71.
49 Stewart-Williams and Thomas (2013a, 2013b).
50 Marlowe (2003a).
51 H. E. Fisher (2016).
52 Boutwell, Barnes, and Beaver (2014); Buss *et al.* (2017).
53 Buckle, Gallup, and Todd (1996).
54 Betzig (1989).
55 Buckle *et al.* (1996).
56 Burns (2000), p. 481.
57 Jankowiak and Fischer (1992).
58 Gottschall and Nordlund (2006).
59 Easton and Hardy (1997); Ryan and Jethá (2010).
60 Mead (1928); Ryan and Jethá (2010).
61 Daly and Wilson (1988); Jankowiak, Nell, and Buckmaster (2002); Saxon (2012).
62 Handy (1923); Stephens (1963), both cited in Daly and Wilson (1988), p. 204.
63 Freeman (1983); Mead (1928).
64 Inuit: Rasmussen (1927). Other societies: Daly and Wilson (1988).
65 C. R. Harris (2013).
66 Petrie and Kempenaers (1998).
67 Sommer and Reichard (2000); Young and Alexander (2012).
68 M. Wilson and Daly (1992), p. 289.
69 Stewart-Williams and Thomas (2013a, 2013b).
70 Buss *et al.* (1992); Daly *et al.* (1982); Symons (1979).
71 Kuhle (2011); Sagarin *et al.* (2012); Scelza (2014); Shackelford *et al.* (2004).
72 Stewart-Williams (in press-b); Stewart-Williams and Thomas (2013a).
73 Buss *et al.* (1992).
74 Sagarin *et al.* (2012).
75 Lishner *et al.* (2008).
76 C. R. Harris (2013); Stewart-Williams (in press-b).
77 Anderson (2006); Larmuseau *et al.* (2013).
78 Stewart-Williams and Thomas (2013b).
79 Diamond (1997); James (1890).
80 Glocker *et al.* (2009); Perrett (2010).
81 Note, though, that baby scorpions are called *scorplings*, which *is* cute.
82 Archer (1998).

83 Harford (2016).
84 Kahneman *et al.* (2004).
85 Daly and Wilson (1988, 1998).
86 Hrdy (1977); Lukas and Huchard (2014).
87 Anderson, Kaplan, Lam, and Lancaster (1999); Case, Lin, and McLanahan (1999); Case and Paxson (2001).
88 Anderson, Kaplan, Lam, and Lancaster (1999); Daly and Wilson (2001); Flinn (1988); Marlowe (1999b).
89 Flinn (1988).
90 Daly and Wilson (2001).
91 Rohwer, Herron, and Daly (1999), pp. 385–386.
92 Daly and Wilson (1998); Rohwer *et al.* (1999).
93 Miller (2000b).
94 R. A. Fisher (1930).
95 Cf. Burnstein, Crandall, and Kitayama (1994).
96 Burnstein *et al.* (1994).
97 Daly and Wilson (1988), pp. 75–77.
98 Burnstein *et al.* (1994).
99 Mann (1992).
100 Daly and Wilson (1988), pp. 72–73.
101 Littlefield and Rushton (1986).
102 Stewart-Williams (2010).
103 Stewart-Williams and Thomas (2013b).
104 See Schmitt (2014), table 1.2, p. 27.
105 Barash (2007), p. 206.
106 Harvey and Bradbury (1991).
107 Murdock (1967).
108 Lassek and Gaulin (2009).
109 Apicella (2014); Washburn and Lancaster (1968).
110 Marlowe (2003b).
111 Stewart-Williams and Thomas (2013b).
112 de Waal (1997a); Woods (2010).
113 Saxon (2016).
114 Ryan and Jethá (2010).
115 H. E. Fisher *et al.* (2002); Schmitt (2005); R. L. Smith (1984).
116 Birkhead and Møller (1998); G. A. Parker (1970); Shackelford and Goetz (2006).
117 A. F. Dixson (1998).
118 de Waal (2005), p. 113.
119 Stewart-Williams and Thomas (2013b), pp. 150–151.
120 Wingfield *et al.* (1990).
121 Archer (2006).
122 Gettler *et al.* (2011); Holmboe *et al.* (2017).
123 Stewart-Williams and Thomas (2013b).
124 Marlowe (2000).
125 Anderson, Kaplan, and Lancaster (1999); Apicella and Marlowe (2004); Daly and Wilson (1987).

126 Marlowe (1999a).
127 Stewart-Williams and Thomas (2013b); A. G. Thomas and Stewart-Williams (2018).
128 Alexander (1974).
129 Marlowe (1999a); Pedersen (1991).
130 Alexander *et al.* (1979).
131 Buss (1996).

Chapter 5: The Altruistic Animal

1 This section draws on Stewart-Williams (2015a).
2 Cheney and Seyfarth (1990); Sherman (1977).
3 Hamilton (1964).
4 Daly, Salmon, and Wilson (1997); Stewart-Williams (2007, 2008).
5 Daly *et al.* (1997).
6 Dawkins (1976/2016).
7 Dawkins (1976/2016).
8 Dawkins (1982).
9 Dawkins (1979).
10 Even some scholars have made this mistake; see, e.g., Washburn (1978), p. 415.
11 Curry (2006).
12 Dawkins (1979).
13 Emlen (1984).
14 Sherman (1977).
15 Spiders: Anthony (2003). Squirrels: Sherman (1980).
16 Holmes and Sherman (1982).
17 Holmes and Sherman (1982).
18 For recent debate on this point, see Nowak, Tarnita, and Wilson (2010), and responses from, e.g., Abbot *et al.* (2011); Liao, Rong, and Queller (2015).
19 Singer (1981).
20 Essock-Vitale and McGuire (1985); Neyer and Lang (2003); M. G. Thomas *et al.* (2018).
21 Burnstein *et al.* (1994).
22 Stewart-Williams (2007, 2008).
23 Fitzgerald, Thomson, and Whittaker (2010); Xue (2013). The cost-of-help effect may also be found in some nonhuman animals. Carter, Wilkinson, and Page (2017) report, for instance, that vampire bats are more kin biased in sharing food when the risks of sharing are greater.
24 M. S. Smith, Kish, and Crawford (1987).
25 Stewart-Williams (2007, 2008).
26 Judge and Hrdy (1992).
27 Stewart-Williams (2008).
28 Daly and Wilson (1982).
29 Apicella and Marlowe (2004); DeBruine (2002); Krupp, DeBruine, and Barclay (2008); D. Lieberman *et al.* (2007).

30 R. C. Johnson (1996).
31 Curry, Mullins, and Whitehouse (in press); Essock-Vitale and McGuire (1980); Madsen *et al.* (2007); Silk (1987).
32 Jankowiak and Diderich (2000).
33 Dudley and File (2007).
34 Griffin, West, and Buckling (2004).
35 Dunn, Aknin, and Norton (2008).
36 Trivers (1971).
37 de Waal (1996).
38 Cosmides and Tooby (1992).
39 Trivers (1971); see also Cosmides and Tooby (1992).
40 Mencken (1916/1982), p. 617.
41 Wright (1994).
42 Axelrod (1984).
43 Hardin (1968).
44 Trivers (1985), p. 392.
45 Nowak and Sigmund (1992).
46 Axelrod (1984); Dawkins (1976/2016).
47 Ridley (1996), p. 84.
48 Curry *et al.* (in press); Essock-Vitale and McGuire (1980).
49 Nietzsche (1886/1966), p. 28.
50 Essock-Vitale and McGuire (1980); M. G. Thomas *et al.* (2018).
51 Hames (1987).
52 Berté (1988).
53 Stewart-Williams (2007).
54 Carter and Wilkinson (2013); Wilkinson (1984).
55 de Waal (1997b); Jaeggi and Gurven (2013).
56 Dolivo and Taborsky (2015); Trivers (1985); Voelkl *et al.* (2015).
57 Massen, Ritter, and Bugnyar (2015).
58 Seyfarth and Cheney (1984).
59 Miller (2007).
60 Miller (2000b), p. 292.
61 Cited in Litchfield (1904), p. 419.
62 Buss (1989).
63 Arnocky *et al.* (2016).
64 Blurton Jones (1987).
65 Hawkes (1991).
66 Gurven and Hill (2009); Marlowe (1999b); Stewart-Williams and Thomas (2013b).
67 Apicella (2014); Hill and Hurtado (1996); Marlowe (2004); E. A. Smith, Bird, and Bird (2003).
68 Miller (2000a), p. 336.
69 Zahavi and Zahavi (1997).
70 Fehr and Gaechter (2002).
71 Gintis (2000).
72 Hrdy (2009); Ridley (1996).

73 Bowles and Gintis (2011); D. S. Wilson (2015); D. S. Wilson and Wilson (2007).
74 D. S. Wilson (2015).
75 Bowles (2009).
76 Haidt (2012).
77 Burnham and Johnson (2005); Hagen and Hammerstein (2006).
78 Pinker (2012); M. E. Price (2012).
79 G. R. Price (1970); D. S. Wilson (2015).
80 van Veelen *et al.* (2012).
81 See, e.g., Nowak *et al.* (2010).
82 Henrich (2004).
83 West, Griffin, and Gardner (2007a, 2007b).
84 Nesse (2006); Ridley (1996).
85 El Mouden *et al.* (2012), p. 42.

Chapter 6: The Cultural Animal

1 Bennett, Li, and Ma (2003), p. 77.
2 Blackmore (1999); Dawkins (1976/2016); Dennett (1995, 2017).
3 Boyd and Richerson (1985); Cavalli-Sforza and Feldman (1981); Henrich (2016); Laland (2017); Mesoudi (2011); Richerson and Boyd (2005); Sperber (1996); Tooby and Cosmides (1992).
4 Cited in Dennett (2017), p. 210.
5 Dennett (2017).
6 Skinner (1953), p. 430.
7 Pinker (1994); Pinker and Bloom (1992).
8 Darwin (1871), pp. 60–61.
9 Kirby, Cornish, and Smith (2008); Pagel (2012).
10 Lee and Hasegawa (2014).
11 Shermer (2009).
12 Hinde and Barden (1985).
13 M. Friedman (1953).
14 Shermer (2009).
15 Hofstede (2001).
16 Tylor (1871), p. 62.
17 Popper (1979), p. 261.
18 Mace and Holden (2005); Mesoudi (2011).
19 See, e.g., Matzke (2015); Prentiss *et al.* (2011); Tehrani and Collard (2002).
20 Goodall (1986).
21 Whiten *et al.* (1999).
22 van Schaik *et al.* (2003).
23 S. Perry (2011).
24 Whitehead and Rendell (2014).
25 Yudkowsky (2006).
26 Henrich (2016); Richerson and Boyd (2005).

27 Legare and Nielsen (2015); Ridley (2010); Tennie, Call, and Tomasello (2009).
28 Boesch (2003); see also Sasaki and Biro (2017).
29 Tomasello (1999).
30 Dennett (2017).
31 Cited in S. Harris (2016).
32 Russell (1926), p. 54.
33 Ridley (2010), p. 8.
34 S. Johnson (2010).
35 Needham (1970), p. 202.
36 Arthur (2009); Basalla (1988).
37 Ridley (2010).
38 Richerson and Boyd (2005).
39 Kellogg and Kellogg (1933).
40 Darwin (1871), p. 39.
41 Lumsden and Wilson (1981).
42 Boyd and Richerson (1985); Henrich and McElreath (2003).
43 Henrich and Gil-White (2001).
44 Kendal *et al.* (2014).
45 Bandura (1986).
46 Asch (1951).
47 Touboul (2014).
48 Henrich and McElreath (2007).
49 Henrich and McElreath (2007).
50 Stack (1987).
51 Henrich (2004); Richerson *et al.* (2016).
52 Cavalli-Sforza, Menozzi, and Piazza (1996).
53 Murdock (1967).
54 Orians (1969).
55 Betzig (2012).
56 Henrich, Boyd, and Richerson (2012).
57 Blume (2009).
58 Kaufmann (2010).
59 Henrich (2016).
60 Daly and Wilson (1988).
61 Henrich (2016).
62 Norenzayan (2013); D. S. Wilson (2002).
63 Dunbar (1993).
64 Richerson *et al.* (2016).
65 Dawkins (1976/2016, 1983).
66 Lumsden and Wilson (1981).
67 Dennett (2009), p. 3.
68 Graham (2010).
69 Blackmore (1999), p. 37.
70 Bentley, Hahn, and Shennan (2004); Newberry *et al.* (2017).
71 Blackmore (1999).
72 Berlin and Kay (1969).

73 Haidt (2012).
74 Stewart-Williams (2015b, in press-a).
75 Zuckerman, Silberman, and Hall (2013).
76 Polderman *et al.* (2015); Turkheimer (2000).
77 Boyd and Richerson (1985).
78 Lynch (1996).
79 Blume (2009); Kaufmann (2010).
80 Lynch (1996).
81 Blackmore (1999); Dennett (2006).
82 Hamer (2004).
83 Stewart-Williams (2010).
84 Dawkins (1976/2016), p. 250.
85 Levitan (2006); Miller (2000a).
86 Changizi (2011); Marcus (2012).
87 Pinker (1994); Pinker and Bloom (1992).
88 Deacon (1997). Note that Deacon has since changed his tune on this issue, and no longer thinks that humans evolved specifically to use language. I agree with 1997 Deacon rather than later Deacon.
89 Dawkins (1993); Ray (2009).
90 Dehaene (2010).
91 Dawkins (2006); Dennett (2006).
92 Ray (2009).
93 Dawkins (1976/2016, 2006).
94 Dawkins (2006); Dennett (2006).
95 Blackmore (1999); Dennett (2017).
96 J. Hughes (2011).
97 Dennett (2006).
98 Boyd and Richerson (1985); Cavalli-Sforza and Feldman (1981); Cochran and Harpending (2009); Durham (1991); Lumsden and Wilson (1981); Richerson, Boyd, and Henrich (2010).
99 Laland and Brown (2002), p. 242.
100 Durham (1991); Simoons (1969).
101 Tishkoff *et al.* (2007).
102 Cochran and Harpending (2009).
103 Harpending (2013).
104 Coyne (2011).
105 Hardy *et al.* (2015).
106 G. H. Perry *et al.* (2007).
107 Axelsson *et al.* (2013).
108 Henrich (2016).
109 Henrich (2016).
110 Wrangham (2009).
111 See, e.g., Henrich (2016); Taylor (2010); Tomasello (1999).
112 Henrich (2016); Kaplan *et al.* (2000).
113 Gottfredson (2007).
114 Deacon (1997); Dennett (2017); Henrich (2016).
115 Deacon (1997); Dunbar (1996); Miller (2000b).

116 Deacon (2013).
117 Blackmore (1999); Haidt (2012).
118 Frost (2015).
119 Banerjee and Bloom (2013).
120 Levitan (2006); Miller (2000a); Pinker (1997).
121 Melfi (2005).
122 Taylor (2010).
123 Dolgin (2015).

Appendix A: How to Win an Argument with a Blank Slater

1 Tybur, Miller, and Gangestad (2007).
2 See Stewart-Williams (2010), chapter 12, for a more detailed discussion of the naturalistic fallacy.
3 Stewart-Williams (2010), chapter 12.
4 See, e.g., Bleier (1984), p. 15.
5 Pinker (2011). See also Pinker (2018).
6 See, e.g., Confer *et al.* (2010).
7 E. O. Wilson (1978).
8 Bobrow and Bailey (2001); although see Vasey, Pocock, and VanderLaan (2007).
9 Buss (1989).
10 Gould and Lewontin (1979).
11 The acronym comes from Henrich, Heine, and Norenzayan (2010).
12 Pinker (2002).
13 Henrich *et al.* (2010).
14 Buss (1989); Schmitt (2005).
15 Daly *et al.* (1997).
16 Gould (2000a).
17 Geary (1995).
18 Female orgasm as adaptation: King and Belsky (2012); Miller (2000b); Puts *et al.* (2012). Female orgasm as by-product: Symons (1979). Homicide as adaptation: Buss (2005). Homicide as by-product: Daly and Wilson (1988). Music as adaptation: Levitan (2006); Miller (2000b). Music as by-product: Pinker (1997). Rape as adaptation *and* rape as by-product: Thornhill and Palmer (2000). Religion as adaptation: D. S. Wilson (2002). Religion as by-product: Atran (2002); Boyer (2001); Stewart-Williams (2015b).
19 Yawning: Massen *et al.* (2014). Dreaming: Revonsuo (2000). PMS: Gillings (2014). Masturbation: Baker and Bellis (1993). Penis as sperm displacement device: Gallup, Burch, and Mitchell (2006). Autism: Reser (2011).
20 Kurzban (2002).

Appendix B: How to Win an Argument with an Anti-Memeticist

1 Benitez-Bribiesca (2001), p. 29.
2 Blackmore (1999).

3 Laland and Brown (2002).
4 Gardner (2000).
5 Stewart-Williams (2015b, in press-a).
6 See, e.g., Sperber (1985).
7 Pagel (2012).
8 Sperber (1996).
9 Boyer (2001).
10 Pinker (1997), p. 209.
11 Warraq (2002).
12 Bateson (1979), p. 182.
13 James (1890).
14 Mach (1896), p. 174.
15 Pagel (2012).
16 Dawkins (1993); Ray (2009).
17 Ewald (1996).

References

Abbot, P., Abe, J., Alcock, J., *et al.* (2011). Inclusive fitness theory and eusociality. *Nature, 471,* E1–E4.

Alexander, R. D. (1974). The evolution of social behavior. *Annual Review of Ecology and Systematics, 5,* 325–383.

(1987). *The biology of moral systems.* Hawthorne, NY: Aldine de Gruyter.

Alexander, R. D., Hoogland, J. L., Howard, R. D., Noonan, K. M., & Sherman, P. W. (1979). Sexual dimorphisms and breeding systems in pinnipeds, ungulates, primates, and humans. In N. A. Chagnon & W. Irons (Eds.), *Evolutionary biology and human social behavior: An anthropological perspective* (pp. 402–435). North Scituate, MA: Duxbury Press.

Allison, R., & Risman, B. J. (2013). A double standard for "hooking up": How far have we come toward gender equality? *Social Science Research, 42,* 1191–1206.

Anderson, K. G. (2006). How well does paternity confidence match actual paternity? Evidence from worldwide nonpaternity rates. *Current Anthropology, 47,* 513–520.

Anderson, K. G., Kaplan, H., Lam, D., & Lancaster, J. B. (1999). Paternal care by genetic fathers and stepfathers II: Reports from Xhosa high school students. *Evolution and Human Behavior, 20,* 433–451.

Anderson, K. G., Kaplan, H., & Lancaster, J. B. (1999). Paternal care by genetic fathers and stepfathers I: Reports from Albuquerque men. *Evolution and Human Behavior, 20,* 405–431.

Andersson, M. (1994). *Sexual selection.* Princeton, NJ: Princeton University Press.

Anthony, C. D. (2003). Kinship influences cannibalism in the wolf spider, *Pardosa milvina. Journal of Insect Behavior, 16,* 23–36.

Apicella, C. L. (2014). Upper-body strength predicts hunting reputation and reproductive success in Hadza hunter-gatherers. *Evolution and Human Behavior, 35,* 508–518.

Apicella, C. L., Feinberg, D. R., & Marlowe, F. W. (2007). Voice pitch predicts reproductive success in male hunter-gatherers. *Evolution and Human Behavior, 3*, 682–684.

Apicella, C. L., & Marlowe, F. W. (2004). Perceived mate fidelity and paternal resemblance predict men's investment in children. *Evolution and Human Behavior, 25*, 371–378.

Archer, J. (1998). Why do people love their pets? *Evolution and Human Behavior, 18*, 237–260.

(2002). Sex differences in physically aggressive acts between heterosexual partners: A meta-analytic review. *Aggression and Violent Behavior, 7*, 313–351.

(2004). Sex differences in aggression in real-world settings: A meta-analytic review. *Review of General Psychology, 8*, 291–322.

(2006). Testosterone and human aggression: An evaluation of the challenge hypothesis. *Neuroscience and Biobehavioral Reviews, 30*, 319–345.

(2009). Does sexual selection explain human sex differences in aggression? *Behavioral and Brain Sciences, 32*, 249–266.

Archie, E. A., Hollister-Smith, J. A., Poole, J. H., et al. (2007). Behavioural inbreeding avoidance in wild African elephants. *Molecular Ecology, 16*, 4138–4148.

Arnocky, S., Piché, T., Albert, G., Ouellette, D., & Barclay, P. (2016). Altruism predicts mating success in humans. *British Journal of Psychology, 108*, 416–435.

Arthur, W. B. (2009). *The nature of technology: What it is and how it evolves.* New York: Free Press.

Asch, S. E. (1951). Effects of group pressure on the modification and distortion of judgments. In H. Guetzkow (Ed.), *Groups, leadership and men* (pp. 177–190). Pittsburgh, PA: Carnegie Press.

Atran, S. (2002). *In Gods we trust: The evolutionary landscape of religion.* Oxford University Press.

Axelrod, R. (1984). *The evolution of cooperation.* New York: Basic Books.

Axelsson, E., Ratnakumar, A., Arendt, M.-L., et al. (2013). The genomic signature of dog domestication reveals adaptation to a starch-rich diet. *Nature, 495*, 360–364.

Badinter, E. (2010/2011). *The conflict: How modern motherhood undermines the status of women* (A. Hunter, Trans.). New York: Henry Holt.

Baker, R. R., & Bellis, M. A. (1993). Human sperm competition: Ejaculate adjustment by males and the function of masturbation. *Animal Behaviour, 46*, 861–885.

Bandura, A. (1986). *Social foundations of thought and action: A social cognitive theory.* Englewood Cliffs, NJ: Prentice-Hall.

Banerjee, K., & Bloom, P. (2013). Would Tarzan believe in God? Conditions for the emergence of religious belief. *Trends in Cognitive Sciences, 17*, 7–8.

Barash, D. P. (2007). *Natural selections: Selfish altruists, honest liars, and other realities of evolution.* New York: Bellevue Literary Press.

Barkow, J. H., Cosmides, L., & Tooby, J. (Eds.). (1992). *The adapted mind: Evolutionary psychology and the generation of culture.* Oxford University Press.

Barrett, L., Dunbar, R. I. M., & Lycett, J. (2002). *Human evolutionary psychology*. Basingstoke, UK: Palgrave.

Basalla, G. (1988). *The evolution of technology*. Cambridge University Press.

Bateson, G. (1979). *Mind and nature: A necessary unity*. New York: Dutton.

Baumeister, R. F., Catanese, K. R., & Vohs, K. D. (2002). Is there a gender difference in strength of sex drive? Theoretical views, conceptual distinctions, and a review of relevant evidence. *Personality and Social Psychology Review*, *5*, 242–273.

Baumeister, R. F., & Vohs, K. D. (2004). Sexual economics: Sex as female resource for social exchange in heterosexual interactions. *Personality and Social Psychology Review*, *8*, 339–363.

Bekoff, M., & Byers, J. A. (Eds.). (1998). *Animal play: Evolutionary, comparative and ecological perspectives*. Cambridge University Press.

Bell, A. P., & Weinberg, M. S. (1978). *Homosexualities: A study of diversity among men and women*. New York: Simon & Schuster.

Bendixen, M., Asao, K., Wyckoff, J. P., Buss, D. M., & Kennair, L. E. O. (2017). Sexual regret in US and Norway: Effects of culture and individual differences in religiosity and mating strategy. *Personality and Individual Differences*, *116*, 246–251.

Benitez-Bribiesca, L. (2001). Memetics: A dangerous idea. *Interciencia*, *26*, 29–31.

Bennett, C. H., Li, M., & Ma, B. (2003). Chain letters and evolutionary histories. *Scientific American*, 76–81.

Bentley, R. A., Hahn, M. W., & Shennan, S. J. (2004). Random drift and culture change. *Proceedings of the Royal Society B*, *271*, 1443–1450.

Berlin, B., & Kay, P. (1969). *Basic color terms: Their universality and evolution*. Berkeley, CA: University of California Press.

Bernard, J. (1974). *Sex differences: An overview*. New York: MSS Modular Publications.

Berté, N. A. (1988). Kékch'i horticultural labor exchange: Productive and reproductive implications. In L. Betzig, M. Borgerhoff Mulder, & P. Turke (Eds.), *Human reproductive behavior* (pp. 83–96). Cambridge University Press.

Betzig, L. (1986). *Despotism and differential reproduction: A Darwinian view of history*. Hawthorne, NY: Aldine.

(1989). Causes of conjugal dissolution: A cross-cultural study. *Current Anthropology*, *30*, 654–676.

(2012). Means, variances, and ranges in reproductive success: Comparative evidence. *Evolution and Human Behavior*, *33*, 309–317.

Birkhead, T. R., & Møller, A. P. (1998). *Sperm competition and sexual selection*. New York: Academic Press.

Bixler, R. H. (1982). Comment on the incidence and purpose of royal sibling incest. *American Ethnologist*, *9*, 580–582.

(1992). Why littermates don't: The avoidance of inbreeding depression. *Annual Review of Sex Research*, *3*, 291–328.

Blackmore, S. (1999). *The meme machine*. Oxford University Press.

Bleier, R. (1984). *Science and gender: A critique of biology and its theories on women*. New York: Pergamon.

Blume, M. (2009). The reproductive benefits of religious affiliation. In E. Voland & W. Schiefenhövel (Eds.), *The biological evolution of religious mind and behavior* (pp. 117–126). Berlin, Germany: Springer-Verlag.

Blurton Jones, N. (1987). Tolerated theft: Suggestions about the ecology and evolution of sharing, hoarding, and scrounging. *Social Science Information*, 26, 31–54.

Bobrow, D., & Bailey, J. M. (2001). Is homosexuality maintained via kin selection? *Evolution and Human Behavior*, 22, 361–368.

Boesch, C. (2003). Is culture a golden barrier between human and chimpanzee? *Evolutionary Anthropology*, 12, 82–91.

Boulet, M., Charpentier, M. J., & Drea, C. M. (2009). Decoding an olfactory mechanism of kin recognition and inbreeding avoidance in a primate. *BMC Evolutionary Biology*, 9, 281.

Boutwell, B. B., Barnes, J. C., & Beaver, K. M. (2014). When love dies: Further elucidating the existence of a mate ejection module. *Review of General Psychology*, 19, 30–38.

Bowles, S. (2009). Did warfare among ancestral hunter-gatherers affect the evolution of human social behaviors? *Science*, 324, 1293–1298.

Bowles, S., & Gintis, H. (2011). *A cooperative species: Human reciprocity and its evolution*. Princeton, NJ: Princeton University Press.

Boyd, R., & Richerson, P. J. (1985). *Culture and the evolutionary process*. Chicago, IL: University of Chicago Press.

Boyer, P. (2001). *Religion explained: The evolutionary origins of religious thought*. New York: Basic Books.

Brown, J. L. (1975). *The evolution of behavior*. New York: Norton.

Buckle, L., Gallup, G. G., & Todd, Z. A. (1996). Marriage as a reproductive contract: Patterns of marriage, divorce, and remarriage. *Ethology and Sociobiology*, 17, 363–377.

Burghardt, G. M. (2005). *Genesis of animal play: Testing the limits*. Cambridge, MA: MIT Press.

Burnham, T. C., & Johnson, D. D. P. (2005). The biological and evolutionary logic of human cooperation. *Analyse and Kritik*, 27, 113–135.

Burns, A. (2000). Looking for love in intimate heterosexual relationships. *Feminism and Psychology*, 10, 481–485.

Burnstein, E., Crandall, C., & Kitayama, S. (1994). Some neo-Darwinian decision rules for altruism: Weighing cues for inclusive fitness as a function of the biological importance of the decision. *Journal of Personality and Social Psychology*, 67, 773–789.

Burt, A., & Trivers, R. L. (2006). *Genes in conflict: The biology of selfish genetic elements*. Cambridge, MA: Harvard University Press.

Buss, D. M. (1988). The evolution of human intrasexual competition: Tactics of mate attraction. *Journal of Personality and Social Psychology*, 54, 616–628.

(1989). Sex differences in human mate preferences: Evolutionary hypotheses tested in 37 cultures. *Behavioral and Brain Sciences*, 12, 1–49.

(1996). Sexual conflict: Evolutionary insights into feminism and the "battle of the sexes". In D. M. Buss & N. M. Malamuth (Eds.), *Sex, power, conflict: Evolutionary and feminist perspectives* (pp. 296–318). Oxford University Press.

(2003). Sexual strategies: A journey into controversy. *Psychological Inquiry*, *14*, 219–226.

(2005). *The murderer next door: Why the mind is designed to kill* (5th ed.). New York: Penguin.

(2014). *Evolutionary psychology: The new science of the mind* (5th ed.). New York: Pearson.

(2016). *The evolution of desire: Strategies of human mating* (rev. ed.). New York: Basic Books.

Buss, D. M., Goetz, C., Duntley, J. D., Asao, K., & Conroy-Beam, D. (2017). The mate switching hypothesis. *Personality and Individual Differences*, *104*, 143–149.

Buss, D. M., Larsen, R. J., Westen, D., & Semmelroth, J. (1992). Sex differences in jealousy: Evolution, physiology, and psychology. *Psychological Science*, *3*, 251–255.

Buss, D. M., & Schmitt, D. P. (1993). Sexual strategies theory: An evolutionary perspective on human mating. *Psychological Review*, *100*, 204–232.

Buss, D. M., Shackelford, T. K., Kirkpatrick, L. A., & Larsen, R. J. (2001). A half century of mate preferences: The cultural evolution of values. *Journal of Marriage and Family*, *63*, 491–503.

Campbell, A. (2002). *A mind of her own: The evolutionary psychology of women*. Oxford University Press.

(2005). Feminism and evolutionary psychology. In J. H. Barkow (Ed.), *Missing the revolution: Darwinism for the social sciences* (pp. 63–99). Oxford University Press.

Camperio-Ciani, A., Corna, F., & Capiluppi, C. (2004). Evidence for maternally inherited factors favouring male homosexuality and promoting female fecundity. *Proceedings of the Royal Society B*, *271*, 2217–2221.

Carter, G. G., & Wilkinson, G. S. (2013). Food sharing in vampire bats: Reciprocal help predicts donations more than relatedness or harassment. *Proceedings of the Royal Society B*, *280*, 20122573.

Carter, G. G., Wilkinson, G. S., & Page, R. A. (2017). Food-sharing vampire bats are more nepotistic under conditions of perceived risk. *Behavioral Ecology*, *28*, 565–569.

Case, A. C., Lin, I.-F., & McLanahan, S. (1999). Household resource allocation in stepfamilies: Darwin reflects on the plight of Cinderella. *American Economic Review*, *89*, 234–238.

Case, A. C., & Paxson, C. (2001). Mothers and others: Who invests in children's health? *Journal of Health Economics*, *20*, 301–328.

Cavalli-Sforza, L. L., & Feldman, M. W. (1981). *Cultural transmission and evolution: A quantitative approach*. Princeton, NJ: Princeton University Press.

Cavalli-Sforza, L. L., Menozzi, P., & Piazza, A. (1996). *The history and geography of human genes*. Princeton, NJ: Princeton University Press.

Changizi, M. (2011). *Harnessed: How language and music mimicked nature and transformed ape to man*. Dallas, TX: BenBella Books.

Chasin, B. (1977). Sociobiology: A sexist synthesis. *Science for the People*, 9, 27–31.

Cheney, D. L., & Seyfarth, R. M. (1990). *How monkeys see the world: Inside the mind of another species*. Chicago, IL: University of Chicago Press.

Clark, R. D., & Hatfield, E. (1989). Gender differences in receptivity to sexual offers. *Journal of Psychology and Human Sexuality*, 2, 39–55.

Cloud, J. M., & Perilloux, C. (2014). Bodily attractiveness as a window to women's fertility and reproductive value. In V. A. Weekes-Shackelford & T. K. Shackelford (Eds.), *Evolutionary perspectives on human sexual psychology and behavior* (pp. 135–152). New York: Springer.

Clutton-Brock, T. H. (1991). *The evolution of parental care*. Princeton, NJ: Princeton University Press.

Clutton-Brock, T. H., & Vincent, A. C. J. (1991). Sexual selection and the potential reproductive rates of males and females. *Nature*, 351, 58–60.

Cochran, G., & Harpending, H. (2009). *The 10,000 year explosion: How civilization accelerated human evolution*. New York: Basic Books.

Collins, K. (2015). America's most prolific wall punchers, charted. *Quartz*. Retrieved February 24, 2017, from https://qz.com/582720/americas-most-prolific-wall-punchers-charted/

Confer, J., Easton, J., Fleischman, D., et al. (2010). Evolutionary psychology: Controversies, questions, prospects, and limitations. *American Psychologist*, 65, 110–126.

Conley, T. D. (2011). Perceived proposer personality characteristics and gender differences in acceptance of casual sex offers. *Journal of Personality and Social Psychology*, 100, 309–329.

Cook, M., & Mineka, S. (1987). Second-order conditioning and overshadowing in the observational conditioning of fear in monkeys. *Behaviour Research and Therapy*, 25, 349–364.

Cosmides, L., & Tooby, J. (1992). Cognitive adaptations for social exchange. In J. H. Barkow, L. Cosmides, & J. Tooby (Eds.), *The adapted mind: Evolutionary psychology and the generation of culture* (pp. 163–228). Oxford University Press.

(1997). Evolutionary psychology: A primer. *Center for Evolutionary Psychology*. Retrieved March 26, 2015, from www.cep.ucsb.edu/primer.html

Coyne, J. A. (2009). *Why evolution is true*. New York: Viking.

(2011). Evolution 2011: Hot research – with cats! *Why Evolution is True*. Retrieved March 30, 2015, from https://whyevolutionistrue.wordpress.com/2011/06/19/evolution-2011-hot-research—with-cats/

Cronin, H. (1991). *The ant and the peacock: Altruism and sexual selection from Darwin to today*. Cambridge University Press.

Cunningham, M. R., Roberts, A. R., Barbee, A. P., Druen, P. B., & Wu, C. H. (1995). "Their ideas of beauty are, on the whole, the same as ours": Consistency and variability in the cross-cultural perception of physical attractiveness. *Journal of Personality and Social Psychology*, 68, 261–279.

Curry, O. (2006). One good deed. *Nature*, 444, 683.

Curry, O. S., Mullins, D. A., & Whitehouse, H. (in press). Is it good to cooperate? Testing the theory of morality-as-cooperation in 60 societies. *Current Anthropology*.

Curtis, V., de Barra, M., & Aunger, R. (2011). Disgust as an adaptive system for disease avoidance behaviour. *Philosophical Transactions of the Royal Society B, 366,* 389–401.

Daly, M., Salmon, C., & Wilson, M. (1997). Kinship: The conceptual hole in psychological studies of social cognition and close relationships. In J. A. Simpson & D. T. Kenrick (Eds.), *Evolutionary social psychology* (pp. 265–296). Mahwah, NJ: Lawrence Erlbaum.

Daly, M., & Wilson, M. (1982). Homicide and kinship. *American Anthropologist, 84,* 372–378.

(1987). The Darwinian psychology of discriminative parental solicitude. *Nebraska Symposium on Motivation, 35,* 91–144.

(1988). *Homicide.* Hawthorne, NY: Aldine de Gruyter.

(1998). *The truth about Cinderella: A Darwinian view of parental love.* New Haven, CT: Yale University Press.

(2001). An assessment of some proposed exceptions to the phenomenon of nepotistic discrimination against stepchildren. *Annales Zoologici Fennici, 36,* 287–296.

Daly, M., Wilson, M., & Weghorst, S. J. (1982). Male sexual jealousy. *Ethology and Sociobiology, 3,* 11–27.

Darwin, C. (1859). *On the origin of the species by means of natural selection.* London: Murray.

(1871). *The descent of man and selection in relation to sex.* London: Murray.

(1872). *The expression of the emotions in man and animals.* London: Murray.

Dawkins, R. (1976/2016). *The selfish gene* (fortieth anniversary ed.). Oxford University Press.

(1979). Twelve misunderstandings of kin selection. *Zeitschrift für Tierpsychologie, 51,* 184–200.

(1982). *The extended phenotype: The long reach of the gene.* Oxford University Press.

(1983). Universal Darwinism. In D. S. Bendall (Ed.), *Evolution from molecules to men* (pp. 403–428). Cambridge University Press.

(1986). *The blind watchmaker: Why the evidence of evolution reveals a universe without design.* London: Penguin.

(1993). Viruses of the mind. In B. Dahlbom (Ed.), *Dennett and his critics: Demystifying mind* (pp. 13–27). Oxford, UK: Blackwell.

(1998). *Unweaving the rainbow: Science, delusion, and the appetite for wonder.* London: Penguin.

(2006). *The God delusion.* New York: Houghton Mifflon.

de Bernières, L. (1994). *Captain Corelli's mandolin.* London: Vintage Books.

de Waal, F. B. M. (1996). *Good natured: The origins of right and wrong in humans and other animals.* Cambridge, MA: Harvard University Press.

(1997a). *Bonobo: The forgotten ape.* Berkeley, CA: University of California Press.

(1997b). The chimpanzee's service economy: Food for grooming. *Evolution and Human Behavior, 18,* 375–386.

(2005). *Our inner ape: A leading primatologist explains why we are who we are.* New York: Riverhead Books.

Deacon, T. W. (1997). *The symbolic species: The co-evolution of language and the brain.* New York: Norton.

(2013). *Incomplete nature: How mind emerged from matter*. New York: Norton.

Deaner, R. O., & Smith, B. A. (2013). Sex differences in sports across 50 societies. *Cross-Cultural Research, 47*, 268–309.

DeBruine, L. M. (2002). Facial resemblance enhances trust. *Proceedings of the Royal Society B, 269*, 1307–1312.

(2005). Trustworthy but not lust-worthy: Context-specific effects of facial resemblance. *Proceedings of the Royal Society B, 272*, 919–922.

Dehaene, S. (2010). *Reading in the brain: The new science of how we read*. New York: Penguin.

Del Giudice, M. (2013). Multivariate misgivings: Is *D* a valid measure of group and sex differences? *Evolutionary Psychology, 11*, 147470491301100511.

Del Giudice, M., Booth, T., & Irwing, P. (2012). The distance between Mars and Venus: Measuring global sex differences in personality. *PLoS ONE, 7*, e29265.

Dennett, D. C. (1995). *Darwin's dangerous idea: Evolution and the meanings of life*. New York: Simon & Schuster.

(2006). *Breaking the spell: Religion as a natural phenomenon*. New York: Viking.

(2009). *The cultural evolution of words and other thinking tools*. Paper presented at the Cold Spring Harbor Symposia on Quantitative Biology. Retrieved March 3, 2018, from http://ase.tufts.edu/cogstud/dennett/papers/coldspring.pdf

(2017). *From bacteria to Bach and back: The evolution of minds*. New York: Norton.

Diamond, J. (1997). *Why is sex fun? The evolution of human sexuality*. London: Weidenfeld & Nicolson.

Dixson, A. F. (1998). *Primate sexuality: Comparative studies of the prosimians, monkeys, apes, and human beings*. Oxford University Press.

Dixson, B. J., Sagata, K., Linklater, W. L., & Dixson, A. F. (2010). Male preferences for female waist-to-hip ratio and body mass index in the highlands of Papua New Guinea. *American Journal of Physical Anthropology, 141*, 620–625.

Dixson, B. J., & Vasey, P. L. (2012). Beards augment perceptions of men's age, social status, and aggressiveness, but not attractiveness. *Behavioral Ecology, 23*, 481–490.

Dolgin, E. (2015). The myopia boom. *Nature, 519*, 276–278.

Dolivo, V., & Taborsky, M. (2015). Norway rats reciprocate help according to the quality of help they received. *Biology Letters, 11*, 20140959.

Dudley, S. A., & File, A. L. (2007). Kin recognition in an annual plant. *Biology Letters, 3*, 435–438.

Dunbar, R. I. M. (1993). Coevolution of neocortical size, group size, and language in humans. *Behavioral and Brain Sciences, 16*, 681–735.

(1996). *Grooming, gossip, and the evolution of language*. London: Faber & Faber.

Dunn, E. W., Aknin, L. B., & Norton, M. I. (2008). Spending money on others promotes happiness. *Science, 319*, 1687–1688.

Durham, W. H. (1991). *Coevolution: Genes, culture, and human diversity*. Stanford, CA: Stanford University Press.

Eagly, A. H. (1995). The science and politics of comparing men and women. *American Psychologist, 50,* 145–158.

Eagly, A. H., & Wood, W. (1999). The origins of sex differences in human behavior: Evolved dispositions versus social roles. *American Psychologist, 54,* 408–423.

Easton, D., & Hardy, J. W. (1997). *The ethical slut: A practical guide to polyamory, open relationships and other adventures.* Berkeley, CA: Celestial Arts.

Eaton, S. B., Konner, M., & Shostak, M. (1988). Stone Agers in the fast lane: Chronic degenerative diseases in evolutionary perspective. *American Journal of Medicine, 84,* 739–749.

Eens, M., & Pinxten, R. (2000). Sex-role reversal in vertebrates: Behavioural and endocrinological accounts. *Behavioural Processes, 51,* 135–147.

El Mouden, C., Burton-Chellew, M., Gardner, A., & West, S. A. (2012). What do humans maximize? In S. Okasha & K. Binmore (Eds.), *Evolution and rationality: Decisions, co-operation and strategic behaviour* (pp. 23–49). Cambridge University Press.

Ellis, B. J., & Symons, D. (1990). Sex differences in sexual fantasy: An evolutionary approach. *Journal of Sex Research, 4,* 527–555.

Emlen, S. T. (1984). Cooperative breeding in birds and mammals. In J. R. Krebs & N. B. Davies (Eds.), *Behavioural ecology: An evolutionary approach* (pp. 305–339). Sunderland, MA: Sinauer.

Emlen, S. T., & Wrege, P. H. (2004). Size dimorphism, intrasexual competition, and sexual selection in Wattled Jacana (*Jacana jacana*), a sex-role-reversed shorebird in Panama. *Auk, 121,* 391–403.

Essock-Vitale, S. M., & McGuire, M. (1980). Predictions derived from the theories of kin selection and reciprocation assessed by anthropological data. *Ethology and Sociobiology, 1,* 233–243.

    (1985). Women's lives viewed from an evolutionary perspective: II. Patterns of helping. *Ethology and Sociobiology, 6,* 155–173.

Ewald, P. W. (1996). *Evolution of infectious disease.* Oxford University Press.

Fausto-Sterling, A. (2000). Beyond difference: Feminism and evolutionary psychology. In H. Rose & S. Rose (Eds.), *Alas poor Darwin: Arguments against evolutionary psychology* (pp. 174–189). London: Jonathan Cape.

Fehr, E., & Gaechter, S. (2002). Altruistic punishment in humans. *Nature, 415,* 137–140.

Feinberg, D. R., Jones, B. C., Little, A. C., Burt, D. M., & Perrett, D. I. (2005). Manipulations of fundamental and formant frequencies influence the attractiveness of human male voices. *Animal Behaviour, 69,* 561–568.

Feingold, A. (1990). Gender differences in effects of physical attractiveness on romantic attraction: A comparison across five research paradigms. *Journal of Personality and Social Psychology, 59,* 981–993.

    (1992). Gender differences in mate selection preferences: A test of the paternal investment model. *Psychological Bulletin, 112,* 125–139.

Feingold, A., & Mazzella, R. (1998). Gender differences in body image are increasing. *Psychological Science, 9,* 190–195.

Fine, C. (2010). *Delusions of gender: The real science behind sex differences.* New York: Norton.

(2017). *Testosterone rex: Myths of sex, science, and society.* New York: Norton.

Fisher, H. E. (2006). Why we love, why we cheat. *Ted.* Retrieved March 3, 2018, from www.ted.com/talks/helen_fisher_tells_us_why_we_love_cheat

(2008). The brain in love. *Ted.* Retrieved March 3, 2018, from www.ted.com/talks/helen_fisher_studies_the_brain_in_love

(2009). *Why him? Why her? How to find and keep lasting love.* New York: Henry Holt.

(2016). *Anatomy of love: A natural history of mating, marriage and why we stray* (2nd ed.). New York: Norton.

Fisher, H. E., Aron, A., Mashek, D., Li, H., & Brown, L. L. (2002). Defining the brain systems of lust, romantic attraction, and attachment. *Archives of Sexual Behavior, 31,* 413–419.

Fisher, R. A. (1930). *The genetical theory of natural selection.* Oxford, UK: Clarendon Press.

Fitzgerald, C. J., Thomson, M. C., & Whittaker, M. B. (2010). Altruism between romantic partners: Biological offspring as a genetic bridge between altruist and recipient. *Evolutionary Psychology, 8,* 462–476.

Fletcher, G., Simpson, J. A., Campbell, L., & Overall, N. (2013). *The science of intimate relationships.* Oxford, UK: Blackwell.

Flinn, M. V. (1988). Step- and genetic parent/offspring relationships in a Caribbean village. *Ethology and Sociobiology, 9,* 1–34.

Ford, C. S., & Beach, F. A. (1951). *Patterns of sexual behavior.* New York: Harper & Row.

Freeman, D. (1983). *Margaret Mead and Samoa: The making and unmaking of an anthropological myth.* Canberra, ACT: Australian National University Press.

Freud, S. (1914). *On narcissism: An introduction.* London: Karnac Books.

Friday, N. (1973). *My secret garden: Women's sexual fantasies.* New York: Simon & Schuster.

Friedman, M. (1953). The methodology of positive economics. In *Essays in positive economics* (pp. 3–43). Chicago, IL: University of Chicago Press.

Friedman, R. A. (2014). A natural fix for A.D.H.D. *The New York Times.* Retrieved March 27, 2015, from www.nytimes.com/2014/11/02/opinion/sunday/a-natural-fix-for-adhd.html

Frost, P. (2015). The fellowship instinct. *Evo and Proud.* Retrieved January 30, 2017, from http://evoandproud.blogspot.my/2015/11/the-fellowship-instinct.html

Gaiman, N. (1996). *The sandman, Vol. IX: The kindly ones.* New York: DC Comics.

Gallup, G. G., Burch, R. L., & Mitchell, T. J. B. (2006). Semen displacement as a sperm competition strategy. *Human Nature, 17,* 253–264.

Gangestad, S. W., & Simpson, J. A. (2000). The evolution of human mating: Trade-offs and strategic pluralism. *Behavioral and Brain Sciences, 23,* 573–587.

Gardner, M. (2000). Kilroy was here. *Los Angeles Times.* Retrieved April 2, 2015, from http://articles.latimes.com/2000/mar/05/books/bk-5402

Geary, D. C. (1995). Reflections of evolution and culture in children's cognition: Implications for mathematical development and instruction. *American Psychologist, 50*, 24–37.

(2010). *Male, female: The evolution of human sex differences* (2nd ed.). Washington, DC: American Psychological Association.

Gettler, L. T., McDade, T. W., Feranil, A. B., & Kuzawa, C. W. (2011). Longitudinal evidence that fatherhood decreases testosterone in human males. *Proceedings of the National Academy of Sciences, 108*, 16194–16199.

Gillings, M. R. (2014). Were there evolutionary advantages to premenstrual syndrome? *Evolutionary Applications, 7*, 897–904.

Gintis, H. (2000). Strong reciprocity and human sociality. *Journal of Theoretical Biology, 206*, 169–179.

Glocker, M. L., Langleben, D. D., Ruparel, K., Loughead, J. W., Gur, R. C., & Sachser, N. (2009). Baby schema in infant faces induces cuteness perception and motivation for caretaking in adults. *Ethology, 115*, 257–263.

Goodall, J. (1986). *The chimpanzees of Gombe: Patterns of behavior*. Cambridge, MA: Harvard University Press.

Gottfredson, L. S. (2007). Innovation, fatal accidents, and the evolution of general intelligence. In M. J. Roberts (Ed.), *Integrating the mind: Domain general versus domain specific processes in higher cognition* (pp. 387–425). Hove, UK: Psychology Press.

Gottschall, J., Martin, J., Quish, H., & Rea, J. (2004). Sex differences in mate choice criteria are reflected in folktales from around the world and in historical European literature. *Evolution and Human Behavior, 25*, 102–112.

Gottschall, J., & Nordlund, M. (2006). Romantic love: A literary universal. *Philosophy and Literature, 30*, 432–452.

Gould, S. J. (1981). *The mismeasure of man*. New York: Norton.

(2000a). More things in heaven and earth. In H. Rose & S. Rose (Eds.), *Alas poor Darwin: Arguments against evolutionary psychology* (pp. 101–126). London: Jonathan Cape.

(2000b). The spice of life: An interview with Stephen Jay Gould. *Leader to Leader, 15*, 14–19.

Gould, S. J., & Lewontin, R. C. (1979). The spandrels of San Marco and the Panglossian program: A critique of the adaptationist programme. *Proceedings of the Royal Society B, 205*, 281–288.

Graham, P. (2010). The acceleration of addictiveness. *PaulGraham.com*. Retrieved March 30, 2015, from www.paulgraham.com/addiction.html

Gray, P. B., McHale, T. S., & Carré, J. M. (2017). A review of human male field studies of hormones and behavioral reproductive effort. *Hormones and Behavior, 91*, 52–67.

Griffin, A. S., West, S. A., & Buckling, A. (2004). Cooperation and competition in pathogenic bacteria. *Nature, 430*, 1024–1027.

Gurven, M., & Hill, K. (2009). Why do men hunt? A reevaluation of "Man the Hunter" and the sexual division of labor. *Current Anthropology, 50*, 51–74.

Gwynne, D. T., & Rentz, D. C. F. (1983). Beetles on the bottle: Male buprestids mistake stubbies for females (*Coleoptera*). *Australian Journal of Entomology, 22*, 79–80.

Haddad, W. A., Reisinger, R. R., Scott, T., Bester, M. N., & de Bruyn, P. J. N. (2015). Multiple occurrences of king penguin (*Aptenodytes patagonicus*) sexual harassment by Antarctic fur seals (*Arctocephalus gazella*). *Polar Biology, 38,* 741–746.

Hagen, E. H. (2005). Controversial issues in evolutionary psychology. In D. M. Buss (Ed.), *The handbook of evolutionary psychology* (pp. 145–176). Hoboken, NJ: Wiley.

Hagen, E. H., & Hammerstein, P. (2006). Game theory and human evolution: A critique of some recent interpretations of experimental games. *Theoretical Population Biology, 69,* 339–348.

Hahn-Holbrook, J., & Haselton, M. (2014). Is postpartum depression a disease of modern civilization? *Current Directions in Psychological Science, 23,* 395–400.

Haidt, J. (2012). *The righteous mind: Why good people are divided by politics and religion.* New York: Random House.

Halpern, D. F. (2012). *Sex differences in cognitive abilities* (4th ed.). New York: Psychology Press.

Hamer, D. H. (2004). *The God gene: How faith is hardwired into our genes.* New York: Random House.

Hames, R. B. (1987). Relatedness and garden labor exchange among the Ye'kwana. *Ethology and Sociobiology, 8,* 354–392.

Hamilton, W. D. (1963). The evolution of altruistic behavior. *American Naturalist, 97,* 354–356.

(1964). The genetical evolution of social behaviour: I & II. *Journal of Theoretical Biology, 7,* 1–52.

(1996). *Narrow roads of gene land: The collected papers of W. D. Hamilton.* Oxford, UK: Freeman/Spektrum.

Hamilton, W. D., & Zuk, M. (1982). Heritable true fitness and bright birds: A role for parasites? *Science, 218,* 384–387.

Hardin, G. (1968). The tragedy of the commons. *Science, 162,* 1243–1248.

Hardy, K., Brand-Miller, J., Brown, K. D., Thomas, M. G., & Copeland, L. (2015). The importance of dietary carbohydrate in human evolution. *Quarterly Review of Biology, 90,* 251–268.

Harford, T. (2016). How to be a happier man. *Men's Health.* Retrieved March 21, 2017, from www.menshealth.co.uk/healthy/stress/how-to-be-a-happier-man

Harpending, H. (2013). *Ongoing evolution in humans.* Paper presented at Culture-Gene Interactions in Human Origins CARTA Symposium. Retrieved March 3, 2018, from www.ucsd.tv/search-details.aspx?showID=24112

Harris, C. R. (2013). Humans, deer, and sea dragons: How evolutionary psychology has misconstrued human sex differences. *Psychological Inquiry, 24,* 195–201.

Harris, F. (1975). *The short stories of Frank Harris: A selection.* Carbondale, IL: Southern Illinois University Press.

Harris, S. (Producer). (2016). Complexity and stupidity: A conversation with David Krakauer. *Waking Up* [Podcast]. Retrieved March 3, 2018, from https://samharris.org/complexity-stupidity/

Harvey, P. H., & Bradbury, J. W. (1991). Sexual selection. In J. R. Krebs & N. B. Davies (Eds.), *Behavioural ecology: An evolutionary approach* (3rd ed., pp. 203–233). Oxford, UK: Blackwell.

Hawkes, K. (1991). Showing off: Tests of an hypothesis about men's foraging goals. *Ethology and Sociobiology, 12*, 29–54.

Hawking, S. (2008). Stephen Hawking asks big questions about the universe. *Ted*. Retrieved March 3, 2018, from www.ted.com/talks/stephen_hawking_asks_big_questions_about_the_universe

Hawks, J., Wang, E. T., Cochran, G. M., Harpending, H. C., & Moyzis, R. K. (2007). Recent acceleration of human adaptive evolution. *Proceedings of the National Academy of Sciences, 104*, 20753–20758.

Henrich, J. (2004). Cultural group selection, coevolutionary processes and large-scale cooperation. *Journal of Economic Behaviour and Organisation, 53*, 3–35.

(2016). *The secret of our success: How culture is driving human evolution, domesticating our species, and making us smarter*. Princeton, NJ: Princeton University Press.

Henrich, J., Boyd, R., & Richerson, P. J. (2012). The puzzle of monogamous marriage. *Philosophical Transactions of the Royal Society B, 367*, 657–669.

Henrich, J., & Gil-White, F. J. (2001). The evolution of prestige: Freely conferred deference as a mechanism for enhancing the benefits of cultural transmission. *Evolution and Human Behavior, 22*, 165–196.

Henrich, J., Heine, S. J., & Norenzayan, A. (2010). The weirdest people in the world? *Behavioral and Brain Sciences, 33*, 61–83.

Henrich, J., & McElreath, R. (2003). The evolution of cultural evolution. *Evolutionary Anthropology, 12*, 123–135.

(2007). Dual-inheritance theory: The evolution of human cultural capacities and cultural evolution. In R. I. M. Dunbar & L. Barrett (Eds.), *Oxford handbook of evolutionary psychology* (pp. 555–570). Oxford University Press.

Hewlett, B. S. (1991). *Intimate fathers: The nature and context of Aka pygmy paternal infant care*. Ann Arbor, MI: University of Michigan Press.

Hill, K., & Hurtado, A. M. (1996). *Aché life history: The ecology and demography of a foraging people*. Hawthorne, NY: Aldine de Gruyter.

Hill, K. R., Walker, R. S., Božičević, M., *et al.* (2011). Co-residence patterns in hunter-gatherer societies show unique human social structure. *Science, 331*, 1286–1289.

Hinde, R. A., & Barden, L. (1985). The evolution of the teddy bear. *Animal Behaviour, 33*, 1371–1373.

Hoffman, J. I., Forcada, J., Trathan, P. N., & Amos, W. (2007). Female fur seals show active choice for males that are heterozygous and unrelated. *Nature, 445*, 912–914.

Hofstede, G. (2001). *Culture's consequences: Comparing values, behaviors, institutions, and organizations across nations*. Thousand Oaks, CA: Sage.

Holmboe, S. A., Priskorn, L., Jørgensen, N., *et al.* (2017). Influence of marital status on testosterone levels: A ten year follow-up of 1113 men. *Psychoneuroendocrinology, 80*, 155–161.

Holmes, W. G., & Sherman, P. W. (1982). The ontogeny of kin recognition in two species of ground squirrels. *American Zoologist, 22,* 491–517.

Hrdy, S. B. (1977). *The langurs of Abu.* Cambridge, MA: Harvard University Press.

(2000). *Mother nature: Maternal instincts and how they shape the human species.* New York: Ballentine Books.

(2009). *Mothers and others: The evolutionary origin of mutual understanding.* Cambridge, MA: Harvard University Press.

Hughes, J. (2011). *On the origin of tepees: The evolution of ideas (and ourselves).* New York: Free Press.

Hughes, S. M., Dispenza, F., & Gallup, G. G. (2003). Sex differences in morphological predictors of sexual behavior: Shoulder to hip and waist to hip ratios. *Evolution and Human Behavior, 24,* 173–178.

Huxley, J. (1957). *New bottles for new wine.* London: Chatto & Windus.

Inge, W. R. (1929). *Labels and libels.* New York: Harper & Brothers.

Jablonski, N. G. (2006). *Skin: A natural history.* Berkeley, CA: University of California Press.

Jaeggi, A. V., & Gurven, M. (2013). Reciprocity explains food sharing in humans and other primates independent of kin selection and tolerated scrounging: A phylogenetic meta-analysis. *Proceedings of the Royal Society B, 280,* 20131615.

Jahr, C. (1976). Elton John: It's lonely at the top. *Rolling Stone, 223,* 11, 16–17.

James, W. (1880). Great men, great thoughts, and the environment. *Atlantic Monthly, 66,* 441–459.

(1890). *Principles of psychology.* New York: Dover.

Janicke, T., Häderer, I. K., Lajeunesse, M. J., & Anthes, N. (2016). Darwinian sex roles confirmed across the animal kingdom. *Science Advances, 2,* e1500983.

Jankowiak, W., & Diderich, M. (2000). Sibling solidarity in a polygamous community in the USA: Unpacking inclusive fitness. *Evolution and Human Behavior, 21,* 125–139.

Jankowiak, W. R., & Fischer, E. F. (1992). A cross-cultural perspective on romantic love. *Ethnology, 31,* 149–156.

Jankowiak, W. R., Nell, M. D., & Buckmaster, A. (2002). Managing infidelity: A cross-cultural perspective. *Ethnology, 41,* 85–101.

Johnson, R. C. (1996). Attributes of Carnegie medalists performing acts of heroism and of the recipients of these acts. *Ethology and Sociobiology, 17,* 355–362.

Johnson, S. (2010). *Where good ideas come from: The natural history of innovation.* New York: Penguin.

Johnston, V. S., & Franklin, M. (1993). Is beauty in the eye of the beholder? *Ethology and Sociobiology, 14,* 183–199.

Jones, B. C., Hahn, A. C., Fisher, C. I., et al. (2017). No evidence that more physically attractive women have higher estradiol or progesterone. *bioRxiv.*

Jordan, B. K., Mohammed, M., Ching, S. T., et al. (2001). Up-regulation of WNT-4 signaling and dosage-sensitive sex reversal in humans. *American Journal of Human Genetics, 68,* 1102–1109.

Joshi, P. K., Esko, T., Mattsson, H., et al. (2015). Directional dominance on stature and cognition in diverse human populations. *Nature, 523,* 459–462.

Judge, D. S., & Hrdy, S. B. (1992). Allocation of accumulated resources among close kin: Inheritance in Sacramento, California, 1890–1984. *Ethology and Sociobiology, 13,* 495–522.

Jussim, L. (2012). *Social perception and social reality: Why accuracy dominates bias and self-fulfilling prophecy.* Oxford University Press.

Kahlenberg, S. M., & Wrangham, R. W. (2010). Sex differences in chimpanzees' use of sticks as play objects resemble those of children. *Current Biology, 20,* R1067–1068.

Kahneman, D., Krueger, A. B., Schkade, D. A., Schwarz, N., & Stone, A. A. (2004). A survey method for characterizing daily life experience: The day reconstruction method. *Science, 306,* 1776–1780.

Kaplan, H. (1994). Evolutionary and wealth flow theories of fertility: Empirical tests and new models. *Population and Development Review, 20,* 753–791.

Kaplan, H., Hill, K., Lancaster, J., & Hurtado, A. M. (2000). A theory of human life history evolution: Diet, intelligence and longevity. *Evolutionary Anthropology, 9,* 156–185.

Kaufmann, E. (2010). *Shall the rich inherit the Earth? Demography and politics in the twenty-first century.* London: Profile Books.

Kellogg, W. N., & Kellogg, L. A. (1933). *The ape and the child: A comparative study of the environmental influence upon early behavior.* New York: Hafner.

Kelly, K. (2010). *What technology wants.* New York: Viking.

Kendal, R., Hopper, L. M., Whiten, A., *et al.* (2014). Chimpanzees copy dominant and knowledgeable individuals: Implications for cultural diversity. *Evolution and Human Behavior, 36,* 65–72.

Kennair, L. E. O., Bendixen, M., & Buss, D. M. (2016). Sexual regret. *Evolutionary Psychology, 14,* 1474704916682903.

Kenrick, D. T., Groth, G., Trost, M. R., & Sadalla, E. K. (1993). Integrating evolutionary and social exchange perspectives on relationships: Effects of gender, self-appraisal, and involvement level on mate selection criteria. *Journal of Personality and Social Psychology, 64,* 951–969.

Kenrick, D. T., & Keefe, R. C. (1992). Age preferences in mates reflect sex differences in reproductive strategies. *Behavioral and Brain Sciences, 15,* 75–133.

Kenrick, D. T., Keefe, R. C., Bryan, A., Barr, A., & Brown, S. (1995). Age preferences and mate choice among homosexuals and heterosexuals: A case for modular psychological mechanisms. *Journal of Personality and Social Psychology, 69,* 1166–1172.

Kenrick, D. T., & Sheets, V. (1993). Homicidal fantasies. *Ethology and Sociobiology, 14,* 231–246.

King, R. (2013). Baby got back: Some brief observations on obesity in ancient female figurines: Limited support for waist to hip ratio constant as a signal of fertility. *Journal of Obesity and Weight Loss Therapy, 3,* 1000159.

King, R., & Belsky, J. (2012). A typological approach to testing the evolutionary functions of human female orgasm. *Archives of Sexual Behavior, 41,* 1145–1160.

Kirby, S., Cornish, H., & Smith, K. (2008). Cumulative cultural evolution in the laboratory: An experimental approach to the origins of structure in human language. *Proceedings of the National Academy of Sciences, 105,* 10681–10686.

Kokko, H., & Ots, I. (2006). When not to avoid inbreeding. *Evolution, 60,* 467–475.

Kordsmeyer, T. L., & Penke, L. (2017). The association of three indicators of developmental instability with mating success in humans. *Evolution and Human Behavior, 38,* 704–713.

Kruger, D. J., & Nesse, R. M. (2006). An evolutionary life-history framework for understanding sex differences in human mortality rates. *Human Nature, 17,* 74–97.

Krupp, D. B., DeBruine, L. M., & Barclay, P. (2008). A cue to kinship promotes cooperation for the public good. *Evolution and Human Behavior, 29,* 49–55.

Kuhle, B. X. (2011). Did you have sex with him? Do you love her? An in vivo test of sex differences in jealous interrogations. *Personality and Individual Differences, 51,* 1044–1047.

Kurzban, R. (2002). Alas poor evolutionary psychology: Unfairly accused, unjustly condemned. *Human Nature Review, 2,* 99–109.

Labuda, D., Lefebvre, J.-F., Nadeau, P., & Roy-Gagnon, M.-H. (2010). Female-to-male breeding ratio in modern humans: An analysis based on historical recombinations. *American Journal of Human Genetics, 86,* 353–363.

Lack, D. (1968). *Ecological adaptations for breeding in birds.* London: Methuen.

Laland, K. N. (2017). *Darwin's unfinished symphony: How culture made the human mind.* Princeton, NJ: Princeton University Press.

Laland, K. N., & Brown, G. R. (2002). *Sense and nonsense: Evolutionary perspectives on human behaviour.* Oxford University Press.

Laland, K. N., Uller, T., Feldman, M., *et al.* (2014). Does evolutionary theory need a rethink? *Nature, 514,* 161–164.

Lamminmäki, A., Hines, M., Kuiri-Hänninen, T., *et al.* (2012). Testosterone measured in infancy predicts subsequent sex-typed behavior in boys and in girls. *Hormones and Behavior, 61,* 611–616.

Langlois, J. H., Roggman, L. A., Casey, R. J., *et al.* (1987). Infant preferences for attractive faces: Rudiments of a stereotype? *Developmental Psychology, 23,* 363–369.

Larmuseau, M. H. D., Vanoverbeke, J., Van Geystelen, A., *et al.* (2013). Low historical rates of cuckoldry in a Western European human population traced by Y-chromosome and genealogical data. *Proceedings of the Royal Society B, 280,* 20132400.

Lassek, W. D., & Gaulin, S. J. C. (2009). Costs and benefits of fat-free muscle mass in men: Relationship to mating success, dietary requirements, and native immunity. *Evolution and Human Behavior, 30,* 322–328.

LeBoeuf, B. J. (1974). The hectic life of the alpha bull: Elephant seal as fighter and lover. *Psychology Today, 8,* 104–108.

Lee, S., & Hasegawa, T. (2014). Oceanic barriers promote language diversification in the Japanese Islands. *Journal of Evolutionary Biology, 27,* 1905–1912.

Legare, C. H., & Nielsen, M. (2015). Imitation and innovation: The dual engines of cultural learning. *Trends in Cognitive Sciences, 19,* 688–699.

LeVay, S. (2011). *Gay, straight, and the reason why: The science of sexual orientation.* Oxford University Press.

Levitan, D. (2006). *This is your brain on music: The science of a human obsession.* New York: Penguin.

Li, G., Kung, K. T., & Hines, M. (2017). Childhood gender-typed behavior and adolescent sexual orientation: A longitudinal population-based study. *Developmental Psychology, 53,* 764–777.

Li, N. P., Bailey, J. M., & Kenrick, D. T. (2002). The necessities and luxuries of mate preferences: Testing the tradeoffs. *Journal of Personality and Social Psychology, 82,* 947–955.

Liao, X., Rong, S., & Queller, D. C. (2015). Relatedness, conflict, and the evolution of eusociality. *PLoS Biology, 13,* e1002098.

Lieberman, D., Tooby, J., & Cosmides, L. (2007). The architecture of human kin detection. *Nature , 445,* 727–731.

Lieberman, D. E. (2013). *The story of the human body: Evolution, health, and disease.* New York: Pantheon.

Lihoreau, M., Zimmer, C., & Rivault, C. (2007). Kin recognition and incest avoidance in a group-living insect. *Behavioral Ecology, 18,* 880–887.

Lippa, R. A. (2009). Sex differences in sex drive, sociosexuality, and height across 53 nations: Testing evolutionary and social structural theories. *Archives of Sexual Behavior, 38,* 631–651.

Lishner, D. A., Nguyen, S., Stocks, E. L., & Zillmer, E. J. (2008). Are sexual and emotional infidelity equally upsetting to men and women? Making sense of forced-choice responses. *Evolutionary Psychology, 6,* 667–675.

Litchfield, H. E. (Ed.). (1904). *Emma Darwin, wife of Charles Darwin: A century of family letters.* Cambridge University Press.

Littlefield, C. H., & Rushton, J. P. (1986). When a child dies: The sociobiology of bereavement. *Journal of Personality and Social Psychology, 51,* 797–802.

Lloyd, E. A. (2006). *The case of the female orgasm: Bias in the science of evolution.* Cambridge, MA: Harvard University Press.

Lonsdorf, E. V. (2017). Sex differences in nonhuman primate behavioral development. *Journal of Neuroscience Research, 95,* 213–221.

López, P., Muñoz, A., & Martín, J. (2002). Symmetry, male dominance and female mate preferences in the Iberian rock lizard, *Lacerta monticola. Behavioral Ecology and Sociobiology, 52,* 342–347.

Lorenz, K. (1943). Die angeborenen formen möglicher erfahrung. *Zeitschrift für Tierpsychologie, 5,* 235–409.

Low, B. S. (1989). Cross-cultural patterns in the training of children: An evolutionary perspective. *Journal of Comparative Psychology, 103,* 311–319.

Lubinski, D., Benbow, C. P., & Kell, H. J. (2014). Life paths and accomplishments of mathematically precocious males and females four decades later. *Psychological Science, 25,* 2217–2232.

Lukas, D., & Huchard, E. (2014). The evolution of infanticide by males in mammalian societies. *Science, 346,* 841–844.

Lumsden, C. J., & Wilson, E. O. (1981). *Genes, mind, and behavior: The coevolutionary process.* Cambridge, MA: Harvard University Press.

Lykken, D. T. (1995). *The antisocial personalities.* Hillsdale, NJ: Lawrence Erlbaum.

Lynch, A. (1996). *Thought contagion: How belief spreads through society.* New York: Basic Books.

Maccoby, E., & Jacklin, C. N. (1974). *The psychology of sex differences.* Stanford, CA: Stanford University Press.

Mace, R., & Holden, C. J. (2005). A phylogenetic approach to cultural evolution. *Trends in Ecology and Evolution, 20,* 116–121.

Mach, E. (1896). On the part played by accident in invention and discovery. *Monist, 6,* 161–175.

Madsen, E., Tunney, R., Fieldman, G., *et al.* (2007). Kinship and altruism: A cross-cultural experimental study. *British Journal of Psychology, 98,* 339–359.

Mann, J. (1992). Nurturance or negligence: Maternal psychology and behavioral preference among preterm twins. In J. H. Barkow, L. Cosmides, & J. Tooby (Eds.), *The adapted mind: Evolutionary psychology and the generation of culture* (pp. 367–390). Oxford University Press.

Marcus, G. (2012). *Guitar zero: The science of becoming musical at any age.* New York: Penguin.

Marks, I. M., & Nesse, R. M. (1994). Fear and fitness: An evolutionary analysis of anxiety disorders. *Ethology and Sociobiology, 15,* 247–261.

Marlowe, F. W. (1999a). Male care and mating effort among Hadza foragers. *Behavioral Ecology and Sociobiology, 46,* 57–64.

   (1999b). Showoffs or providers? The parenting effort of Hadza men. *Evolution and Human Behavior, 20,* 391–404.

   (2000). Paternal investment and the human mating system. *Behavioural Processes, 51,* 45–61.

   (2003a). A critical period for provisioning by Hadza men: Implications for pair bonding. *Evolution and Human Behavior, 24,* 217–229.

   (2003b). The mating system of foragers in the Standard Cross-Cultural Sample. *Cross-Cultural Research, 37,* 282–306.

   (2004). Mate preferences among Hadza hunter-gatherers. *Human Nature, 15,* 365–376.

Massen, J. J. M., Dusch, K., Eldakar, O. T., & Gallup, A. C. (2014). A thermal window for yawning in humans: Yawning as a brain cooling mechanism. *Physiology and Behavior, 130,* 145–148.

Massen, J. J. M., Ritter, C., & Bugnyar, T. (2015). Tolerance and reward equity predict cooperation in ravens (*Corvus corax*). *Scientific Reports, 5,* 15021.

Matzke, N. J. (2015). The evolution of antievolution policies after Kitzmiller v. Dover. *Science, 38,* 28–30.

Maugham, W. S. (1938/2001). *The summing up.* London: Vintage Books.

McClintock, E. A. (2011). Handsome wants as handsome does: Physical attractiveness and gender differences in revealed sexual preferences. *Biodemography and Social Biology, 57,* 221–257.

McClure, B. (2004). Reproductive biology: Pillow talk in plants. *Nature, 429,* 249–250.

McLuhan, M. (1964). *Understanding media: The extensions of man.* New York: McGraw-Hill.

Mead, M. (1928). *Coming of age in Samoa: A psychological study of primitive youth for Western civilization.* New York: Morrow.

Melfi, V. (2005). Red baboon bottoms as sexual traffic lights. *The Naked Scientists* [Podcast]. Retrieved January 26, 2015, from www.thenakedscientists.com/podcasts/naked-scientists/animal-communication-sexual-signalling-and-emotions

Mencken, H. L. (1916/1982). *A Mencken chrestomathy: His own selection of his choicest writings*. New York: Vintage Books.

Mesoudi, A. (2011). *Cultural evolution: How Darwinian theory can explain human culture and unify the social sciences*. Chicago, IL: University of Chicago Press.

Miller, G. F. (2000a). Evolution of human music through sexual selection. In N. L. Wallin, B. Merker, & S. Brown (Eds.), *The origins of music* (pp. 329–360). Harvard, MA: MIT Press.

  (2000b). *The mating mind: How sexual choice shaped the evolution of human nature*. New York: Vintage Books.

  (2007). Sexual selection for moral virtues. *Quarterly Review of Biology, 82,* 97–125.

  (2009). *Spent: Sex, evolution, and consumer behavior*. New York: Viking.

  (2013). Mutual mate choice models as the red pill in evolutionary psychology: Long delayed, much needed, ideologically challenging, and hard to swallow. *Psychological Inquiry, 24,* 207–210.

Millward, J. (2013). Deep inside: A study of 10,000 porn stars and their careers. *Jon Millward Data Journalist*. Retrieved February 24, 2017, from http://jonmillward.com/blog/studies/deep-inside-a-study-of-10000-porn-stars/

Moeliker, C. W. (2001). The first case of homosexual necrophilia in the mallard *Anas platyrhynchos* (Aves: Anatidae). *Deinsea, 8,* 243–247.

Møller, A. P. (1993). Female preference for apparently symmetrical male sexual ornaments in the barn swallow *Hirundo rustica*. *Behavioral Ecology and Sociobiology, 32,* 371–376.

  (1997). Developmental stability and fitness: A review. *American Naturalist, 149,* 916–932.

Møller, A. P., & Birkhead, T. R. (1993). Certainty of paternity covaries with paternal care in birds. *Behavioral Ecology and Sociobiology, 33,* 261–268.

Montagu, A. (1973). *Man and aggression* (2nd ed.). Oxford University Press.

Morris, C. (2016). Porn's dirtiest secret: What everyone gets paid. *CNBC*. Retrieved February 24, 2017, from www.cnbc.com/2016/01/20/porns-dirtiest-secret-what-everyone-gets-paid.html

Muller, M. N., Thompson, M. E., & Wrangham, R. W. (2006). Male chimpanzees prefer mating with old females. *Current Biology, 16,* 2234–2238.

Muniz, L., Perry, S., Manson, J. H., *et al.* (2006). Father-daughter avoidance in a wild primate population. *Current Biology, 16,* R156–157.

Murdock, G. P. (1967). *Ethnographic atlas*. Pittsburgh, PA: University of Pittsburgh Press.

Murdock, G. P., & Provost, C. (1973). Factors in the division of labor by sex: A cross-cultural analysis. *Ethnology, 12,* 203–225.

Nedelec, J. L., & Beaver, K. M. (2014). Physical attractiveness as a phenotypic marker of health: An assessment using a nationally representative sample of American adults. *Evolution and Human Behavior, 35,* 456–463.

Needham, J. (1970). *Clerks and craftsmen in China and the West: Lectures and addresses on the history of science and technology.* Cambridge University Press.

Nesse, R. M. (1990). Evolutionary explanations of emotions. *Human Nature, 1,* 261–289.

(2006). Why a lot of people with selfish genes are pretty nice except for their hatred of *The Selfish Gene*. In A. Grafen & M. Ridley (Eds.), *Richard Dawkins: How a scientist changed the way we think: Reflections by scientists, writers, and philosophers* (pp. 203–212). Oxford University Press.

Nesse, R. M., & Williams, G. C. (1994). *Why we get sick: The new science of Darwinian medicine.* New York: Vintage Books.

Newberry, M. G., Ahern, C. A., Clark, R., & Plotkin, J. B. (2017). Detecting evolutionary forces in language change. *Nature, 551,* 223–226.

Neyer, F. J., & Lang, F. R. (2003). Blood is thicker than water: Kinship orientation across adulthood. *Journal of Personality and Social Psychology, 84,* 310–321.

Nietzsche, F. (1886/1966). *Beyond good and evil: Prelude to a philosophy of the future* (W. Kaufmann, Trans.). New York: Vintage Books.

Norenzayan, A. (2013). *Big gods: How religion transformed cooperation and conflict.* Princeton, NJ: Princeton University Press.

Nowak, M. A., & Sigmund, K. (1992). Tit for tat in heterogeneous populations. *Nature, 355,* 250–253.

Nowak, M. A., Tarnita, C. E., & Wilson, E. O. (2010). The evolution of eusociality. *Nature, 466,* 1057–1062.

Oberzaucher, E., & Grammer, K. (2014). The case of Moulay Ismael – fact or fancy? *PLoS ONE, 9,* e85292.

Ogas, O., & Gaddam, S. (2011). *A billion wicked thoughts: What the Internet tells us about sexual relationships.* New York: Penguin.

Öhman, A., & Mineka, S. (2001). Fears, phobias and preparedness: Toward an evolved module of fear and fear learning. *Psychological Review, 108,* 483–522.

Orians, G. H. (1969). On the evolution of mating systems in birds and mammals. *American Naturalist, 28,* 1–16.

Paczolt, K. A., & Jones, A. G. (2010). Post-copulatory sexual selection and sexual conflict in the evolution of male pregnancy. *Nature, 464,* 401–404.

Pagel, M. (2012). *Wired for culture: Origins of the human social mind.* New York: Norton.

Parker, G. A. (1970). Sperm competition and its evolutionary consequences in the insects. *Biological Reviews, 45,* 525–567.

Parker, K., & Wang, W. (2013). *Modern parenthood: Roles of moms and dads converge as they balance work and family.* Washington, DC: Pew Research Center.

Pasterski, V. L., Geffner, M. E., Brain, C., *et al.* (2005). Prenatal hormones and postnatal socialization by parents as determinants of male-typical toy play in girls with congenital adrenal hyperplasia. *Child Development, 76,* 264–278.

Pasterski, V., Hindmarsh, P., Geffner, M., *et al.* (2007). Increased aggression and activity level in 3- to 11-year-old girls with congenital adrenal hyperplasia (CAH). *Hormones and Behavior, 52,* 368–374.

Pedersen, F. A. (1991). Secular trends in human sex ratios: Their influence on individual and family behavior. *Human Nature*, 2, 271–291.

Pelé, M., Bonnefoy, A., Shimada, M., & Sueur, C. (2017). Interspecies sexual behaviour between a male Japanese macaque and female sika deer. *Primates*, 58, 275–278.

Penn, D., & Potts, W. (1998). MHC-disassortative mating preferences reversed by cross-fostering. *Proceedings of the Royal Society B*, 265, 1299–1306.

Perrett, D. (2010). *In your face: The new science of human attraction*. Basingstoke, UK: Palgrave Macmillan.

Perry, G. H., Dominy, N. J., Claw, K. G., et al. (2007). Diet and the evolution of human amylase gene copy number variation. *Nature Genetics*, 39, 1256–1260.

Perry, S. (2011). Social traditions and social learning in capuchin monkeys (*Cebus*). *Philosophical Transactions of the Royal Society B*, 366, 988–996.

Petrie, M., & Kempenaers, B. (1998). Extra-pair paternity in birds: Explaining variation between species and populations. *Trends in Ecology and Evolution*, 13, 52–58.

Pinker, S. (1994). *The language instinct: The new science of language and mind*. New York: Penguin.

(1997). *How the mind works*. New York: Norton.

(2002). *The blank slate: The modern denial of human nature*. New York: Penguin.

(2011). *The better angels of our nature: Why violence has declined*. New York: Penguin.

(2012). The false allure of group selection. *Edge.org*. Retrieved May 20, 2013, from http://edge.org/conversation/the-false-allure-of-group-selection

(2018). *Enlightenment now: The case for reason, science, humanism, and progress*. New York: Viking.

Pinker, S., & Bloom, P. (1992). Natural language and natural selection. In J. H. Barkow, L. Cosmides, & J. Tooby (Eds.), *The adapted mind: Evolutionary psychology and the generation of culture* (pp. 451–493). Oxford University Press.

Pinker, Susan. (2008). *The sexual paradox: Troubled boys, gifted girls, and the real difference between the sexes*. New York: Scribner.

Polderman, T. J. C., Benyamin, B., de Leeuw, C. A., et al. (2015). Meta-analysis of the heritability of human traits based on fifty years of twin studies. *Nature Genetics*, 47, 702–709.

Pollan, M. (1990). *The botany of desire: A plant's-eye view of the world*. New York: Random House.

Popper, K. R. (1979). *Objective knowledge: An evolutionary approach*. Oxford, UK: Clarendon.

Pound, N., Lawson, D. W., Toma, A. M., et al. (2014). Facial fluctuating asymmetry is not associated with childhood ill-health in a large British cohort study. *Proceedings of the Royal Society B*, 281, 20141639.

Power, M., & Schulkin, J. (2009). *The evolution of obesity*. Baltimore, MD: Johns Hopkins University Press.

Prentiss, A. M., Skelton, R. R., Eldredge, N., & Quinn, C. (2011). Get rad! The evolution of the skateboard deck. *Evolution: Education and Outreach*, 4, 379–389.

Price, G. R. (1970). Selection and covariance. *Nature*, 227, 520–521.

Price, M. E. (2012). Group selection theories are now more sophisticated, but are they more predictive? *Evolutionary Psychology*, 10, 45–49.

Principe, G. (2012). Orangutans on ritalin: An evolutionary developmental psychology perspective on ADHD. *This View of Life*. Retrieved March 27, 2015, from https://evolution-institute.org/article/orangutans-on-ritalin-an-evolutionary-developmental-psychology-perspective/

Pusey, A. (2004). Inbreeding avoidance in primates. In A. P. Wolf & D. H. Durham (Eds.), *Inbreeding, incest, and the incest taboo*. Stanford, CA: Stanford University Press.

Puts, D. A. (2010). Beauty and the beast: Mechanisms of sexual selection in humans. *Evolution and Human Behavior*, 31, 157–175.

Puts, D. A., Dawood, K., & Welling, L. L. M. (2012). Why women have orgasms: An evolutionary analysis. *Archives of Sexual Behavior*, 41, 1127–1143.

Rasmussen, K. (1927). *Across Arctic America: Narrative of the Fifth Thüle Expedition*. Fairbanks, AK: University of Alaska Press.

Ray, D. W. (2009). *The God virus: How religion infects our lives and culture*. Bonner Springs, KS: IPC Press.

Rees, M. (2004). *Our final century*. London: Heinemann.

Reser, J. E. (2011). Conceptualizing the autism spectrum in terms of natural selection and behavioral ecology: The solitary forager hypothesis. *Evolutionary Psychology*, 9, 207–238.

Revonsuo, A. (2000). The reinterpretation of dreams: An evolutionary hypothesis of the function of dreaming. *Behavioral and Brain Sciences*, 23, 877–901.

Richerson, P. J., Baldini, R., Bell, A., et al. (2016). Cultural group selection plays an essential role in explaining human cooperation: A sketch of the evidence. *Behavioral and Brain Sciences*, 39, e30.

Richerson, P. J., & Boyd, R. (2005). *Not by genes alone: How culture transformed human evolution*. Chicago, IL: University of Chicago Press.

Richerson, P. J., Boyd, R., & Henrich, J. (2010). Gene–culture coevolution in the age of genomics. *Proceedings of the National Academy of Sciences*, 7, 8985–8992.

Ridley, M. (1993). *The red queen: Sex and the evolution of human nature*. London: Penguin.

(1996). *The origins of virtue: Human instincts and the evolution of cooperation*. London: Penguin.

(2010). *The rational optimist: How prosperity evolves*. New York: HarperCollins.

Rohwer, S., Herron, J. C., & Daly, M. (1999). Stepparental behavior as mating effort in birds and other animals. *Evolution and Human Behavior*, 20, 367–390.

Rozin, P., Haidt, J., & McCauley, C. R. (1993). Disgust. In M. Lewis & J. M. Haviland (Eds.), *Handbook of emotions* (pp. 575–594). New York: Guilford Press.

Russell, B. (1926). *On education: Especially in early childhood*. London: Routledge.

(1961). *Has man a future?* New York: Simon & Schuster.

Ryan, C., & Jethá, C. (2010). *Sex at dawn: The prehistoric origins of modern sexuality*. New York: HarperCollins.

Saad, G., & Gill, T. (2003). An evolutionary psychology perspective on gift giving among young adults. *Psychology and Marketing*, 20, 765–784.

Sagarin, B. J., Martin, A. L., Coutinho, S. A., *et al.* (2012). Sex differences in jealousy: A meta-analytic examination. *Evolution and Human Behavior, 33,* 595–614.

Salmon, C. A. (2012). The pop culture of sex: An evolutionary window on the worlds of pornography and romance. *Review of General Psychology, 16,* 152–160.

Sasaki, T., & Biro, D. (2017). Cumulative culture can emerge from collective intelligence in animal groups. *Nature Communications, 8,* 15049.

Saxon, L. (2012). *Sex at dusk: Lifting the shiny wrapping from* Sex at Dawn. Lexington, KY: Createspace.

(2016). *The naked bonobo.* Lexington, KY: Createspace.

Scelza, B. A. (2014). Jealousy in a small-scale, natural fertility population: The roles of paternity, investment and love in jealous response. *Evolution and Human Behavior, 35,* 103–108.

Schärer, L., Rowe, L., & Arnqvist, G. (2012). Anisogamy, chance and the evolution of sex roles. *Trends in Ecology and Evolution, 27,* 260–264.

Scheidel, W. (2004). Ancient Egyptian sibling marriage and the Westermarck effect. In A. P. Wolf & W. H. Durham (Eds.), *Inbreeding, incest, and the incest taboo: The state of knowledge at the turn of the century* (pp. 93–108). Stanford, CA: Stanford University Press.

Schmitt, D. P. (2005). Sociosexuality from Argentina to Zimbabwe: A 48-nation study of sex, culture, and strategies of human mating. *Behavioral and Brain Sciences, 28,* 247–275.

(2014). Evaluating evidence of mate preference adaptations: How do we really know what *Homo sapiens sapiens* really want? In V. A. Weekes-Shackelford & T. K. Shackelford (Eds.), *Evolutionary perspectives on human sexual psychology and behavior* (pp. 3–39). New York: Springer.

Schmitt, D. P., & Pilcher, J. J. (2004). Evaluating evidence of psychological adaptation: How do we know one when we see one? *Psychological Science, 15,* 643–649.

Schmitt, D. P., & 118 Members of the International Sexuality Description Project. (2003). Universal sex differences in the desire for sexual variety: Tests from 52 nations, 6 continents, and 13 islands. *Journal of Personality and Social Psychology, 85,* 85–104.

Schopenhauer, A. (1818/1966). *The world as will and representation, Vol. II: Supplements to the fourth book* (E. F. J. Payne, Trans.). New York: Dover Publications.

Scott, I. M. L., Clark, A. P., Boothroyd, L. G., & Penton-Voak, I. S. (2013). Do men's faces really signal heritable immunocompetence? *Behavioral Ecology, 24,* 579–589.

Sear, R., & Mace, R. (2008). Who keeps children alive? A review of the effects of kin on child survival. *Evolution and Human Behavior, 29,* 1–18.

Segerstråle, U. (2000). *Defenders of the truth: The sociobiology debate.* Oxford University Press.

Seligman, M. E. P., & Hager, J. L. (Eds.). (1990). *Biological boundaries of learning.* New York: Appleton-Century-Crofts.

Sell, A., Lukazsweski, A. W., & Townsley, M. (2017). Cues of upper body strength account for most of the variance in men's bodily attractiveness. *Proceedings of the Royal Society B, 284,* 20171819.

Seyfarth, R. M., & Cheney, D. L. (1984). Grooming alliances and reciprocal alliances in vervet monkeys. *Nature, 308,* 541–543.

Shackelford, T. K., & Goetz, A. T. (2006). Comparative evolutionary psychology of sperm competition. *Journal of Comparative Psychology, 120,* 139–146.

Shackelford, T. K., & Larsen, R. J. (1997). Facial asymmetry as an indicator of psychological, emotional, and physiological distress. *Journal of Personality and Social Psychology, 72,* 456–466.

Shackelford, T. K., Voracek, M., Schmitt, D. P., *et al.* (2004). Romantic jealousy in early adulthood and in later life. *Human Nature, 15,* 283–300.

Shepher, J. (1971). Mate selection among second generation kibbutz adolescents and adults: Incest avoidance and negative imprinting. *Archives of Sexual Behavior, 1,* 293–307.

Sherman, P. W. (1977). Nepotism and the evolution of alarm calls. *Science, 197,* 1246–1253.

(1980). The limits of ground squirrel nepotism. In G. B. Barlow & J. Silverberg (Eds.), *Sociobiology: Beyond nature/nurture? AAAS Selected Symposium 35* (pp. 505–544). Boulder, CO: Westview Press.

Shermer, M. (2004). *The science of good and evil: Why people cheat, gossip, care, share, and follow the Golden Rule.* New York: Henry Holt.

(2009). *The mind of the market: How biology and psychology shape our economic lives.* New York: Henry Holt.

Shostak, M. (1981). *Nisa.* Cambridge, MA: Harvard University Press.

Shubin, N. (2008). *Your inner fish: A journey into the 3.5 billion-year history of the human body.* New York: Pantheon.

Silk, J. B. (1987). Adoption among the Inuit. *Ethos, 15,* 320–330.

Simoons, F. J. (1969). Primary adult lactose intolerance and the milking habit: A problem in biological and cultural interrelations. *American Journal of Digestive Diseases, 14,* 819–836.

Simpson, J. A., & Gangestad, S. W. (1991). Individual differences in sociosexuality: Evidence for convergent and discriminant validity. *Journal of Personality and Social Psychology, 60,* 870–883.

Singer, P. (1981). *The expanding circle: Ethics and sociobiology.* New York: Farrar, Strauss, & Giroux.

Singh, D. (1993). Adaptive significance of female physical attractiveness: Role of waist-to-hip ratio. *Journal of Personality and Social Psychology, 65,* 293–307.

(1995). Female judgment of male attractiveness and desirability for relationships: Role of waist-to-hip ratio and financial status. *Journal of Personality and Social Psychology, 69,* 1089–1101.

Singh, D., Dixson, B. J., Jessop, T. S., Morgan, B., & Dixson, A. F. (2010). Cross-cultural consensus for waist–hip ratio and women's attractiveness. *Evolution and Human Behavior, 31,* 176–181.

Singh, D., & Luis, S. (1995). Ethnic and gender consensus for the effect of waist-to-hip ratio on judgement of women's attractiveness. *Human Nature, 6*, 51–65.

Skinner, B. F. (1953). *Science and human behavior*. New York: Free Press.

Slater, A., Quinn, P. C., Hayes, R., & Brown, E. (2000). The role of facial orientation in newborn infants' preference for attractive faces. *Developmental Science, 3*, 181–185.

Smith, E. A., Bird, R. B., & Bird, D. W. (2003). The benefits of costly signaling: Meriam turtle hunters. *Behavioral Ecology, 14*, 116–126.

Smith, E. A., Hill, K., Marlowe, F., *et al.* (2010). Wealth transmission and inequality among hunter-gatherers. *Current Anthropology, 51*, 19–34.

Smith, M. S., Kish, B. J., & Crawford, C. B. (1987). Inheritance of wealth as human kin investment. *Ethology and Sociobiology, 8*, 171–182.

Smith, R. L. (1984). Human sperm competition. In R. L. Smith (Ed.), *Sperm competition and the evolution of animal mating systems* (pp. 601–659). London: Academic Press.

Sober, E., & Wilson, D. S. (1998). *Unto others: The evolution and psychology of unselfish behavior*. Cambridge, MA: Harvard University Press.

Sohn, K. (2016). Men's revealed preferences regarding women's ages: Evidence from prostitution. *Evolution and Human Behavior, 37*, 272–280.

  (2017). Men's revealed preference for their mates' ages. *Evolution and Human Behavior, 38*, 58–62.

Sommer, V., & Reichard, U. (2000). Rethinking monogamy: The gibbon case. In P. Kappeler (Ed.), *Primate males* (pp. 159–168). Cambridge University Press.

Sommers, C. H. (2013a). *Freedom feminism: Its surprising history and why it matters today*. Washington, DC: AEI Press.

  (2013b). *The war against boys: How misguided policies are harming our young men* (rev. ed.). New York: Simon & Schuster.

  (2013c). What "lean in" misunderstands about gender differences. *The Atlantic*. Retrieved January 12, 2017, from www.theatlantic.com/sexes/archive/2013/03/what-lean-in-misunderstands-about-gender-differences/274138/

Sperber, D. (1985). Anthropology and psychology: Towards an epidemiology of representations. *Man, 20*, 73–89.

  (1996). *Explaining culture: A naturalistic approach*. Oxford, UK: Blackwell.

Stack, S. (1987). Celebrities and suicide: A taxonomy and analysis, 1948–1983. *American Sociological Review, 52*, 401–412.

Starr, S. B. (2015). Estimating gender disparities in federal criminal cases. *American Law and Economics Review, 17*, 127–159.

Stephen, I. D., Hiew, V., Coetzee, V., Tiddeman, B. P., & Perrett, D. I. (2017). Facial shape analysis identifies valid cues to aspects of physiological health in Caucasian, Asian, and African populations. *Frontiers in Psychology, 8*, 1883.

Sternberg, R. J. (1986). A triangular theory of love. *Psychological Review, 93*, 119–135.

Stewart-Williams, S. (2002). Gender, the perception of aggression, and the overestimation of gender bias. *Sex Roles, 46*, 177–189.

  (2007). Altruism among kin vs. nonkin: Effects of cost of help and reciprocal exchange. *Evolution and Human Behavior, 28*, 193–198.

(2008). Human beings as evolved nepotists: Exceptions to the rule and effects of cost of help. *Human Nature, 19,* 414–425.

(2010). *Darwin, God, and the meaning of life: How evolutionary theory undermines everything you thought you knew.* Cambridge University Press.

(2014). The sticking point: Why men still outnumber women in science. *Psychology Today.* Retrieved January 12, 2017, from www.psychologytoday.com/us/blog/the-nature-nurture-nietzsche-blog/201402/the-sticking-point-why-men-still-outnumber-women-in

(2015a). Evolution and morality. In J. D. Wright (Ed.), *International encyclopedia of the social and behavioral sciences* (2nd ed., Vol. XV, pp. 811–818). Oxford, UK: Elsevier.

(2015b). Foreword: On the origin of afterlife beliefs by means of memetic selection. In M. Martin & K. Augustine (Eds.), *The myth of afterlife: Essays on the case against life after death* (pp. xiii–xxv). Jefferson, NC: McFarland.

(2017). The science of human sex differences: Implications for policy on gender equity. *Gifted Women, Fragile Men.* Retrieved March 21, 2017, from http://euromind.global/steve-stewart-williams/

(in press-a). Afterlife beliefs: An evolutionary approach. In D. M. Wulff (Ed.), *Oxford handbook of the psychology of religion* (2nd ed.). Oxford University Press.

(in press-b). Are humans peacocks or robins? In L. Workman, W. Reader, & J. Barkow (Eds.), *Cambridge handbook of evolutionary perspectives on human behavior* (2nd ed.). Cambridge University Press.

Stewart-Williams, S., Butler, C. A., & Thomas, A. G. (2017). Sexual history and present attractiveness: People want a mate with a bit of a past, but not too much. *Journal of Sex Research, 54,* 1097–1105.

Stewart-Williams, S., & Thomas, A. G. (2013a). The ape that kicked the hornet's nest: Response to commentaries on "The ape that thought it was a peacock". *Psychological Inquiry, 24,* 248–271.

(2013b). The ape that thought it was a peacock: Does evolutionary psychology exaggerate human sex differences? *Psychological Inquiry, 24,* 137–168.

Strassmann, B. I. (1997). The biology of menstruation in *Homo sapiens*: Total lifetime menses, fecundity, and nonsynchrony in a natural fertility population. *Current Anthropology, 38,* 123–129.

Swaddle, J. P., & Cuthill, I. C. (1994). Preference for symmetric males by female zebra finches. *Nature, 367,* 165–166.

Symons, D. (1979). *The evolution of human sexuality.* Oxford University Press.

(1989). The psychology of human mate preferences. *Behavioral and Brain Sciences, 12,* 34–35.

Tallis, F. (2005). *Love sick: Love as a mental illness.* New York: Avalon.

Taylor, T. (2010). *The artificial ape: How technology changed the course of human evolution.* New York: Palgrave Macmillan.

Tehrani, J., & Collard, M. (2002). Investigating cultural evolution through biological phylogenetic analyses of Turkmen textiles. *Journal of Anthropological Archaeology, 21,* 443–463.

Tennie, C., Call, J., & Tomasello, M. (2009). Ratcheting up the ratchet: On the evolution of cumulative culture. *Philosophical Transactions of the Royal Society B, 364,* 2405–2415.

Tennov, D. (1979). *Love and limerence: The experience of being in love.* Lanham, ML: Scarborough House.

Thomas, A. G., & Stewart-Williams, S. (2018). Mating strategy flexibility in the laboratory: Preferences for long- and short-term mating change in response to evolutionarily relevant variables. *Evolution and Human Behavior, 39,* 82–93.

Thomas, M. G., Ji, T., Wu, J., *et al.* (2018). Kinship underlies costly cooperation in Mosuo villages. *Royal Society Open Science, 5,* 171535.

Thornhill, R. (1976). Sexual selection and nuptial feeding behavior in *Bittacus apicalis* (Insecta: Mecoptera). *American Naturalist, 110,* 529–548.

  (1992). Female preference for the pheromone of males with low fluctuating asymmetry in the Japanese scorpionfly (*Panorpa japonica*: Mecoptera). *Behavioral Ecology, 3,* 277–283.

Thornhill, R., & Gangestad, S. W. (1993). Human facial beauty: Averageness, symmetry, and parasite resistance. *Human Nature, 4,* 237–269.

  (1994). Human fluctuating asymmetry and sexual behavior. *Psychological Science, 5,* 297–302.

Thornhill, R., & Palmer, C. T. (2000). *A natural history of rape: Biological bases of sexual coercion.* Cambridge, MA: MIT Press.

Tidière, M., Gaillard, J.-M., Müller, D. W. H., *et al.* (2015). Does sexual selection shape sex differences in longevity and senescence patterns across vertebrates? A review and new insights from captive ruminants. *Evolution, 69,* 3123–3140.

Tiger, L., & Shepher, J. (1975). *Women in the kibbutz.* New York: Harcourt, Brace, Jovanovich.

Tishkoff, S. A., Reed, F. A., Ranciaro, A., *et al.* (2007). Convergent adaptation of human lactase persistence in Africa and Europe. *Nature Genetics, 39,* 31–40.

Tomasello, M. (1999). *The cultural origins of human cognition.* Cambridge, MA: Harvard University Press.

Tooby, J., & Cosmides, L. (1992). The psychological foundations of culture. In J. H. Barkow, L. Cosmides, & J. Tooby (Eds.), *The adapted mind: Evolutionary psychology and the generation of culture* (pp. 19–136). Oxford University Press.

Touboul, J. (2014). The hipster effect: When anticonformists all look the same. *arXiv,* 1410.8001.

Trivers, R. L. (1971). The evolution of reciprocal altruism. *Quarterly Review of Biology, 46,* 35–57.

  (1972). Parental investment and sexual selection. In B. Campbell (Ed.), *Sexual selection and the descent of man: 1871–1971* (pp. 136–179). Chicago, IL: Aldine Press.

  (1985). *Social evolution.* Menlo Park, CA: Benjamin/Cummings.

  (2002). *Natural selection and social theory: Selected papers of Robert Trivers.* Oxford University Press.

Turkheimer, E. (2000). Three laws of behavior genetics and what they mean. *Current Directions in Psychological Science, 9,* 160–164.

Tybur, J. M., Miller, G. F., & Gangestad, S. W. (2007). Testing the controversy: An empirical examination of adaptationists' attitudes towards politics and science. *Human Nature, 18,* 313–328.

Tylor, E. B. (1871). *Primitive culture: Researches into the development of mythology, philosophy, religion, art and customs.* New York: Henry Holt.

van den Berghe, P. L. (1979). *Human family systems: An evolutionary view.* New York: Elsevier.

Van Dongen, S., & Gangestad, S. W. (2011). Human fluctuating asymmetry in relation to health and quality: A meta-analysis. *Evolution and Human Behavior, 32,* 380–398.

van Schaik, C. P., Ancrenaz, M., Borgen, G., *et al.* (2003). Orangutan cultures and the evolution of material culture. *Science, 299,* 102–105.

van Veelen, M., García, J., Sabelis, M. W., & Egas, M. (2012). Group selection and inclusive fitness are *not* equivalent; the Price equation vs. models and statistics. *Journal of Theoretical Biology, 299,* 64–80.

Vasey, P. L., Pocock, D. S., & VanderLaan, D. P. (2007). Kin selection and male androphilia in Samoan fa'afafine. *Evolution and Human Behavior, 28,* 159–167.

Voelkl, B., Portugal, S. J., Unsöld, M., Usherwood, J. R., Wilson, A. M., & Fritz, J. (2015). Matching times of leading and following suggest cooperation through direct reciprocity during V-formation flight in ibis. *Proceedings of the National Academy of Sciences, 112,* 2115–2120.

Waitt, C., & Little, A. C. (2006). Preferences for symmetry in conspecific facial shape among *Macaca mulatta. International Journal of Primatology, 27,* 133–145.

Wallen, K. (2005). Hormonal influences on sexually differentiated behavior in nonhuman primates. *Frontiers in Neuroendocrinology, 26,* 7–26.

Walpole, S. C., Prieto-Merino, D., Edwards, P., *et al.* (2012). The weight of nations: An estimation of adult human biomass. *BMC Public Health, 12,* 439.

Warraq, I. (2002). Virgins? What virgins? *The Guardian.* Retrieved March 30, 2015, from www.theguardian.com/books/2002/jan/12/books.guardianreview5

Washburn, S., & Lancaster, C. (1968). The evolution of hunting. In R. B. Lee & I. DeVore (Eds.), *Man the hunter* (pp. 193–303). Chicago, IL: Aldine.

Washburn, S. L. (1978). Human behavior and the behavior of other animals. *American Psychologist, 33,* 405–418.

Weinberg, K. S. (1963). *Incest behavior.* New York: Citadel Press.

West, S. A., Griffin, A. S., & Gardner, A. (2007a). Social semantics: Altruism, cooperation, mutualism, strong reciprocity and group selection. *Journal of Evolutionary Biology, 20,* 415–432.

(2007b). Social semantics: How useful has group selection been? *Journal of Evolutionary Biology, 21,* 374–385.

Westermarck, E. A. (1891). *The history of human marriage.* London: Macmillan.

Whitehead, H., & Rendell, L. (2014). *The cultural lives of whales and dolphins.* Chicago, IL: University of Chicago Press.

Whiten, A., Goodall, J., McGrew, W. C., *et al.* (1999). Cultures in chimpanzees. *Nature, 399,* 682–685.

Wilkinson, G. S. (1984). Reciprocal food sharing in the vampire bat. *Nature, 308,* 181–184.

Williams, G. C. (1966). *Adaptation and natural selection: A critique of some current evolutionary thought*. Princeton, NJ: Princeton University Press.

Wilson, D. S. (2002). *Darwin's cathedral: Evolution, religion, and the nature of society*. Chicago, IL: University of Chicago Press.

(2015). *Does altruism exist? Culture, genes, and the welfare of others*. New Haven, CT: Yale University Press.

Wilson, D. S., & Sober, E. (1994). Reintroducing group selection to the human behavioral sciences. *Behavioral and Brain Sciences*, *17*, 585–654.

Wilson, D. S., & Wilson, E. O. (2007). Rethinking the theoretical foundation of sociobiology. *Quarterly Review of Biology*, *82*, 327–348.

Wilson, E. O. (1975). *Sociobiology: The new synthesis*. Cambridge, MA: Harvard University Press.

(1978). *On human nature*. Cambridge, MA: Harvard University Press.

Wilson, M., & Daly, M. (1985). Competitiveness, risk taking, and violence: The young male syndrome. *Ethology and Sociobiology*, *6*, 59–73.

(1992). The man who mistook his wife for a chattel. In J. H. Barkow, L. Cosmides, & J. Tooby (Eds.), *The adapted mind: Evolutionary psychology and the generation of culture* (pp. 289–322). Oxford University Press.

Wilson, M. L., Boesch, C., Fruth, B., *et al.* (2014). Lethal aggression in *Pan* is better explained by adaptive strategies than human impacts. *Nature*, *513*, 414–417.

Wingfield, J. C., Hegner, R. E., Dufty, A. M. J., & Ball, G. F. (1990). The "challenge hypothesis": Theoretical implications for patterns of testosterone secretion, mating systems, and breeding strategies. *American Naturalist*, *136*, 829–846.

Wishnia, K. (2015). *Soft money: A Filomena Buscarsela mystery*. Oakland, CA: PM Press.

Wolf, A. P. (1966). Childhood association, sexual attraction, and the incest taboo. *American Anthropologist*, *68*, 883–898.

(1970). Childhood association and sexual attraction: A further test of the Westermarck hypothesis. *American Anthropologist*, *72*, 530–515.

Wolf, N. (1991). *The beauty myth: How images of beauty are used against women*. New York: Morrow.

Wood, W., & Eagly, A. H. (2002). A cross-cultural analysis of the behavior of women and men: Implications for the origins of sex differences. *Psychological Bulletin*, *128*, 699–727.

(2012). Biosocial construction of sex differences and similarities in behavior. In J. M. Olson & M. P. Zanna (Eds.), *Advances in experimental social psychology, Vol. XLVI* (pp. 55–123). London: Elsevier.

Woods, V. (2010). *Bonobo handshake: A memoir of love and adventure in the Congo*. New York: Penguin.

Workman, L., & Reader, W. (2014). *Evolutionary psychology: An introduction* (3rd ed.). Cambridge University Press.

Wrangham, R. (2009). *Catching fire: How cooking made us human*. London: Profile.

Wright, R. (1994). *The moral animal: The new science of evolutionary psychology*. New York: Vintage Books.

Wynne-Edwards, V. C. (1962). *Animal dispersion in relation to social behaviour.* Edinburgh, UK: Oliver & Boyd.

Xue, M. (2013). Altruism and reciprocity among friends and kin in a Tibetan village. *Evolution and Human Behavior, 34,* 323–329.

Young, L., & Alexander, B. (2012). *The chemistry between us: Love, sex, and the science of attraction.* New York: Penguin.

Yudkowsky, E. S. (2006). *The human importance of the intelligence explosion.* Paper presented at the Singularity Summit 2006, Stanford, CA. Retrieved March 3, 2018, from www.scribd.com/document/2327576/The-Human-Importance-of-the-Intelligence-Explosion-Powperpoint-Presentation-Handout

Zahavi, A. (1975). Mate selection: A selection for a handicap. *Journal of Theoretical Biology, 53,* 205–214.

Zahavi, A., & Zahavi, A. (1997). *The handicap principle: A missing piece of Darwin's puzzle.* Oxford University Press.

Zuckerman, M., Silberman, J., & Hall, J. A. (2013). The relation between intelligence and religiosity: A meta-analysis and some proposed explanations. *Personality and Social Psychology Review, 17,* 325–354.

Zuk, M. (2013). *Paleofantasy: What evolution really tells us about sex, diet, and how we live.* New York: Norton.

Index